New Insights in GNSS Remote Sensing for Ionosphere Monitoring and Modeling

New Insights in GNSS Remote Sensing for Ionosphere Monitoring and Modeling

Guest Editor

Angela Aragon-Angel

MDPI

Basel • Beijing • Wuhan • Barcelona • Belgrade • Novi Sad • Cluj • Manchester

Guest Editor
Angela Aragon-Angel
Mathematics Department
Universitat Politècnica de
Catalunya
Barcelona
Spain

Editorial Office
MDPI AG
Grosspeteranlage 5
4052 Basel, Switzerland

This is a reprint of the Special Issue, published open access by the journal *Remote Sensing* (ISSN 2072-4292), freely accessible at: https://www.mdpi.com/journal/remotesensing/special_issues/9QX7A1N97W.

For citation purposes, cite each article independently as indicated on the article page online and as indicated below:

Lastname, A.A.; Lastname, B.B. Article Title. *Journal Name* **Year**, *Volume Number*, Page Range.

ISBN 978-3-7258-6099-9 (Hbk)
ISBN 978-3-7258-6100-2 (PDF)
https://doi.org/10.3390/books978-3-7258-6100-2

Contents

About the Editor

Angela Aragon-Angel

Angela Aragon-Angel is a Senior Researcher with the research group of Astronomy and Geomatics and an Associate Professor at the Universitat Politècnica de Catalunya (UPC). Her primary research interests lie in satellite navigation, with a strong focus on signal processing, multi-frequency and multi-constellation Global Navigation Satellite Systems (GNSS), and the development of robust positioning algorithms for challenging environments. A significant thread of her work explores modeling the Earth's ionosphere to enhance environmental monitoring and geodetic applications. She has been actively involved in numerous European projects, such as those funded by the Horizon 2020 framework and the European Space Agency (ESA), where she has contributed to advancing high-precision GNSS technologies. Among her key achievements, Dr. Aragon-Angel has authored over 60 peer-reviewed scientific publications, serves as a regular reviewer for leading IEEE and navigation journals, and plays an active role in international scientific working groups dedicated to the future of satellite navigation and its applications. Her work bridges the gap between fundamental signal processing and practical, multi-disciplinary remote sensing solutions.

Preface

The ionosphere, a dynamic and ionized region of Earth's upper atmosphere, has evolved from a subject of pure scientific curiosity to a critical infrastructure layer for our modern technological society. It directly influences the propagation of radio signals, impacting global navigation satellite systems (GNSS), satellite communications, and even power grids. The central subject of this Special Issue is the advanced application of GNSS as a powerful remote sensing tool to monitor, model, and fundamentally understand this complex atmospheric region.

The scope of this collection is intentionally broad and interdisciplinary. It encompasses studies that leverage diverse GNSS data sources—from ground-based networks to space-based platforms like COSMIC—and focuses on both methodological innovation and applied science. Contributions span the development of novel algorithms for ionospheric determination, multi-scale monitoring (from local to global), and the advancement of space weather services. One of the key aims is to explore how the integration of multisource data and multiscale approaches can consolidate our knowledge of the underlying physics governing ionospheric behavior.

The primary motivation for compiling this work is twofold. First, we live in an era of unprecedented data availability from current and planned GNSS observing systems, coupled with the proliferation of distributed sensor arrays. This presents a novel opportunity to deepen research with richer datasets. Second, as we progress through the rising phase of Solar Cycle 25, the frequency and intensity of ionospheric disturbances are increasing. There is a pressing, practical need to enhance the resilience of GNSS-dependent technologies against these disruptions, which is vividly demonstrated by research in this issue on the degradation of precise point positioning accuracy during strong geomagnetic storms.

Therefore, the core purpose of this Special Issue is to highlight and disseminate cutting-edge contributions that bridge theoretical advancement with practical application. We seek to provide a platform for research that not only refines our observational capabilities—such as new models for plasmaspheric electron content or real-time error correction for smartphone GNSS—but also embraces emerging methodologies like transformer-based deep learning for prediction and explainability analysis, pushing the boundaries of ionospheric science.

This volume is addressed to a wide audience within the geoscience and engineering communities. It is intended for researchers and scientists specializing in atmospheric remote sensing, space weather, and geodesy. It is equally pertinent to engineers and technologists working on high-accuracy GNSS positioning, navigation systems, and satellite communications, who require the latest models and mitigation strategies for ionospheric effects. Finally, it serves as a valuable reference for graduate students entering these dynamic and critically important fields.

Angela Aragon-Angel
Guest Editor

remote sensing

MDPI

Article

Transformer-Based Ionospheric Prediction and Explainability Analysis for Enhanced GNSS Positioning

He-Sheng Wang, Dah-Jing Jwo * and Yu-Hsuan Lee

Department of Communications, Navigation, and Control Engineering, National Taiwan Ocean University, Keelung 202, Taiwan; hswang@email.ntou.edu.tw (H.-S.W.); 11167019@mail.ntou.edu.tw (Y.-H.L.)
* Correspondence: djjwo@mail.ntou.edu.tw

Abstract: This study aims to investigate the impact of ionospheric models on Global Navigation Satellite System (GNSS) positioning and proposes an ionospheric prediction method based on a Transformer deep learning model. We construct a Transformer-based deep learning model that utilizes global ionospheric maps as input to achieve spatiotemporal prediction of Total Electron Content (TEC). To gain a deeper understanding of the model's prediction mechanism, we employ integrated gradients for explainability analysis. The results reveal the key ionospheric features that the model focuses on during prediction, providing guidance for further model optimization. This study demonstrates the efficacy of a Transformer-based model in predicting Vertical Total Electron Content (VTEC), achieving comparable accuracy to traditional methods while offering enhanced adaptability to spatial and temporal variations in ionospheric behavior. Furthermore, the application of advanced explainability techniques, particularly the Integrated Decision Gradient (IDG) method, provides unprecedented insights into the model's decision-making process, revealing complex feature interactions and spatial dependencies in VTEC prediction, thus bridging the gap between deep learning capabilities and explainable scientific modeling in geophysical applications. The model achieved positioning accuracies of −1.775 m, −2.5720 m, and 2.6240 m in the East, North, and Up directions respectively, with standard deviations of 0.3399 m, 0.2971 m, and 1.3876 m. For VTEC prediction, the model successfully captured the diurnal variations of the Equatorial Ionization Anomaly (EIA), with differences between predicted and CORG VTEC values typically ranging from −6 to 6 TECU across the study region. The gradient score analysis revealed that solar activity indicators (F10.7 and sunspot number) showed the strongest correlations (0.7–0.8) with VTEC variations, while geomagnetic indices exhibited more localized impacts. The IDG method effectively identified feature importance variations across different spatial locations, demonstrating the model's ability to adapt to regional ionospheric characteristics.

Keywords: GNSS; transformer; ionospheric effect; explainability; integrated gradient; integrated decision gradient

1. Introduction

1.1. Introduction to GNSS and XAI

Global Navigation Satellite Systems (GNSSs) have become an integral part of modern life, providing precise positioning and navigation capabilities for a wide range of military, civilian, and commercial applications. Since the launch of the first Global Positioning System (GPS) satellite in 1978 and the completion of the GPS network in 1993, other countries and regions have developed their own systems, including Russia's GLONASS,

https://doi.org/10.3390/rs17010081

the European Union's Galileo, and China's BeiDou. These systems have enabled numerous applications, from vehicle navigation and mobile phone positioning to autonomous vehicles [1].

However, GNSS accuracy is significantly affected by ionospheric delay, which remains a critical challenge in satellite positioning [2,3]. Recent studies have further highlighted these challenges, particularly in the context of precise positioning applications. Lyu et al. [4] investigated the uncertainties in interpolating satellite-specific slant ionospheric delays and their impacts on PPP-RTK, while Tang et al. [5] examined how equatorial plasma bubbles and associated ionospheric gradients affect GNSS PPP-RTK performance. The ionosphere, a layer of the Earth's upper atmosphere, contains free electrons that can alter the speed and direction of radio signals passing through it [6]. This delay varies with factors such as solar activity, geomagnetic activity, and atmospheric disturbances, leading to reduced positioning accuracy.

In recent years, researchers have turned to deep learning methods to predict ionospheric Total Electron Content (TEC) and correct for navigation errors [7,8]. The rise of artificial intelligence, particularly with breakthroughs like OpenAI's GPT models, has further fueled interest in applying these technologies to various domains, including ionospheric prediction [9–14]. Traditional approaches to ionospheric prediction can be broadly categorized into three groups:

- Empirical Models: These include the International Reference Ionosphere (IRI), NeQuick, and various regional models. They rely on historical data and statistical analysis to predict ionospheric behavior.
- Physical Models: These are based on solving continuity, momentum, and energy equations for the ionospheric plasma.
- Data Assimilation Methods: These combine observations with physical or empirical models to provide more accurate predictions. The Global Assimilative Ionospheric Model (GAIM) is a prominent example.

Recent years have seen significant advances in deep learning approaches for ionospheric prediction. These include the following:

- Convolutional Neural Networks (CNNs) for spatial feature extraction from global ionospheric maps;
- Long Short-Term Memory (LSTM) networks for capturing temporal dependencies in TEC variations;
- Hybrid architectures, combining different neural network types for improved prediction accuracy;
- Attention-based models that can capture both local and global ionospheric patterns.

Each approach has its strengths and limitations. Deep learning methods have shown particular promise in capturing complex, non-linear relationships in ionospheric behavior and adapting to varying conditions.

Various artificial intelligence approaches have been applied to ionospheric prediction in recent years. Early efforts focused on traditional neural networks, with Tulunay et al. [7,8] demonstrating the feasibility of neural network techniques for ionospheric forecasting. The field has since evolved to incorporate more sophisticated architectures. Tulasi Ram et al. [7] developed an improved two-dimensional artificial neural network model (ANNIM) for ionospheric modeling. Chen et al. [10] applied deep learning techniques to global ionospheric TEC prediction, while Luo et al. [11] introduced the SAM-ConvLSTM model for improved accuracy. Recurrent neural network architectures have also shown promise, with Sun et al. [12] implementing a bidirectional LSTM approach for vertical TEC forecasting. More recent work by Wang et al. [13] has explored automated deep learning frameworks, combining multiple techniques, including residual networks and attention mechanisms. Ren et al. [14] specifically addressed TEC prediction during storm periods using a mixed CNN-BiLSTM approach. While these studies have demonstrated the potential of AI in ionospheric prediction, they typically lack explainability, making it difficult to understand how these models arrive at their predictions. This gap motivates our current work combining Transformer architecture with explainable AI techniques.

As AI systems become more prevalent in critical applications, there is a growing demand for transparency and trustworthiness in their decision-making processes. This has led to the emergence of eXplainable Artificial Intelligence (XAI), which aims to make AI systems more interpretable and their predictions more understandable to humans. XAI is an emerging field that seeks to make AI models more transparent. It focuses on developing methods to create models that are not only highly accurate but also interpretable to humans. XAI aims to elucidate how AI systems arrive at their decisions, thereby fostering trust among stakeholders and supporting the development of responsible AI [15,16]. By enhancing the comprehensibility of complex AI models, XAI plays a crucial role in bridging the gap between advanced machine learning techniques and human understanding.

This paper presents a novel approach that combines a Transformer-based model for regional ionospheric mapping with XAI techniques. Our work builds upon previous research in ionospheric prediction using neural networks, such as the time-based input features proposed by McKinnell and Poole [17] and the incorporation of geomagnetic indices and solar activity by Sivavaraprasad et al. [18]. In our study, we employ the Transformer deep learning architecture to construct an ionospheric prediction model. To analyze the interpretability of this model, we utilize the Integrated Decision Gradient (IDG) method, an advanced technique in the field of XAI.

The IDG method is an enhancement of the widely used Integrated Gradients (IG) approach, first introduced by Sundararajan et al. in 2017 [19]. IG attributes the prediction of a deep network to its input features by accumulating gradients along the straight-line path from a baseline input to the actual input. While IG has been successful in many applications, it can sometimes produce noisy attributions due to the gradient saturation problem, particularly in deep networks with non-linear activations [20]. To address these limitations, Walker et al. proposed the Integrated Decision Gradient method in 2023 [21]. IDG builds upon the foundational principles of IG but introduces key improvements to enhance the quality and reliability of feature attributions. The primary innovation of IDG lies in its approach to gradient calculation.

Instead of computing gradients with respect to the input features directly, IDG calculates gradients with respect to the decision boundary of the model. This subtle but crucial difference allows IDG to capture more meaningful attributions, especially in cases where the model's decision is primarily influenced by features close to the decision boundary.

In the context of our ionospheric prediction model, IDG allows us to achieve the following:

- Identify the most influential input features for each prediction, providing insights into the model's decision-making process.
- Understand how different ionospheric parameters contribute to the model's predictions across various spatial and temporal scales.
- Detect potential biases or unexpected behaviors in the model, which could lead to improvements in model architecture or training procedures.
- Enhance the interpretability of our Transformer-based model, making it more transparent and trustworthy for stakeholders in the GNSS community.

By employing IDG, we not only improve upon the interpretability offered by traditional methods like IG but also gain a more nuanced understanding of our model's behavior. This approach aligns with the growing emphasis on responsible AI development in critical applications such as ionospheric modeling for GNSS systems [22,23].

The integration of advanced deep learning techniques like Transformers with state-of-the-art interpretability methods such as IDG represents a significant step forward in the field of ionospheric prediction. It allows us to leverage the power of complex neural networks while maintaining a high degree of model transparency and interpretability, crucial for the continued advancement and acceptance of AI-driven solutions in scientific and engineering domains. By combining advanced deep learning techniques with explainable AI methods, we aim to enhance the accuracy and reliability of GNSS positioning while providing insights into the model's decision-making process. This approach not only contributes to the field of ionospheric prediction but also addresses the broader need for transparent and trustworthy AI systems in critical applications.

1.2. Ionospheric Effects on Global Navigation Satellite Systems

The ionosphere, a critical layer of the Earth's atmosphere located between the troposphere and the exosphere, plays a significant role in GNSS positioning accuracy. Composed primarily of ionized gases, including oxygen, hydrogen, and other molecules, the ionosphere's electron density, thickness, and height vary with seasonal changes, diurnal cycles, and solar radiation. This variability introduces challenges in precise GNSS positioning, necessitating accurate monitoring and modeling of ionospheric behavior. The ionosphere is formed when high-energy radiation from the sun, particularly ultraviolet and X-ray, ionizes atmospheric molecules and atoms. This ionization process creates free electrons that affect satellite signals, causing variations in signal propagation speed and path delays [24]. The ionosphere's activity is closely correlated with solar and geomagnetic activities, making these features crucial for predicting the Vertical Total Electron Content (VTEC).

Total Electron Content is a key parameter in describing ionospheric behavior. It represents the total number of electrons present along the path from the satellite to the receiver, typically measured in TEC Units (TECU), where 1 TECU equals 10^{16} electrons/m^2 [25]. TEC can be categorized into Slant TEC (STEC) and Vertical TEC (VTEC), with the relationship between them described by a mapping function. The single-layer ionosphere model simplifies the ionosphere as a thin shell concentrated at a specific height, typically between 350 and 450 km. This model, while neglecting the actual structure and thickness variations of the ionosphere, provides a useful approximation for many applications [2].

The influencing factors can be divided into three groups:

- Solar Activity: Solar activity follows an approximate 11-year cycle, characterized by variations in sunspot numbers and solar flux. During periods of high solar activity, phenomena such as solar flares and coronal mass ejections (CMEs) increase, leading to higher levels of ionization in the Earth's upper atmosphere [26]. The F10.7 index, which measures solar radio flux at a wavelength of 10.7 cm, serves as a proxy for solar activity and closely correlates with sunspot numbers.
- Geomagnetic Activity: The Earth's magnetic field significantly influences ionospheric plasma behavior. At the geomagnetic equator, the horizontal magnetic field creates an eastward electric current in the ionosphere. Additionally, the interplanetary magnetic field (IMF) interacts with the Earth's magnetosphere, particularly affecting high-latitude regions and contributing to auroral phenomena [27].
- Anthropogenic Factors: Human activities, such as nuclear tests and rocket launches, can also impact ionospheric structure. Nuclear tests have been observed to affect D-layer ionization, while rocket launches can create ionospheric holes due to the combustion of rocket fuel in the ionosphere [28,29].

The ionosphere significantly affects GNSS positioning by causing signal delays and bending. This effect, known as ionospheric delay, impacts the accuracy of GNSS signals. To compensate for these effects, researchers have developed various mathematical models to describe ionospheric behavior. Some well-known models include the Klobuchar model, NeQuick model, BeiDou Global Ionospheric delay correction Model (BDGIM), International Reference Ionosphere (IRI) model, and Global Ionosphere Map (GIM). Developed by the U.S. Department of Defense, the Klobuchar model is widely used in GPS systems. It employs eight parameters broadcast by GPS satellites to estimate ionospheric delay. While simple and efficient, its accuracy is limited, making it suitable primarily for single-frequency GPS receivers [2]. The NeQuick model is a global ionospheric model based on radiative transfer theory. It simulates electron density distribution and is commonly used in the Galileo navigation system. NeQuick provides more accurate modeling of electron density at different altitudes, thereby improving compensation for ionospheric delay [30]. The BDGIM [31–33] is designed specifically for the BeiDou Navigation Satellite System (BDS). Like other GNSS ionospheric models, it aims to mitigate ionospheric delay effects on positioning accuracy. BDGIM uses a series of coefficients broadcast in the BDS navigation message to calculate ionospheric delays. The model employs a spherical harmonic function to represent global TEC distribution and has demonstrated good performance in ionospheric delay correction, particularly in the Asia-Pacific region where BDS has stronger coverage. Compared to the Klobuchar model, BDGIM generally achieves better accuracy in ionospheric delay estimation, especially in low and middle latitude regions. The IRI model is an internationally recognized standard ionospheric model based on global measurement data and statistical analysis. Updated annually, it is applicable to a wide range of scenarios, including GPS positioning, communication, and radar systems [34].

GIMs, unlike the previous three models, are generated using data from GNSS networks. They rely on measurements from a global network of ground-based GNSS receivers. GIMs are produced by analyzing the ionospheric TEC obtained from GNSS signals. These TEC values are then interpolated over a global grid to create a map of the ionosphere's electron content [35]. Organizations like the Center for Orbit Determination in Europe (CODE) and the International GNSS Service (IGS) routinely produce GIMs using their own algorithms and interpolation techniques. GIMs provide near-real-time or post-processed global maps of the ionosphere's TEC. They are crucial for applications that require current ionospheric conditions, such as GNSS positioning and navigation, where ionospheric delays can significantly impact accuracy. Recent studies have evaluated GIM performance under varying conditions. Wielgosz et al. [36] conducted comprehensive validations of GIM accuracy across different solar activity levels, comparing performance during solar minimum (2014) and maximum (2018) periods. Tang et al. [37] specifically investigated GIM-TEC errors during equatorial plasma bubble events in high solar activity conditions, revealing important limitations in GIM accuracy during ionospheric disturbances. GIMs can provide high temporal resolution (e.g., hourly updates), reflecting the ionosphere's current state, including short-term variations and disturbances like geomagnetic storms. They are ideal for real-time applications requiring current ionospheric information, such as precise GNSS positioning, satellite communication adjustments, and space weather monitoring.

Understanding and modeling ionospheric effects are crucial for improving GNSS positioning accuracy. The choice of model depends on the specific application scenario and required precision. The Taiwan region presents particular challenges for VTEC prediction due to its location near the northern crest of the Equatorial Ionization Anomaly (EIA) [38]. The EIA is characterized by two bands of enhanced ionization approximately 15–20 degrees north and south of the magnetic equator, with a trough at the equator [39]. Over Taiwan, the northern EIA crest creates dynamic ionospheric conditions with significant VTEC variations, even during magnetically quiet periods. This feature typically begins developing during morning hours as solar heating increases, reaches maximum development in the afternoon (around 14:00–16:00 local time), and breaks down during evening hours. The process creates substantial spatial and temporal VTEC gradients that pose significant challenges for prediction. During magnetically disturbed conditions, the EIA becomes even more active and irregular, with enhanced VTEC variations and more unpredictable behavior. These characteristics make our study region particularly suitable for testing ionospheric prediction models, as successful prediction here requires handling both regular daily EIA variations and irregular disturbance effects [40,41].

Our experimental framework consists of several key components. We utilized CORG data spanning from January 2022 to December 2023 for model training and validation, focusing on a regional map around Taiwan (115°E–130°E, 17.5°N–30°N). The study involves three main phases: (1) development and training of the Transformer model using eight input features, including solar activity indicators, geomagnetic indices, and time functions; (2) application of the IDG method to analyze feature importance and model behavior; and (3) validation of the model's performance through VTEC prediction accuracy assessment and GNSS positioning experiments. We evaluate our results using multiple metrics including VTEC prediction errors, positioning accuracy in East–North–Up coordinates, and feature contribution analysis through gradient scores. The positioning experiments were conducted using single-frequency observations, with results compared against dual-frequency solutions and CORG-based corrections to assess practical improvements in GNSS positioning accuracy.

2. Transformer Model

The Transformer model, introduced by Vaswani et al. [42] in their 2017 paper "Attention Is All You Need", revolutionized the field of natural language processing and has since been adapted for various other domains. Unlike previous sequence-to-sequence models that relied on recurrent or convolutional neural networks, the Transformer is based entirely on attention mechanisms, allowing for more parallel processing and better handling of long-range dependencies.

The Transformer architecture comprises an encoder and a decoder, each built from layers that include multi-head self-attention mechanisms and position-wise feed-forward networks.

A. Encoder

The encoder transforms the input sequence into a continuous representation. It begins with input embeddings, where each element of the input sequence is mapped into a high-dimensional space. Positional encodings are added to these embeddings to retain information about the sequence order, as the model does not inherently capture positional information. The core of the encoder consists of multi-head self-attention layers, allowing the model to focus on different positions within the input sequence to capture various types of dependencies. Following the attention layers are feed-forward networks applied to each position separately, introducing non-linearity and enhancing the model's capacity.

B. Decoder

The decoder generates the output sequence by attending to both the encoder's output and the previously generated outputs. It includes masked multi-head self-attention layers to prevent the model from attending to future positions, preserving the auto-regressive property necessary for sequence generation. The decoder also incorporates encoder–decoder attention layers, enabling it to focus on relevant parts of the input sequence. The Dst index measures the intensity of the global geomagnetic storm level. The Kp index quantifies disturbances in the Earth's magnetic field. The Ap index is a daily index derived from the Kp values.

C. Self-Attention Mechanism

At the heart of the Transformer is the self-attention mechanism, which computes a weighted sum of input representations, where weights are determined by the similarity between elements. For each position in the sequence, the model calculates query, key, and value vectors. The attention scores are computed using scaled dot-product attention:

$$\text{Attention}(Q, K, V) = \text{softmax}\left(\frac{QK^T}{\sqrt{d_k}}\right)V, \tag{1}$$

where Q, K, and V represent the query, key, and value matrices, respectively, d_k is the dimension of the key vectors, and softmax is defined as follows:

$$\text{softmax}(x_i) = \frac{\exp(x_i)}{\sum_j \exp(x_j)}$$

where x_i represents the i-th element of the input vector x, and the sum in the denominator is taken over all elements j.

D. Multi-Head Attention

To capture different types of dependencies, the Transformer uses multi-head attention, which allows the model to jointly attend to information from different representation subspaces. Each attention head performs the self-attention mechanism independently, and the outputs are concatenated and linearly transformed:

$$\text{Multihead}(Q, K, V) = \text{Concat}(\text{head}_1, \ldots, \text{head}_h)W^O \tag{2}$$

where Concat(head_1,...,head_h) denotes the concatenation operation that combines multiple head outputs into a single matrix by joining them along their last dimension. For example, if each head_i is a matrix of shape ($N \times D$), then Concat(head_1,...,head_h) produces a matrix of shape ($N \times hD$), where h is the number of attention heads. Each head is calculated as follows:

$$\text{head}_i = \text{Attention}(QW_i^Q, KW_i^K, VW_i^V), \tag{3}$$

where W^O is the weighting matrix for concatenated output, and W^Q, W^K, and W^V are the weighting matrices for query, key, and value, respectively.

E. Positional Encoding

Since the Transformer does not inherently account for the order of the sequence, positional encodings are added to the input embeddings to introduce sequence order information. Commonly, sinusoidal functions are used for positional encodings:

$$\text{PE}_{(pos,2i)} = \sin\left(\frac{pos}{1000^{2i/d_{\text{model}}}}\right), \text{PE}_{(pos,2i+1)} = \cos\left(\frac{pos}{1000^{2i/d_{\text{model}}}}\right) \tag{4}$$

where pos is the position and i is the dimension.

The ionosphere exhibits complex spatial-temporal behaviors influenced by solar activity, geomagnetic conditions, and atmospheric disturbances. Traditional models often fail to capture these complexities due to the following:

- Spatial-Temporal Complexity: The ionosphere's behavior varies significantly across different locations and times.

- Long-Range Dependencies: Past ionospheric conditions can influence future states over extended periods and distances.
- Data Representation: GIM data provides TEC values over a global grid, requiring the effective handling of spatial data in conjunction with temporal sequences.

To apply the Transformer model to ionospheric prediction using GIM data, several adaptations are necessary. The GIM provides TEC values on a global grid defined by latitude and longitude. In our study, we focus on a regional map around Taiwan to capture local ionospheric variations pertinent to GNSS applications in this area. Each TEC measurement corresponds to a specific longitude and latitude grid point. The selected features represent the primary physical factors known to influence ionospheric TEC variations:

- Solar Activity Indicators (Sunspot number and F10.7 index): These directly measure solar radiation levels, which are the primary source of ionospheric ionization. The F10.7 index strongly correlates with solar EUV radiation, which drives ionospheric electron production. Using both indicators provides redundancy and robustness in capturing solar activity effects.
- Geomagnetic Indices (Dst and Kp): Dst index captures global-scale magnetic disturbances and storm effects. The Kp index reflects broader geomagnetic activity, which affects electron density distribution. Together, they account for both sudden and gradual geomagnetic influences on the ionosphere.
- Time Functions (DNS, DNC, HRS, HRC): DNS and DNC capture seasonal variations in solar zenith angle and day length. HRS and HRC represent diurnal variations due to the Earth's rotation. These sinusoidal functions effectively model the periodic nature of solar illumination patterns.

While additional parameters like neutral winds or electric fields also affect TEC, their effects are indirectly reflected in the geomagnetic indices, and direct measurements are not typically available in real time for operational GNSS applications. Our experimental results demonstrate that these selected features capture the main drivers of TEC variations sufficiently well for practical prediction purposes.

Here, the time functions DNS and DNC are defined as follows:

$$\mathrm{DNS} = \sin\left(\frac{2\pi \times \text{Day Number}}{365}\right), \mathrm{DNC} = \cos\left(\frac{2\pi \times \text{Day Number}}{365}\right), \tag{5}$$

representing seasonal variations. HRS and HRC are calculated as follows:

$$\mathrm{HRS} = \sin\left(\frac{2\pi \times \text{Hour}}{365}\right), \mathrm{HRC} = \cos\left(\frac{2\pi \times \text{Hour}}{365}\right), \tag{6}$$

representing diurnal variations. These time functions capture the periodic nature of the ionosphere influenced by the Earth's rotation and orbit. The sunspot number is updated every 24 h, reflecting solar activity levels. The F10.7 Index is the solar radio flux at 10.7 cm wavelength, an indicator of solar EUV emissions affecting the ionosphere. The Dst index measures the intensity of the global geomagnetic storm level. The Kp Index quantifies disturbances in the Earth's magnetic field. The Ap index is a daily index derived from the Kp values. By incorporating these features, the model gains a comprehensive view of factors influencing ionospheric conditions.

The Transformer model implementation utilized open-source frameworks, as the focus of our work was not on developing novel neural architectures but rather on understanding and improving model behavior through explainability analysis. The IDG analysis framework, however, was custom-developed for this specific application to ensure proper handling of ionospheric data characteristics. The Transformer architecture has been adapted for VTEC prediction as follows: For input encoding, each VTEC measurement point is encoded with its corresponding features (solar activity indicators, geomagnetic indices, time functions). Positional encoding represents the spatial relationships between grid points in our Taiwan region (115°E–130°E, 17.5°N–30°N). The Self-Attention Mechanism captures spatial dependencies between different grid points in our region, allowing the model to weigh the importance of neighboring points differently based on ionospheric conditions. This is particularly important for tracking EIA development and movement across our grid. For multi-head attention, different attention heads capture various types of relationships. This is crucial for simultaneously tracking the different ionospheric phenomena affecting our region. For the encoder–decoder structure, the encoder processes input features and their relationships, and the decoder generates VTEC predictions while considering all these relationships. This architecture is particularly suited for VTEC prediction because it can handle multiple input features simultaneously and capture both spatial and temporal dependencies.

3. The Need for Explainable AI

Black-box models, such as deep neural networks, have achieved remarkable success across various domains due to their ability to model complex, non-linear relationships. However, their opacity presents significant challenges. Firstly, the internal decision-making processes are often inaccessible or incomprehensible, making it difficult to understand how inputs are transformed into outputs. Moreover, without interpretability, models may inadvertently perpetuate or amplify biases present in the training data, leading to unfair or erroneous outcomes. In critical applications like GNSS, where decisions can have significant consequences, stakeholders may be hesitant to rely on models they do not understand.

Explainable AI seeks to address these limitations by making AI systems more transparent and interpretable. The motivations for XAI include the following:

- System Verification: Ensuring that AI models make decisions based on relevant and appropriate factors is essential for validating model behavior and performance.
- System Improvement: Understanding the model's reasoning can highlight weaknesses or errors, guiding targeted improvements and refinements.
- Learning from the System: Interpretable models can provide novel insights or patterns not immediately apparent to human experts, advancing domain knowledge.
- Legal and Regulatory Compliance: Emerging regulations, such as the European Union's General Data Protection Regulation (GDPR), mandate explainability in AI systems to protect user rights and promote ethical practices.

Our Transformer-based model predicts ionospheric TEC using GIM data, focusing on a regional map around Taiwan. The model incorporates several input features as described in the previous section. The model architecture includes an encoder and a decoder with multi-head attention layers, residual connections, layer normalization, and feed-forward networks. Key parameters include eight input features, one output feature, 512 neurons in the embedding layers, a dropout rate of 0.1, and the use of the Adam optimizer with a Mean Squared Error (MSE) loss function. The architecture of the model will be given in the experimental section.

Despite its predictive success (the results are given in the next section), the model's complexity poses challenges for interpretability. First of all, the use of high-dimensional embeddings and multiple layers makes it difficult to trace the influence of individual input features. Next, the model captures complex non-linear relationships, complicating efforts to understand how specific inputs affect outputs. Finally, while attention weights provide some insight, they do not always directly correlate with feature importance or causal influence.

To address these challenges, we turn to gradient-based explanation methods, which have shown promise in interpreting deep learning models. These methods analyze the gradients of the output with respect to the input features, identifying how changes in inputs affect the output.

Integrated Gradients, introduced by Sundararajan et al. [19], compute the average gradients along the path from a baseline input to the actual input. The integrated gradient along the ith dimension for an input x and baseline x' (usually a zero vector) is defined as follows:

$$\mathrm{IG}_i(x) = (x_i - x_i') \times \int_{\alpha=0}^{1} \frac{\partial F(x\prime + \alpha(x - x\prime))}{\partial x_i} d\alpha, \tag{7}$$

where x_i denotes the ith dimension, F is the model's output function (the model to be explained), α is the interpolation constant, and i indexes the input features. In practice, the integral is approximated using a summation:

$$\mathrm{IG}_i^{\mathrm{approx}}(x) = (x_i - x_i') \times \sum_{k=1}^{m} \frac{\partial F(x\prime + \frac{k}{m}(x - x\prime))}{\partial x_i} \times \frac{1}{m} \tag{8}$$

While IG provides valuable insights, it can suffer from issues such as gradient saturation and sensitivity to the choice of baseline, potentially leading to noisy or misleading attributions. To overcome these limitations, we employ the Integrated Decision Gradient (IDG) method, which enhances IG by focusing on gradients near the decision boundary. IDG calculates gradients with respect to the model's decision function, capturing more meaningful attributions, especially in deep models with non-linear activations.

IDG introduces an additional weighting factor based on the gradient with respect to the interpolation constant:

$$\mathrm{ImportanceFactor}(\alpha) = \frac{\partial F(x\prime + \alpha(x - x\prime))}{\partial \alpha} \tag{9}$$

$$\mathrm{IDG}_i(x) = \int_{\alpha=0}^{1} \frac{\partial F(x\prime + \alpha(x - x\prime))}{\partial x_i} \frac{\partial F(x\prime + \alpha(x - x\prime))}{\partial \alpha} d\alpha, \tag{10}$$

The approximation for IDG is as follows:

$$\mathrm{IG}_i^{\mathrm{approx}}(x) = (x_i - x_i') \times \sum_{k=1}^{m} \frac{\partial F(x\prime + \frac{k}{m}(x - x\prime))}{\partial x_i} \times \frac{\partial F(x\prime + \frac{k}{m}(x - x\prime))}{\partial \alpha} \times \frac{1}{m} \tag{11}$$

By integrating gradient-based explanation techniques like IDG into our Transformer model, we can identify which input features contribute most to the model's predictions, aiding in understanding and validating the model's behavior. It is also possible to uncover potential biases or unintended influences in the model, facilitating corrective measures. The method may gain insights into the model's weaknesses, guiding targeted refinements and enhancements.

IDG is generally considered a model-agnostic method. It is built upon the principles of gradient-based attribution methods, which are typically applicable to a wide range of differentiable models. This includes most neural network architectures. IDG is an extension of the IG method, which is itself model-agnostic. IG can be applied to any differentiable model, and IDG inherits this property. Like other gradient-based methods, IDG primarily relies on analyzing the relationship between inputs and outputs of the model, rather than requiring access to specific internal structures or parameters. However, it is important to note that IDG, like other gradient-based methods, requires that the model be differentiable. This means it is not applicable to non-differentiable models like decision trees or random forests. While theoretically applicable to many models, the computational efficiency of IDG might vary depending on the model architecture and complexity. In practice, some adaptations might be necessary when applying IDG to different types of models or data, especially when moving beyond the image processing domain, where it has shown clear benefits.

In our VTEC prediction application, IDG reveals how the model processes different input features to make its predictions. The method helps us understand why the model predicts certain VTEC values by showing the relative importance of solar activity indicators, geomagnetic indices, and time functions at each grid point. For example, at the northern grid points (25°N), the interpolation weights show different patterns compared to southern points (22.5°N), suggesting the model adapts its prediction strategy based on geographic location. The gradient scores reveal how the model responds to changes in solar and geomagnetic conditions, while cumulative gradients demonstrate the long-term impact of each feature. These insights not only validate that our model learns physically meaningful relationships but also identify which features drive predictions under different conditions, providing guidance for potential model improvements.

4. Experimental Results

The flow of the experiment is shown in Figure 1, while the flow of the transformer model is shown in Figure 2. In order to analyze the explainability of deep learning models in the context of ionospheric predictions, we first employed a Transformer model to predict VTEC. This approach serves a dual purpose: firstly, to demonstrate the efficacy of the Transformer architecture in capturing the complex spatio-temporal dynamics of the ionosphere, and secondly, to provide a robust foundation for subsequent explainability analyses. By first validating the accuracy of our VTEC predictions, we ensure that the model we seek to interpret is indeed capturing meaningful patterns in the data. This step is crucial, as interpreting a poorly performing model would yield little insight into the underlying physical processes. Once the model's predictive capabilities are established, we can then confidently apply interpretability techniques to unravel the decision-making processes within the model. This approach allows us to not only improve VTEC predictions but also to gain deeper insights into the factors influencing ionospheric behavior, potentially advancing our understanding of space weather phenomena.

Input features of the model are as described in Section 3. The study utilized the rapid ionosphere map product (CODE Orbit Rapid GIM, abbreviated as CORG) retrieved from the CODE website. The prediction area covered longitudes from 115°E to 130°E and latitudes from 17.5°N to 30°N, encompassing 24 grid points (See Figure 3).

The model was trained using data from the following sources and periods. Some key parameters are summarized as follows:

(a) Training Data:
- Time Period: January 2022 to December 2023
- Temporal Resolution: 24 data points per day (hourly samples)
- Spatial Coverage: Regional map around Taiwan (115°E–130°E, 17.5°N–30°N)
- Number of Grid Points: 24 points total (4×6 grid)

(b) Input Features (eight features total):
- Solar Activity Indicators
- Sunspot Number
- F10.7 index
- Geomagnetic Indices
- Dst index

- Kp index

(c) Time Functions

- DNS (Day Number Sine)
- DNC (Day Number Cosine)
- HRS (Hour Sine)
- HRC (Hour Cosine)

(d) Target Variable: VTEC values from CODE's rapid ionosphere map product (CORG)
(e) Training Parameters:

- Training/Validation Split: 80%/20%
- Batch Size: 64
- Learning Rate: 0.001
- Number of Epochs: 20
- Optimizer: Adam
- Loss Function: Mean Squared Error (MSE)

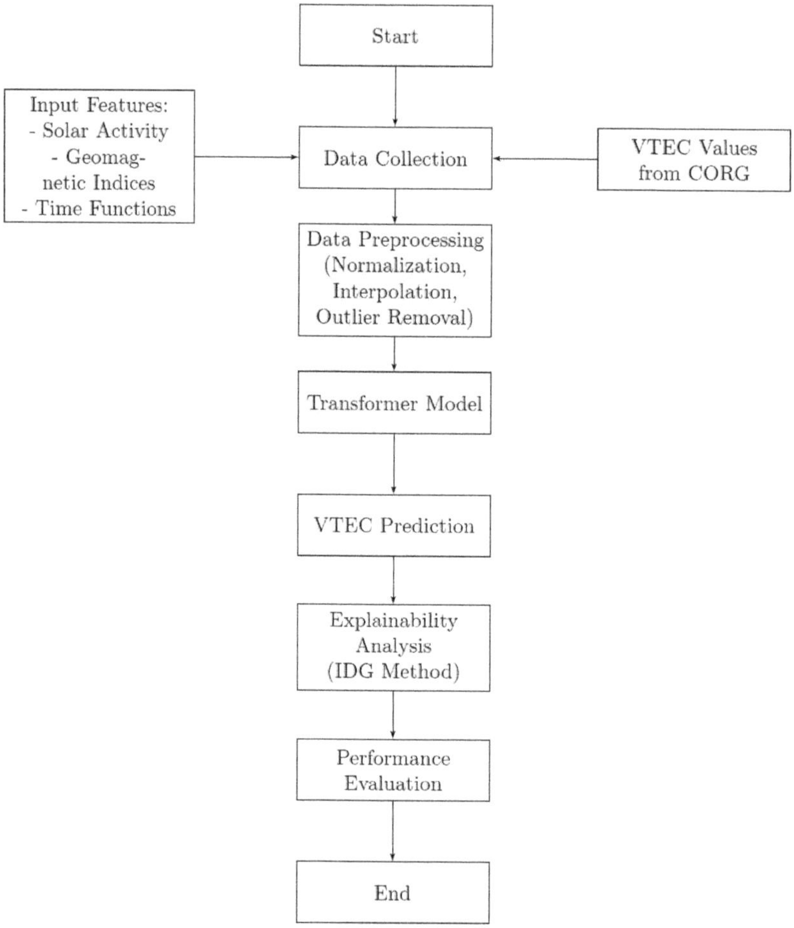

Figure 1. Flowchart of the proposed VTEC prediction algorithm using the Transformer model with explainability analysis.

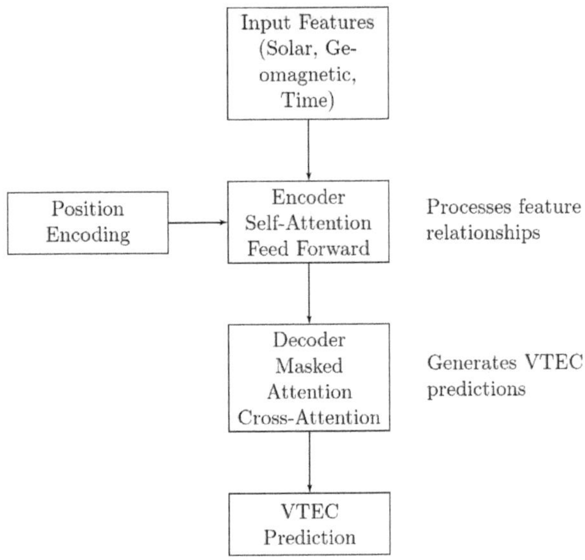

Figure 2. Transformer model architecture for VTEC prediction, showing the flow from input features through encoder–decoder structure to final VTEC predictions. The model incorporates positional encoding and multiple attention mechanisms to capture spatial-temporal relationships in ionospheric behavior.

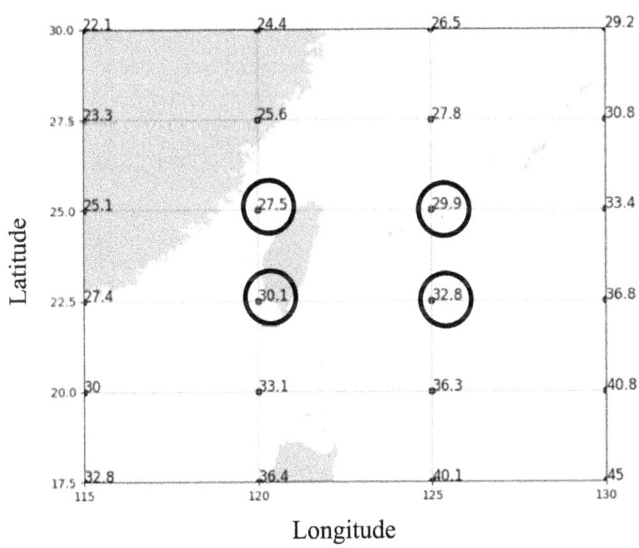

Figure 3. Example of 24 grid points covering the area around Taiwan. The values next to the grid points show VTEC values retrieved from the CODE website at 2022 11/25 00:00. The four grids enclosed with small circles will be used to compute the correlations later on.

The model was evaluated on test data from January 2024, which was completely separate from the training/validation period. The model architecture includes an encoder and a decoder with multi-head attention layers, residual connections, layer normalization, and feed-forward networks. Key parameters include eight input features (four GIM parameters and four time functions. For GIM parameters, the Ap index is not used as an input feature, since it is a daily average of the geomagnetic activity, providing a linear measure of the overall level of geomagnetic disturbances. It is derived from the Kp index by converting each three-hour Kp value into a linear scale. In this case, the Kp index should have enough information for daily geomagnetic activity), one output feature (VTEC), 512 neurons in the embedding layers, a dropout rate of 0.1, and the use of the Adam optimizer with a Mean Squared Error (MSE) loss function. Model parameters are summarized in Table 1.

Table 1. Key parameters of the Transformer model.

Transformer Parameters and Information			
Input Feature	9	Batch Size	64
Output Feature	1	Activation	ReLU
Optimizer	Adam	Loss Function	MSE
Epoch	20	Time Step	24
Neuron Number	512	Dropout	0.1

The choice of 512 neurons in the embedding layers was determined through empirical testing and consideration of several factors:

- Model Capacity: A total of 512 dimensions provide sufficient representational capacity to capture the complex relationships in our input features while avoiding excessive parameters that could lead to overfitting. We experimented with different dimensions (128, 256, 512, and 1024) and found that 512 offered the best balance between model performance and computational efficiency.
- Common Practice: This dimension is a standard choice in Transformer architectures, as established in the original 'Attention is All You Need' paper by Vaswani et al. [31]. It provides adequate capacity for most sequence modeling tasks while maintaining reasonable computational requirements.
- Memory Constraints: While larger dimensions (e.g., 1024) might theoretically capture more subtle patterns, they significantly increase memory usage and computation time without providing substantial improvements in our prediction accuracy.

The model predicted VTEC values for the 24 grid points in the region, with bilinear interpolation used to fill in the remaining areas. We present the results for 25 November 2022, at different times to demonstrate the model's performance. The formula for bilinear interpolation is as follows (See Figure 4):

$$\begin{aligned}\overline{VTEC}(\lambda,\psi,t) = (1-p)(1-q)\overline{VTEC}(\lambda_1,\psi_1,t) + (1-p)q\overline{VTEC}(\lambda_1,\psi_2,t) \\ +p(1-q)\overline{VTEC}(\lambda_2,\psi_1,t) + pq\overline{VTEC}(\lambda_2,\psi_2,t)\end{aligned} \tag{12}$$

where t is the time of observation; λ and ψ are longitude and latitude, respectively; and p and q are defined as follows:

$$p = \frac{\lambda - \lambda_i}{\lambda_j - \lambda_i}, q = \frac{\psi - \psi_i}{\psi_j - \psi_i}$$

Figure 5a shows the interpolated regional ionospheric map using the data taken from the CORG file. Figure 5b shows the model predictions. The predicted VTEC values closely match the CORG file data, with slight variations in some areas. The interpolated maps show similar patterns of VTEC distribution across the region. The result reflects the ionospheric effect at midnight local time, which usually has a lower VTEC value in the Taiwan region. The 24 h difference averages between the CORG and predicted VTEC values (CORG–Predicted) in TECU at 24 grid points are shown in Table 2.

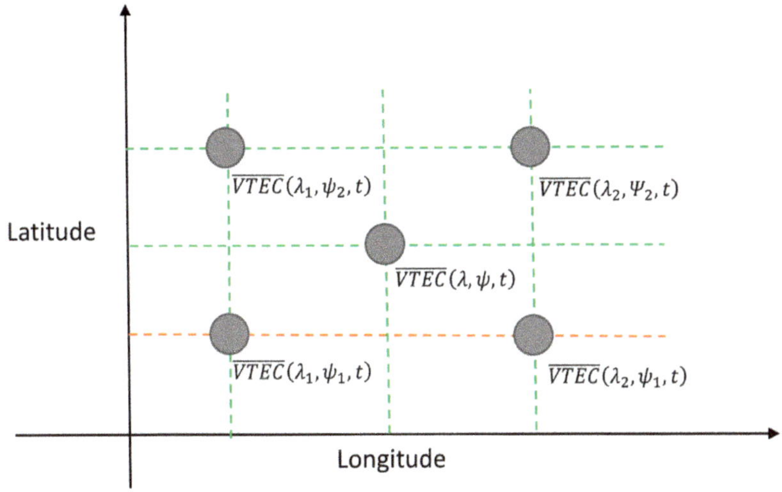

Figure 4. Bilinear interpolation using four grid points for interpolation of VTECs inside the rectangle.

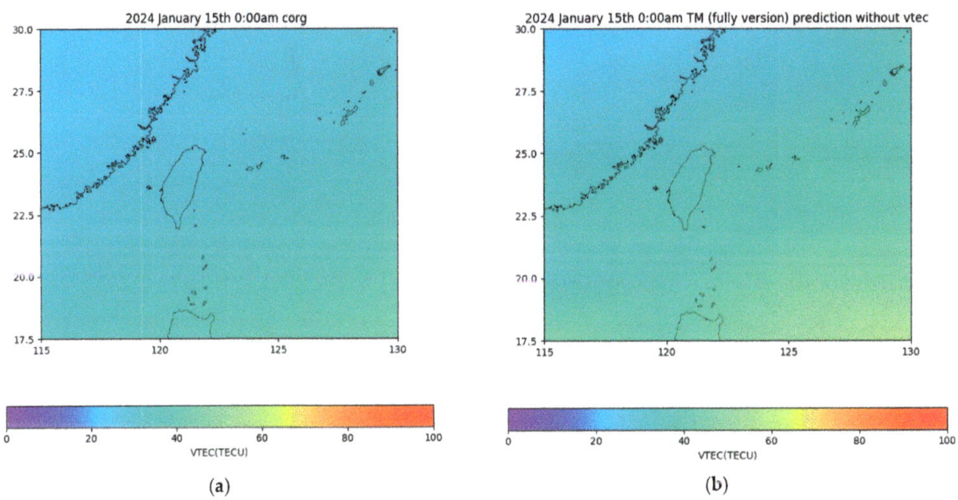

Figure 5. Local ionospheric map with lower VTEC value. (**a**) CORG data. (**b**) Prediction results.

Table 2. Differences between CORG and predicted VTEC values (CORG–Predicted) in TECU at 24 grid points.

Lat/Long	115°E	120°E	125°E	130°E
30.0°N	6.04	2.33	3.70	0.12
27.5°N	−1.37	2.35	1.96	−1.39
25.0°N	−2.07	−5.01	−3.00	−1.09
22.5°N	−6.19	−1.52	−1.28	−0.27
20.0°N	−2.74	−3.48	−1.95	−2.42
17.5°N	−6.45	0.61	−4.99	−1.45

To demonstrate our model's ability to capture the dynamic evolution of ionospheric conditions, particularly the development of the EIA over Taiwan, we present a series of comparisons between CORG data and model predictions throughout key periods on 15 January 2024. Figure 6 shows the progression from the early morning through the afternoon hours. Throughout this sequence, our model successfully reproduces both the magnitude and spatial distribution of VTEC variations, demonstrating its capability to track the complex temporal evolution of ionospheric conditions, particularly during the crucial period of EIA development and intensification. This sequence provides strong evidence that our model can effectively capture not just static conditions but also the dynamic processes driving ionospheric variability over our region of interest.

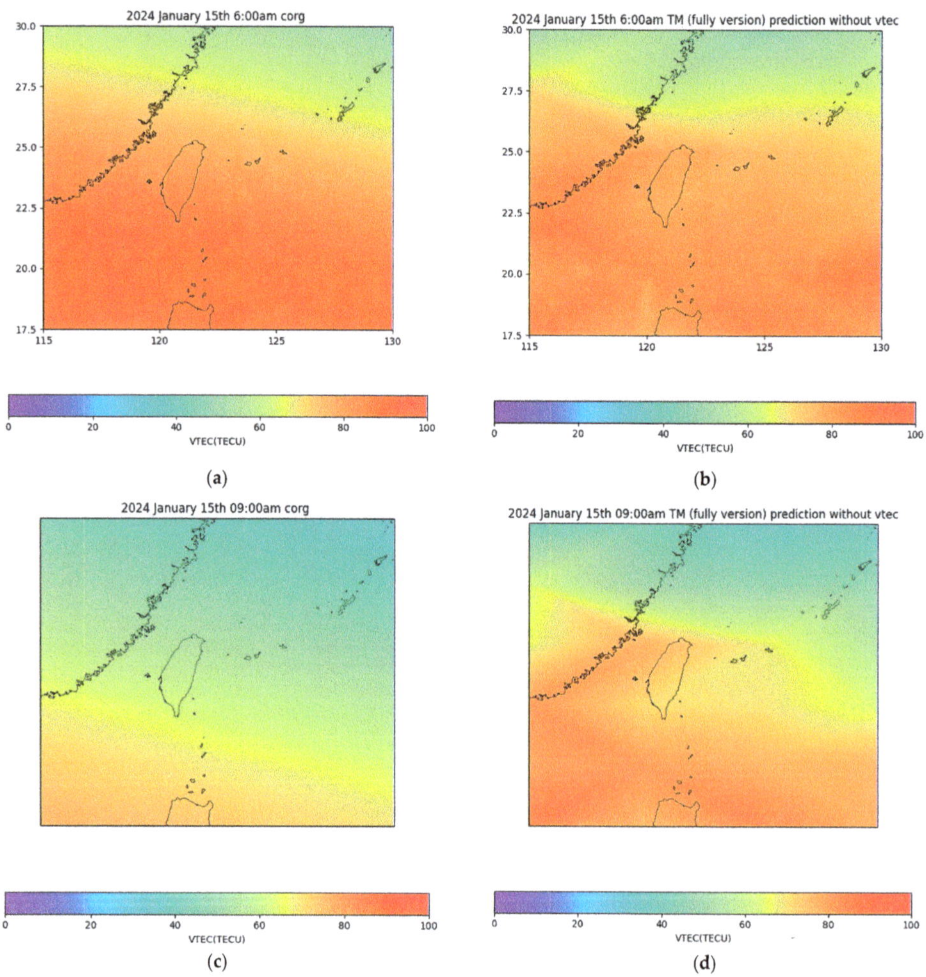

Figure 6. *Cont.*

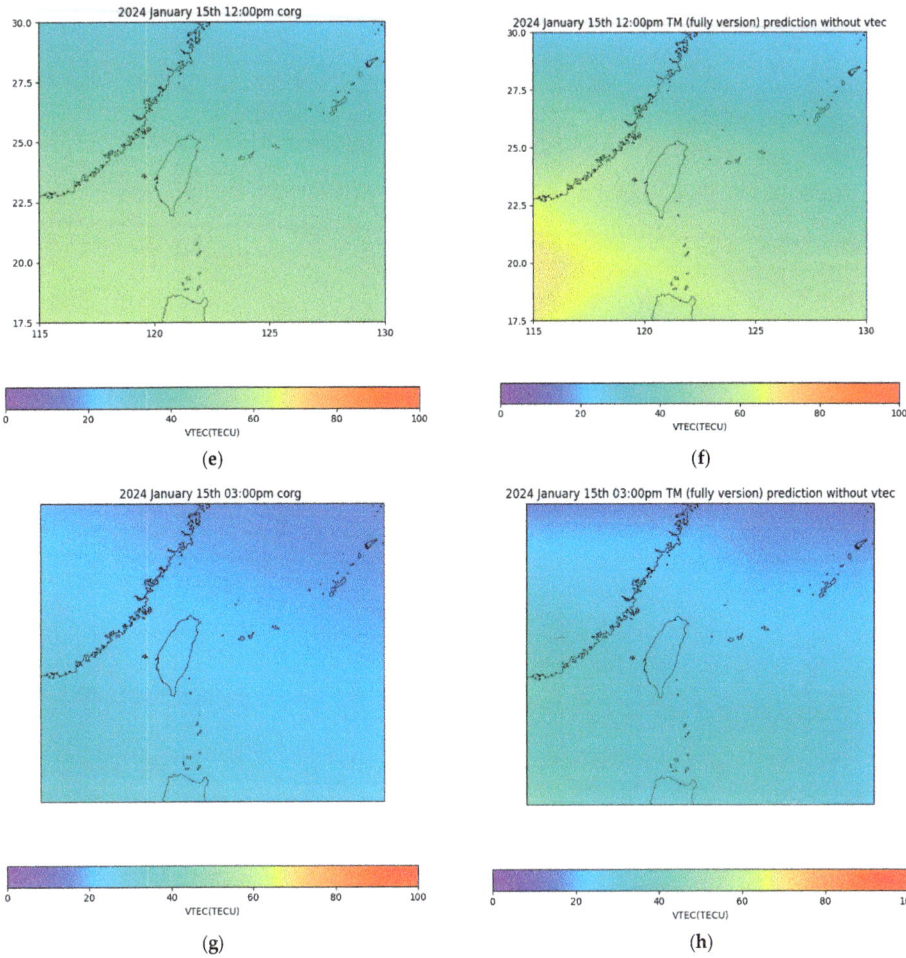

Figure 6. Diurnal variation of VTEC over the Taiwan region on 15 January 2024, showing the development and peak of the EIA. Left column: CORG VTEC maps. Right column: Transformer model predictions. From top to bottom: (**a**,**b**) 06:00 a.m. local time, before EIA development; (**c**,**d**) 09:00 a.m. local time, during EIA buildup; (**e**,**f**) 12:00 p.m. local time, approaching peak EIA intensity; (**g**,**h**) 15:00 local time, EIA transition. Color scale indicates VTEC values in TECU. The sequence demonstrates the model's capability to capture the temporal evolution of the EIA, from minimal VTEC values in early morning to the formation of the characteristic enhanced VTEC band during peak hours.

To evaluate the practical impact of the VTEC predictions, positioning experiments were conducted using data from 21 February 2023, between 04:20 and 04:30 UTC. VTEC maps for this time period are shown in Figure 7. The study compared three methods for ionospheric delay correction:

I. Dual-frequency receiver solution;
II. CORG file VTEC values;
III. Transformer model predictions.

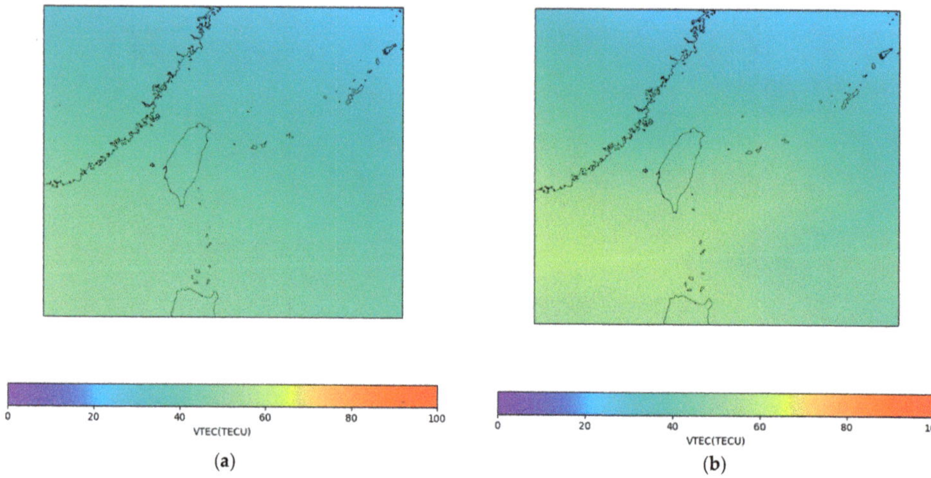

Figure 7. VTEC maps over Taiwan region during local midday (12:20–12:30 LT, 04:20–04:30 UT) on 21 February 2023, showing the northern EIA crest. (**a**) CORG VTEC map showing actual values. (**b**) Predicted VTEC map from our Transformer model. The band of enhanced VTEC values visible in both maps demonstrates the model's ability to capture the EIA's northern crest during its peak development period. Color scale indicates VTEC values in TECU.

In all cases, we only used standard single-frequency code-phase measurements for position calculation. For the dual-frequency case, the L1–L2 combination is utilized only to eliminate the ionospheric effect. For the positioning experiments, we employed the Single Point Positioning (SPP) model using single-frequency code observations. We used broadcast ephemeris data to calculate satellite positions and clock corrections. We chose SPP with broadcast ephemeris to evaluate our model's performance in a common user scenario, as this represents the most widely used positioning approach in mass-market GNSS receivers. While precise products could potentially provide better positioning accuracy, our focus was on comparing the relative performance of different ionospheric correction methods under typical user conditions. We emphasize that, to evaluate the practical impact of the VTEC predictions, our primary focus is on comparing the positioning results obtained using VTEC values from the CORG file and those predicted by our Transformer model. The dual-frequency receiver solution serves as a baseline for comparison but is not the main focus of our analysis. The positioning experiments were conducted under the following conditions:

- Test Site Details
  Location: NTOU GNSS base station (25°09′00″N, 121°46′48″E)
  Antenna Type: Trimble Zephyr Geodetic 2
  Receiver Type: Trimble NetR9
  Height: 32.5 m above sea level
- Data Collection Period
  Date: 21 February 2023
  Time: 04:20–04:30 UTC (12:20–12:30 Local Time)
  Total Duration: 10 min
  Satellite Configuration
- Number of visible satellites: 8–10 GPS satellites
  Elevation cutoff angle: 10 degrees
  PDOP range: 2.1–2.8

The positioning results shown in Figures 8 and 9 and the statistics presented in Tables 2 and 3 were computed using all available epochs during this period. The experiment used a single station to ensure consistent conditions for comparing different ionospheric correction methods. Figure 8 shows the 2D scatter plot of the GPS positioning. Figure 8b uses the Transformer model's predicted VTEC values. For comparison, Figure 8a presents the corresponding results using CORG file VTEC values. Figure 9 shows East–North–Up (ENU) error plots using the Transformer model's predicted VTEC values and the corresponding results using CORG file VTEC values. Tables 3 and 4 summarize the Root Mean Square (RMS) errors and standard deviations (STD) of the positioning errors for all methods, including the dual-frequency solution as a baseline.

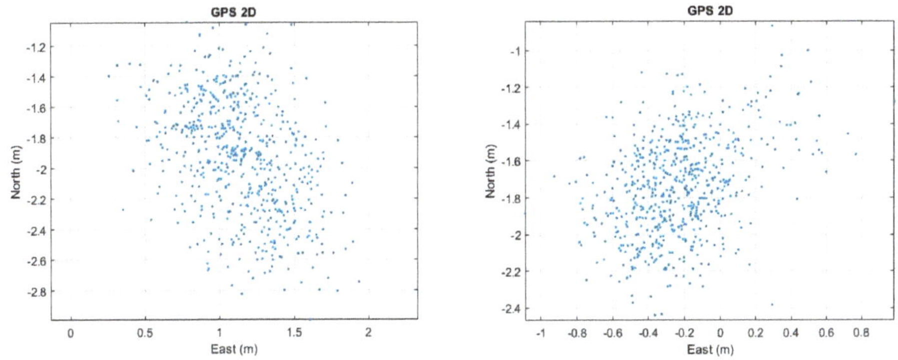

(a) (b)

Figure 8. The 2D scatter plot for GPS positioning. (**a**) Positioning results with the ionospheric effect calculated using CORG VTEC value. (**b**) Positioning results with the ionospheric effect calculated using the Transformer model's predicted VTEC value.

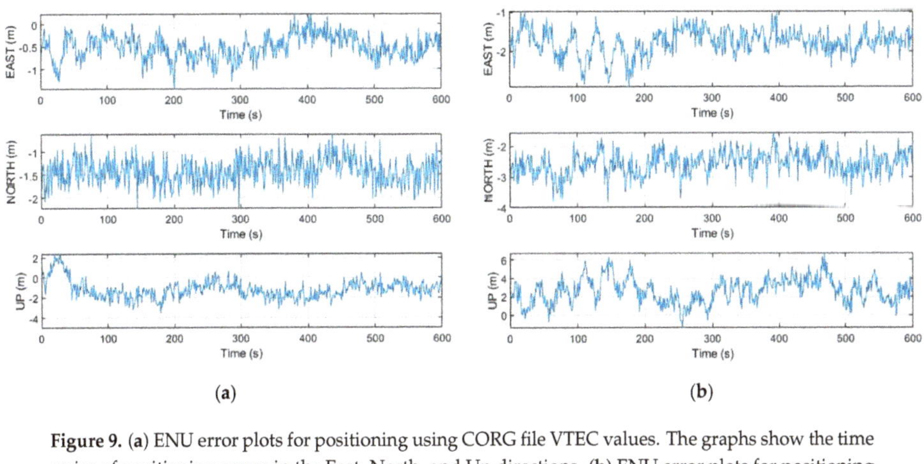

(a) (b)

Figure 9. (**a**) ENU error plots for positioning using CORG file VTEC values. The graphs show the time series of positioning errors in the East, North, and Up directions. (**b**) ENU error plots for positioning using the Transformer model's predicted VTEC values.

Table 3. Positioning results—RMS (m).

Method	E	N	U
Dual-Frequency	0.4406	−1.1318	0.1653
CORG	−0.4008	−1.4018	−1.5746
Transformer	−1.775	−2.5720	2.6240

Table 4. Positioning results—STD (m).

Method	E	N	U
Dual-Frequency	0.2703	0.2659	0.5482
CORG	0.2442	0.3060	0.8494
Transformer	0.3399	0.2971	1.3876

The positioning errors show a bias rather than fluctuating around zero. After careful analysis, we believe this systematic error can be attributed to several factors:

- Use of Broadcast Ephemeris: The broadcast ephemeris contains residual orbit and clock errors that can introduce systematic biases in the position solution. While these errors are typically at the meter level, they do not average to zero over short time spans.
- Code Multipath Effects: In our experimental setup, code measurements are susceptible to multipath effects. Without employing multipath mitigation techniques, these effects can manifest as systematic errors in the position solution.
- Hardware Biases: Code measurements contain various hardware-related biases (both receiver and satellite) that were not explicitly accounted for in our processing strategy.

The experimental results show that while the Transformer model's predictions for VTEC lead to positioning accuracy at the meter level, they do not yet achieve the same level of accuracy as the CORG data. The positioning errors using our model are approximately 1–2 m larger than those obtained using CORG data. This difference suggests that while our approach demonstrates the feasibility of using Transformer-based models for VTEC prediction, there is still room for improvement. Potential enhancements could include expanding the training dataset, incorporating additional input features, fine-tuning the model architecture, and improving the handling of extreme ionospheric conditions.

The current results should be viewed as a proof of concept, demonstrating that deep learning models can provide reasonable ionospheric corrections for GNSS positioning, while acknowledging the need for further refinement to match the performance of established methods. The Transformer model's ability to learn from historical data implies that its predictive capabilities could be extended to longer time intervals. This scalability is a significant advantage over some traditional prediction methods. If provided with a more extensive training dataset covering a longer historical period, the Transformer model could potentially extend its prediction horizon. This could lead to more accurate long-term VTEC forecasts, which would be invaluable for various GNSS applications and space weather prediction.

Next, we would like to investigate the explainability of the proposed model. Prior to conducting the main experiment, we performed a correlation analysis on the input features using Spearman's rank correlation coefficient. This analysis serves as a reference point for interpreting the explainability results of our model. Spearman's rank correlation coefficient is a measure of the strength and direction of association between two variables. It is defined as the Pearson correlation coefficient between the rank variables. The coefficient ranges from −1 to +1, where +1 indicates a perfect positive correlation, −1 indicates a perfect negative correlation, and 0 indicates no correlation. The formula for Spearman's rank correlation coefficient (SRCC) is as follows:

$$\text{SRCC} = \text{Pearson}(S(X), S(Y)) = \frac{\text{cov}(S(X), S(Y))}{\sigma(S(X))\sigma(S(Y))}$$

where $S(X)$ and $S(Y)$ are the ranked versions of the original variables X and Y, cov is the covariance, and σ is the standard deviation.

The data used for training span from 2022 to 2023, while the prediction data are from January 2024, comprising 24 data points per day. In this study, we focus exclusively on four grid points near Taiwan. These points are arranged from left to right and top to bottom, as follows (See Figure 1):

Grid Point 10: 120°E, 25°N
Grid Point 11: 125°E, 25°N
Grid Point 14: 120°E, 22.5°N
Grid Point 15: 125°E, 22.5°N

Figure 10a–d illustrates the rank correlation coefficients among these four grid points. In all cases, the last row indicates the correlation coefficients with VTEC of the present grid. These figures serve as references and comparisons for interpreting the experimental results. After reviewing the Spearman's rank correlation coefficients for the four grid points near Taiwan, it can be seen that the correlation patterns are largely consistent across all four grid points, suggesting that the relationships between variables are relatively stable in the region around Taiwan. Some specific correlations are summarized as follows:

(a) VTEC correlations:
- Strong positive correlation with F10.7 and Sunspot Number (around 0.7 to 0.8), confirming the known relationship between solar activity and ionospheric electron content.
- Moderate positive correlation with HRS (sine of hour, around 0.4 to 0.5), indicating a clear diurnal pattern in VTEC values.
- Weak to moderate negative correlation with HRC (cosine of hour, around −0.2 to −0.3), further supporting the diurnal variation.
- Very weak correlations with DNS and DNC (day number sine and cosine), suggesting that seasonal variations might be less prominent than daily variations in this dataset.

(b) Solar activity indicator:
- Very strong correlation between F10.7 and Sunspot Number (>0.9), as expected, since both are indicators of solar activity.
- These solar activity indicators show similar correlation patterns with other variables, reinforcing their interchangeability in many applications.

(c) Geomagnetic indices:
- Ap and Kp indices show very strong correlation (>0.9), which is expected, as they are closely related measures of geomagnetic activity.
- Dst index shows moderate negative correlations with Ap and Kp (around −0.5 to −0.6), aligning with the understanding that strong geomagnetic storms (indicated by negative Dst) often correspond to higher Ap and Kp values.
- Interestingly, the geomagnetic indices (Ap, Kp, Dst) show very weak correlations with VTEC in this dataset, which might be unexpected given the known influence of geomagnetic activity on the ionosphere.

(d) Time functions:

- HRS and HRC show the expected orthogonal relationship (correlation close to 0), validating their use as independent components of the diurnal cycle.
- Similarly, DNS and DNC show near-zero correlation, correctly representing independent components of the annual cycle.

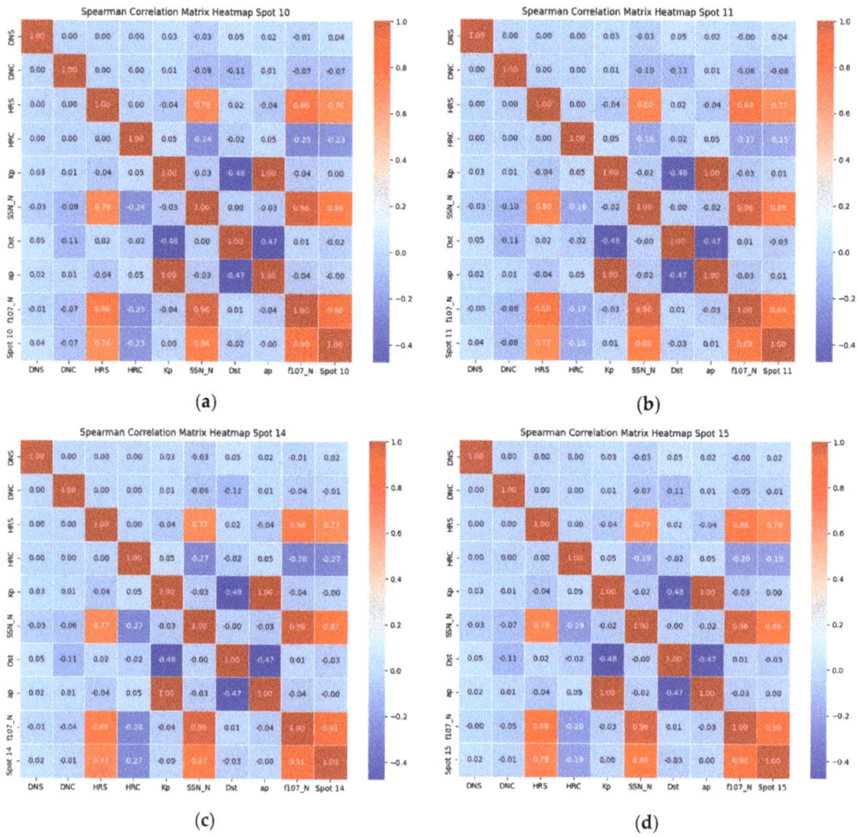

Figure 10. Spearman's rank correlation coefficient matrix for input features and VTEC at four grid points. The matrix displays the strength and direction of correlations between VTEC, solar activity indicators, geomagnetic indices, and time functions. Color intensity represents the magnitude of correlation, with red indicating positive correlations and blue indicating negative correlations. (**a**) Grid point 10. (**b**) Grid point 11. (**c**) Grid point 14. (**d**) Grid point 15.

These observations provide a valuable baseline for interpreting the Transformer model's behavior and feature importance. Any significant deviations from these correlation patterns in the model's learned representations would be particularly interesting to investigate further.

We are now ready to employ the IDG method. This approach calculates an additional gradient of the interpolation constant with respect to the sample to obtain a weight for each interpolated sample. By multiplying the interpolation contribution with this weight, we effectively reduce the impact of noise on the explanations. The IDG method integrates the entire gradient path and additionally integrates each interpolation sample step to minimize noise, resulting in cleaner explanation results. In our experiment, we used 300 interpolation samples. The interpolation samples are defined using the integral gradient to determine their contribution. The IDG method uses important factors to eliminate the saturation effect that usually occurs in the integrated gradient approach.

Figure 11 presents the results of the IDG method as applied to the VTEC prediction model for four specific grid points. These figures illustrate the standardized interpolation sample weights, providing insights into the model's decision-making process and the relative importance of different interpolation steps in VTEC prediction. The differences in weight distributions across grid points indicate that the model adapts its decision-making process based on geographical location.

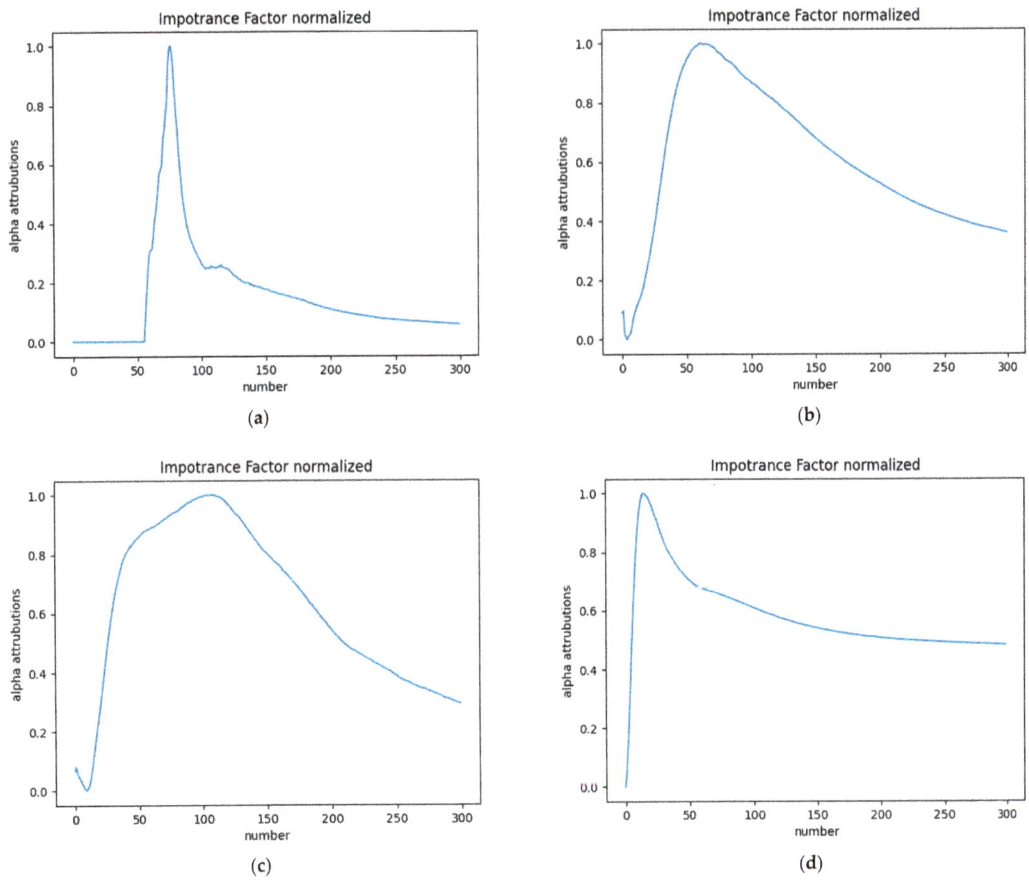

Figure 11. Standardized interpolation sample weights derived from IDG method for four grid points in the VTEC prediction model. (**a**) Grid point 10 (120°E, 25°N); (**b**) Grid point 11 (125°E, 25°N); (**c**) Grid point 14 (120°E, 22.5°N); (**d**) Grid point 15 (125°E, 22.5°N). The x-axis represents 300 interpolation steps from the baseline (step 0) to the actual input (step 300). The y-axis shows the standardized weight (0–1), indicating the relative importance of each interpolation step in the model's decision-making process. These plots illustrate the spatial variability in feature importance and the model's adaptive behavior across different geographical locations. Note the varying patterns between northern (**a**,**b**) and southern (**c**,**d**) grid points, suggesting latitude-dependent prediction strategies in the VTEC model.

All four figures show very low weights for the initial interpolation steps, suggesting that baseline values have minimal impact on the model's predictions. The generally increasing trends in all figures imply that the model considers cumulative changes in features rather than relying on specific threshold values.

The southern grid points (14 and 15) show more linear weight increases, while the northern points (10 and 11) exhibit more pronounced peaks, suggesting latitude-dependent prediction strategies. The varied patterns across grid points highlight the complexity of the model and its ability to capture location-specific ionospheric behaviors. These figures provide valuable insights into the inner workings of the VTEC prediction model, demonstrating its adaptive nature and the complex spatial dependencies in ionospheric modeling.

The next experimental results show the gradient scores (Figure 12) and cumulative gradient scores (Figure 13) for each feature at four grid points over a five-day period. These figures provide insights into the importance and impact of different features on the VTEC prediction model. Each figure contains multiple lines, representing different features: Sunspot Number, F10.7, Dst, Ap, DNS, DNC, HRS, and HRC.

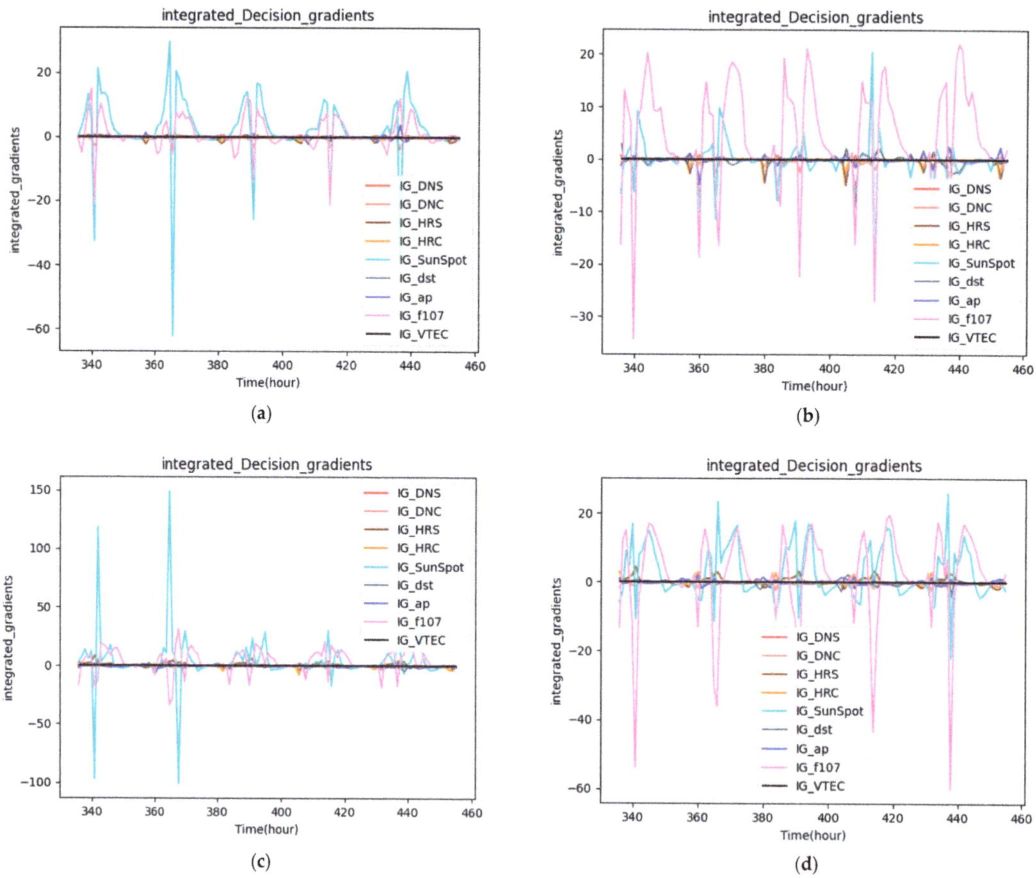

Figure 12. Gradient scores for VTEC prediction features across four grid points. (**a**) Grid point 10 (120°E, 25°N); (**b**) Grid point 11 (125°E, 25°N); (**c**) Grid point 14 (120°E, 22.5°N); (**d**) Grid point 15 (125°E, 22.5°N). Each line represents a different feature: Sunspot Number, F10.7, Dst, Ap, DNS, DNC, HRS, and HRC. The y-axis shows the gradient score, indicating the instantaneous impact of each feature on the VTEC prediction. Positive values suggest an increase in VTEC, while negative values indicate a decrease. Note the varying patterns and magnitudes across different grid points, revealing the spatial dependency of feature importance in the model.

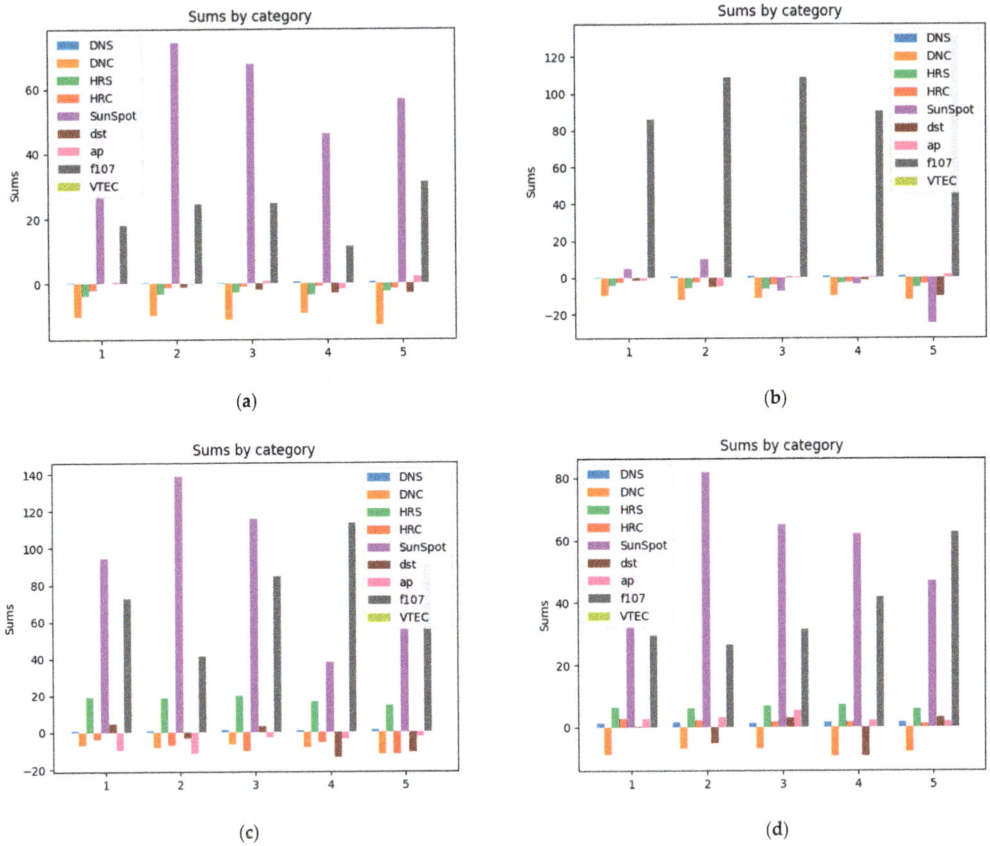

Figure 13. Cumulative gradient scores for VTEC prediction features across four grid points. (**a**) Grid point 10 (120°E, 25°N); (**b**) Grid point 11 (125°E, 25°N); (**c**) Grid point 14 (120°E, 22.5°N); (**d**) Grid point 15 (125°E, 22.5°N). Each line represents the cumulative effect of a different feature: Sunspot Number, F10.7, Dst, Ap, DNS, DNC, HRS, and HRC. The y-axis shows the cumulative gradient score, illustrating the overall impact of each feature on the VTEC prediction over time. The diverging lines highlight the relative importance and long-term influence of different features. Observe the dominance of solar activity indicators (Sunspot Number and F10.7) and the consistent negative contribution of DNC across all grid points.

In these figures, solar activity indicators show similar patterns across all grid points. They often have opposite contributions (one positive, one negative) in the gradient scores. In corresponding cumulative plots, they tend to have the largest magnitudes, indicating their significant impact on VTEC predictions.

Geomagnetic indices generally show smaller contributions compared to solar activity indicators. The Dst index often displays negative contributions, while the Ap index contributions vary. Their impact becomes more noticeable in the cumulative plots, especially during periods of geomagnetic disturbance.

Time functions DNS and DNC show relatively small but consistent contributions. HRS and HRC display more pronounced oscillations in gradient scores. DNC consistently shows a notable negative contribution across all grid points. Clear daily cycles are visible in the gradient scores, particularly for HRS and HRC. The cumulative plots show how these daily variations accumulate over time.

While overall patterns are similar, there are noticeable differences between grid points. Grid points 14 and 15 (southern points) often show more pronounced variations in gradient scores compared to grid points 10 and 11 (northern points).

5. Discussion

This study has explored the application of a Transformer-based model for VTEC prediction, coupled with advanced explainability techniques to understand the model's decision-making process. The research has yielded plausible insights into both the effectiveness of deep learning for ionospheric modeling and the interpretability of complex AI models in geophysical applications. The Transformer model demonstrated impressive capabilities in VTEC prediction, showing comparable performance to traditional methods. The model showed the ability to capture spatial variations in VTEC, adapting its predictions across different grid points. This adaptability is crucial for accurate regional ionospheric mapping. While already competitive, the results suggest that with more extensive training data and further refinement, the Transformer model could potentially outperform traditional methods, especially in capturing complex, non-linear ionospheric behaviors.

The IDG analysis offered deep insights into the model's internal decision-making process. The analysis revealed how the importance of different features varied across the interpolation steps, providing a nuanced view of feature contributions. Differences in IDG patterns across grid points demonstrated the model's ability to adapt its decision-making process based on geographical location. The IDG method effectively reduced noise in the explanations, providing clearer insights into feature contributions compared to traditional gradient-based methods.

The gradient score analysis further elucidated the model's behavior and feature utilization. Consistent with correlation analysis, solar activity indicators emerged as the most influential features, but the analysis also revealed the significant contributions of other features like DNC, which were not apparent from simple correlation studies. The analysis captured both the instantaneous (gradient scores) and cumulative (cumulative gradient scores) impacts of features, providing a comprehensive view of their influence over time. Variations in gradient scores across grid points further confirmed the model's ability to adapt to different geographical locations, potentially capturing localized ionospheric phenomena. The often opposing contributions of related features (e.g., Sunspot Number and F10.7) highlighted the model's ability to capture subtle, complex interactions between features.

An interesting observation from the gradient score analysis is the relatively small variations and contributions shown by the DNS and DNC feature lines (Figure 9). While it might be tempting to consider removing these seasonal indicators from the model, several factors support their retention. First, these features capture fundamental physical processes—specifically, seasonal variations in solar zenith angle and day length that are known to influence ionospheric behavior. Their smaller contributions in our study might be particularly related to our focus on the region around Taiwan, where seasonal effects are less prominent than diurnal variations. Second, while their direct contributions appear small, these features provide stable baseline information that helps the model maintain awareness of the longer-term seasonal context. Finally, the gradient scores only show direct contributions and may not fully capture how these features interact with other inputs to improve overall prediction accuracy. A comprehensive ablation study would be necessary to definitively determine the impact of these seasonal indicators on model performance.

6. Conclusions

In this paper, we demonstrate the potential of deep learning, specifically Transformer models, in advancing the field of ionospheric modeling and VTEC prediction. The application of explainability techniques (IDG and gradient score analysis) to a complex geophysical model may open new research, interesting for interpreting AI models in earth and space sciences. The combination of deep learning and explainable AI techniques in this geophysical application shows the potential for similar approaches in other scientific domains dealing with complex, multi-variate time series data. Future work should include comparative analysis with other deep learning architectures such as LSTM, CNN-LSTM hybrid, and BiLSTM models. Such comparisons would provide quantitative insights into the relative strengths and limitations of different approaches for VTEC prediction. Performance metrics like RMSE, MAE, and MAPE could be used to evaluate prediction accuracy, while also considering computational efficiency and model complexity. This would help establish more comprehensive guidelines for choosing appropriate deep learning architectures for ionospheric prediction tasks.

Additionally, future work might include comprehensive ablation studies to quantify the impact of XAI on model performance. Such studies would compare different model configurations: a base Transformer model without XAI guidance, a model with basic gradient-based analysis, and our full model with IDG-based XAI. This would help isolate the effects of different XAI components and their contributions to prediction accuracy. The ablation analysis could examine how XAI influences feature selection, model refinement, and the handling of edge cases. Additionally, it could investigate whether insights from XAI can guide architectural improvements and hyperparameter optimization. Such studies would contribute to our understanding of how explainability techniques can be leveraged to enhance deep learning models for ionospheric prediction.

Author Contributions: Conceptualization, H.-S.W. and D.-J.J.; methodology, H.-S.W. and Y.-H.L.; software, Y.-H.L. and H.-S.W.; validation, H.-S.W. and D.-J.J.; writing—original draft preparation, H.-S.W.; writing—review and editing, H.-S.W. and D.-J.J.; supervision, H.-S.W. and D.-J.J. All authors have read and agreed to the published version of the manuscript.

Funding: The authors gratefully acknowledge the support of the National Science and Technology Council, Taiwan, under grant number NSTC 113-2221-E-019-059.

Data Availability Statement: The original contributions presented in this study are included in the article. Further inquiries can be directed to the corresponding author.

Conflicts of Interest: The authors declare no conflicts of interest.

References

1. Kaplan, E.; Hegarty, C. *Understanding GPS/GNSS: Principles and Applications*, 3rd ed; Artech House: Boston, MA, USA, 2017.
2. Klobuchar, J.A. Ionospheric time-delay algorithm for single-frequency GPS users. *IEEE Trans. Aerosp. Electron. Syst.* **1987**, *3*, 325–331. [CrossRef]
3. Dubey, S.; Wahi, R.; Gwal, A.K. Ionospheric Effects on GPS Positioning. *Adv. Space Res.* **2006**, *38*, 2478–2484. [CrossRef]
4. Lyu, S.; Xiang, Y.; Soja, B.; Wang, N.; Yu, W.; Truong, T.-K. Uncertainties of interpolating satellite-specific slant ionospheric delays and impacts on PPP-RTK. *IEEE Trans. Aerosp. Electron. Syst.* **2023**, *60*, 490–505. [CrossRef]
5. Tang, L.; Zhang, F.; Li, P.; Deng, Y.; Chen, W. Effects of equatorial plasma bubble-induced ionospheric gradients on GNSS PPP-RTK. *GPS Solut.* **2024**, *28*, 124. [CrossRef]
6. Hagfors, T. 2.1. The Ionosphere. In *Methods in Experimental Physics*; Meeks, M.L., Ed.; Astrophysics; Academic Press: Cambridge, MA, USA, 1976; Volume 12, pp. 119–135.
7. Tulunay, Y.; Tulunay, E.; Senalp, E.T. The neural network technique—1: A general exposition. *Adv. Space Res.* **2004**, *33*, 983–987. [CrossRef]
8. Tulunay, Y.; Tulunay, E.; Senalp, E.T. The neural network technique—2: An ionospheric example illustrating its application. *Adv. Space Res.* **2004**, *33*, 988–992. [CrossRef]

9. Tulasi Ram, S.; Sai Gowtam, V.; Mitra, A.; Reinisch, B. The Improved Two-Dimensional Artificial Neural Network-Based Ionospheric Model (ANNIM). *J. Geophys. Res. Space Phys.* **2018**, *123*, 5807–5820. [CrossRef]
10. Chen, Z.; Liao, W.; Li, H.; Wang, J.; Deng, X.; Hong, S. Prediction of Global Ionospheric TEC Based on Deep Learning. *Space Weather* **2022**, *20*, e2021SW002854. [CrossRef]
11. Luo, H.; Gong, Y.; Chen, S.; Yu, C.; Yang, G.; Yu, F.; Hu, Z.; Tian, X. Prediction of Global Ionospheric Total Electron Content (TEC) Based on SAM-ConvLSTM Model. *Space Weather* **2023**, *21*, e2023SW003707. [CrossRef]
12. Sun, W.; Xu, L.; Huang, X.; Zhang, W.; Yuan, T.; Yan, Y. Bidirectional LSTM for ionospheric vertical Total Electron Content (TEC) forecasting. In Proceedings of the 2017 IEEE Visual Communications and Image Processing (VCIP), St. Petersburg, FL, USA, 10–13 December 2017; pp. 1–4.
13. Wang, H.; Liu, H.; Yuan, J.; Le, H.; Shan, W.; Li, L. MAOOA-Residual-Attention-BiConvLSTM: An Automated Deep Learning Framework for Global TEC Map Prediction. *Space Weather* **2024**, *22*, e2024SW003954. [CrossRef]
14. Ren, X.; Zhao, B.; Ren, Z.; Xiong, B. Ionospheric TEC Prediction in China during Storm Periods Based on Deep Learning: Mixed CNN-BiLSTM Method. *Remote Sens.* **2024**, *16*, 3160. [CrossRef]
15. Ali, S.; Abuhmed, T.; El-Sappagh, S.; Muhammad, K.; Alonso-Moral, J.M.; Confalonieri, R.; Guidotti, R.; Del Ser, J.; Díaz-Rodríguez, N.; Herrera, F. Explainable Artificial Intelligence (XAI): What we know and what is left to attain Trustworthy Artificial Intelligence. *Inf. Fusion* **2023**, *99*, 101805. [CrossRef]
16. Markus, A.F.; Kors, J.A.; Rijnbeek, P.R. The role of explainability in creating trustworthy artificial intelligence for health care: A comprehensive survey of the terminology, design choices, and evaluation strategies. *J. Biomed. Inform.* **2021**, *113*, 103655. [CrossRef] [PubMed]
17. McKinnell, L.A.; Poole, A.W.V. Predicting the ionospheric *F* layer using neural networks. *J. Geophys. Res.* **2004**, *109*, A08308. [CrossRef]
18. Sivavaraprasad, G.; Deepika, V.S.; SreenivasaRao, D.; Ravi Kumar, M.; Sridhar, M. Performance evaluation of neural network TEC forecasting models over equatorial low-latitude Indian GNSS station. *Geod. Geodyn.* **2020**, *11*, 192–201. [CrossRef]
19. Sundararajan, M.; Taly, A.; Yan, Q. Axiomatic Attribution for Deep Networks. In Proceedings of the International Conference on Machine Learning, Sydney, Australia, 6–11 August 2017.
20. Smilkov, D.; Thorat, N.; Kim, B.; Viégas, F.; Wattenberg, M. Smoothgrad: Removing noise by adding noise. *arXiv* **2024**, arXiv:1706.03825.
21. Walker, C.; Jha, S.K.; Chen, K.; Ewetz, R. Integrated Decision Gradients: Compute Your Attributions Where the Model Makes Its Decision. In Proceedings of the AAAI Conference on Artificial Intelligence, Vancouver, BC, Canada, 20–27 February 2024.
22. Arrieta, A.B.; Díaz-Rodríguez, N.; Del Ser, J.; Bennetot, A.; Tabik, S.; Barbado, A.; Garcia, S.; Gil-Lopez, S.; Molina, D.; Benjamins, R.; et al. Explainable Artificial Intelligence (XAI): Concepts, taxonomies, opportunities and challenges toward responsible AI. *Inf. Fusion* **2020**, *58*, 82–115. [CrossRef]
23. Samek, W.; Montavon, G.; Lapuschkin, S.; Anders, C.J.; Müller, K.R. Explaining Deep Neural Networks and Beyond: A Review of Methods and Applications. *Proc. IEEE* **2021**, *109*, 247–278. [CrossRef]
24. Kelley, M.C. *The Earth's Ionosphere: Plasma Physics and Electrodynamics*, 2nd ed.; Academic Press: San Diego, CA, USA, 2009.
25. Mannucci, A.J.; Wilson, B.D.; Yuan, D.N.; Ho, C.H.; Lindqwister, U.J.; Runge, T.F. A global mapping technique for GPS-derived ionospheric total electron content measurements. *Radio Sci.* **1998**, *33*, 565–582. [CrossRef]
26. Tsurutani, B.T.; Gonzalez, W.D.; Tang, F.; Akasofu, S.I.; Smith, E.J. Origin of interplanetary southward magnetic fields responsible for major magnetic storms near solar maximum (1978–1979). *J. Geophy. Res.* **1988**, *93*, 8519–8531. [CrossRef]
27. Basu, S.; Basu, S.; Groves, K.M.; Yeh, H.-C.; Su, S.-Y.; Rich, F.J.; Sultan, P.J.; Keskinen, M.J. Response of the equatorial ionosphere in the South Atlantic Region to the Great Magnetic Storm of July 15, 2000. *Geophys. Res. Lett.* **2001**, *28*, 3577–3580. [CrossRef]
28. Willard, H.R.; Kenney, J.K. Ionospheric effects of high-altitude nuclear tests. *J. Geophys. Res.* **1963**, *68*, 2053–2056. [CrossRef]
29. Mendillo, M.; Hawkins, G.S.; Klobuchar, J.A. A sudden vanishing of the ionospheric F region due to the launch of Skylab. *J. Geophys. Res.* **1975**, *80*, 2217–2228. [CrossRef]
30. Nava, B.; Coïsson, P.; Radicella, S.M. A new version of the NeQuick ionosphere electron density model. *J. Atmos. Sol. Terr. Phys.* **2005**, *70*, 1856–1862. [CrossRef]
31. Yuan, Y.; Wang, N.; Li, Z.; Huo, X. The BeiDou global broadcast ionospheric delay correction model (BDGIM) and its preliminary performance evaluation. *Navigation* **2019**, *66*, 55–69. [CrossRef]
32. Li, Z.; Yuan, Y.; Wang, N.; Hernandez-Pajares, M.; Huo, X. SHPTS: Towards a new method for generating precise global ionospheric TEC map based on spherical harmonic and generalized trigonometric series functions. *J. Geod.* **2015**, *89*, 331–345. [CrossRef]
33. Chen, J.; Zhang, Y.; Wang, J.; Yang, S.; Dong, D.; Zhou, J.; Wang, M.; Wang, J. A simplified and unified model of multi-GNSS precise point positioning. *Adv. Space Res.* **2015**, *55*, 125–134. [CrossRef]
34. Bilitza, D. IRI the International Standard for the Ionosphere. *Adv. Radio Sci.* **2018**, *16*, 1–11. [CrossRef]

35. Hernández-Pajares, M.; Juan, J.M.; Sanz, J. GPS ionospheric tomography: A review of real time implementation and challenges. *GPS Solut.* **2009**, *13*, 103–112.
36. Wielgosz, P.; Milanowska, B.; Krypiak-Gregorczyk, A.; Jarmołowski, W. Validation of GNSS-derived global ionosphere maps for different solar activity levels: Case studies for years 2014 and 2018. *GPS Solut.* **2021**, *25*, 103. [CrossRef]
37. Tang, L.; Zhang, F.; Chen, W. The Error of Global Ionospheric Map-TEC During Equatorial Plasma Bubble Event in the High Solar Activity Year. *Space Weather* **2024**, *22*, e2023SW003714. [CrossRef]
38. Huang, Y.N.; Cheng, K. Solar cycle variations of the equatorial ionospheric anomaly in total electron content in the Asian region. *J. Geophys. Res. Space Phys.* **1996**, *101*, 24513–24520. [CrossRef]
39. Anderson, D.N. A theoretical study of the ionospheric F region equatorial anomaly—I. Theory. *Planet. Space Sci.* **1973**, *21*, 409–419. [CrossRef]
40. Lin, C.H.; Liu, J.Y.; Tsai, H.F.; Cheng, C.Z. Variations in the equatorial ionization anomaly peaks in the Western Pacific region during the geomagnetic storms of April 6 and July 15, 2000. *Earth Planets Space* **2007**, *59*, 401–405. [CrossRef]
41. Mo, X.; Zhang, D. A comparative study of the northern and southern equatorial ionization anomaly crests in the East-Asian sector during 2006–2015. *Adv. Space Res.* **2021**, *68*, 1461–1472. [CrossRef]
42. Vaswani, A.; Shazeer, N.; Parmar, N.; Uszkoreit, J.; Jones, L.; Gomez, A.N.; Kaiser, L.; Polosukhin, I. Attention is all you need. In Proceedings of the 31st International Conference on Neural Information Processing Systems (NIPS'17), Long Beach, CA, USA, 4–9 December 2017.

remote sensing

MDPI

Article

Real-Time Detection and Correction of Abnormal Errors in GNSS Observations on Smartphones

Hongbo Mu [1], Xianwen Yu [1,*], Angela Aragon-Angel [2], Jiafu Wang [1] and Yanze Wu [1]

[1] School of Transportation, Southeast University, Nanjing 211189, China; 220233280@seu.edu.cn (H.M.); seuwjf0507@163.com (J.W.); 220213554.seu@vip.163.com (Y.W.)

[2] Research Group of Astronomy and Geomatics (gAGE), Universitat Politecnica de Catalunya (UPC), 08034 Barcelona, Spain; angela.aragon@upc.edu

* Correspondence: yuxianwen@seu.edu.cn

Abstract: Smartphones, due to the integration of low-cost GNSS chips and linearly polarized antennas, frequently experience abnormal errors in their observations, particularly during positioning on water surfaces. In response to this issue, this paper proposes a method for detecting and correcting abnormal errors in GNSS observations on smartphones. Firstly, the state and observation equations of the Kalman filter are formulated based on the continuous and smooth characteristics of pseudorange and carrier observations. Secondly, real-time detection of abnormal error occurrence in observations is performed by assessing whether the difference between the predicted and observed values computed by the Kalman filter exceeds a specified threshold. Finally, depending on abnormal errors within the epoch, different strategies are applied for real-time reparation of observations containing anomalies. Two smartphones have been used for static tests on land and kinematic tests on water. Results show that under various environmental conditions, the proposed method effectively enhances the quality of observations on smartphones. Specifically, the method achieved a maximum improvement of 86.03% in pseudorange quality and 84.31% in carrier phase quality. The method proposed in this paper outperformed the State-Based method by approximately 10% on land and by 10–35% on water. It also shows high stability and reliability, particularly in complex environments such as navigation on water.

Keywords: smartphone; GNSS observation quality; abnormal error correction; real-time kinematic

1. Introduction

In recent years, smartphones have become ubiquitous in people's lives. As navigation and positioning technologies in smartphones continue to advance, their potential for achieving higher accuracy in navigation and positioning has become increasingly promising [1–8]. Leveraging advantages such as low cost, high precision, and portability, smartphones have wide-ranging applications in areas such as smart driving, maritime positioning, and emergency rescue [9]. However, the use of low-cost GNSS chips and linearly polarized antennas in smartphones leads to poor quality observations, making them prone to frequent abnormal errors [10]. These abnormal errors significantly affect the quality of smartphone GNSS observations, making it difficult to meet the demands of high-precision positioning [11,12].

Currently, to mitigate the impact of abnormal errors and enhance the quality of smartphone observations, some scholars have conducted a series of studies focusing on three aspects: hardware devices, environmental conditions, and positioning algorithms. Regarding smartphone hardware specifically, prior research has demonstrated that the GNSS observation data from smartphones has serious quality issues and is significantly different from that of geodetic receivers, mainly in signal strength, measurement noise, and resistance to multipath effects [13,14].

Pesyna et al. [15] analyzed smartphone GNSS observations, emphasizing that the performance gap between smartphones and geodetic receivers is primarily due to the poorer multipath suppression capability of smartphone antennas compared to measurement-grade antennas. Geng et al. [16] analyzed the impact of smartphone antennas on observations. The results showed that the signal-to-noise ratio of smartphones is about 10 Db-Hz lower than that of geodetic receivers equipped with measurement-grade GNSS antennas. They believed that the linearly polarized smartphone antenna caused its observation data to be seriously affected by multipath effects. Kirkko-Jaakkola et al. [17] analyzed the observation quality of Nokia Lumia 1520 smartphones and U-Blox low-cost receivers. The results indicated that smartphone observations exhibited significant noise and numerous outliers, resulting in meter-level positioning being only achievable through post-processing. For smartphones, hardware such as GNSS chips and antennas are the main factors limiting the quality of their observations [18]. Thus, high-quality hardware could significantly improve the quality of observations, but this option is totally out of our reach and unsuitable for application and promotion.

In terms of usage environments, some studies have highlighted the close relationship between smartphone positioning accuracy and specific environmental conditions. Scholars have analyzed the positioning accuracy of smartphones when located in open spaces against areas with complex environments such as cities, forests, and severe occlusion [19–22]. Findings indicate that smartphone positioning accuracy is profoundly influenced by environmental factors. In urban and forested areas, the positioning accuracy of a smartphone may become dozens of times worse than in the open sky. Shinghal et al. [8] collected single-point positioning (SPP) data of pedestrians running in an area with many tall buildings in downtown Toronto, Canada. The positioning results exhibited deviations ranging from several meters to several tens of meters, with irregular large jumps observed. Weng et al. [9] conducted experiments in urban areas of Hong Kong and analyzed multipath effects on smartphones. The analysis indicated that in urban environments, the quality of GNSS pseudorange measurements on smartphones significantly deteriorates due to the presence of multipath effects and non-line-of-sight (NLOS) effects, leading to a large number of outliers. In the actual environments where these smartphones are frequently employed, positioning often suffers due to complex surroundings.

Unlike urban environments populated with dense buildings, there are fewer obstacles obstructing satellite signals on the ocean, lakes, and waterways. Nevertheless, multipath effects caused by water surface reflections can severely impact the quality of observations gathered by smartphones [23,24]. Multipath effects [25,26] are closely related to factors such as the surrounding environment of the receiver, more precisely to the reflectivity of surrounding surfaces, and satellite elevation angles. Combined effects of various reflected, scattered, and diffracted signals around the antenna cannot be eliminated through modeling and differencing methods. Specht et al. [27] analyzed the accuracy of kinematic positioning using six Samsung Galaxy smartphones during ship navigation and compared the results with land outcomes. The findings indicated that the positioning accuracy of smartphones on water was lower compared to that on land. In 2022, Bai et al. [28] conducted an analysis of GNSS observation data from smartphones at sea, focusing on four aspects: visible satellite count, signal-to-noise ratio, pseudorange noise, and multipath effects. The results indicated that smartphones exhibited significant carrier phase jitter and that poor stability of most positioning errors is within 10 m.

A significant amount of research studies tends to focus on algorithmic improvements not specifically targeting smartphones but rather, GNSS receivers in general [29,30]. Anomaly detection and repair methods in GNSS navigation can be broadly categorized into three main types: algorithms based on linear combinations of observations, TurboEdit-like algorithms, and time-correlation-based algorithms. The linear combination method is widely used due to its simplicity and convenience, such as the GF (Geometry-Free), HMW (Hatch–Melbourne–Wübbena) combination [31], and its related extensions and improvements. However, the GF method requires high precision in pseudorange, which may be difficult for smartphones to achieve. The HMW combination [32], as a dual-frequency pseudorange carrier phase combination, effectively removes most interference by differencing wide-lane carrier phase observations and narrow-lane pseudorange observations.

This algorithm introduces pseudorange noise and pseudorange multipath into cycle slip detection values, making it challenging to detect small cycle slips.

Some researchers have proposed an anomaly detection and correction method based on TurboEdit [33]. This approach requires observations from at least two or more frequencies to be viable. Due to the limited GNSS frequencies received by smartphones, such methods that rely solely on multi-frequency data combinations cannot be directly utilized for error detection and correction in smartphone observations at present. Li et al. [34] proposed a Time-Correlated Pseudorange Correction (TCPC) method. This approach utilizes trend analysis and breakpoint detection methods to handle data anomalies and employs an adaptive smoothing technique to estimate pseudorange errors. In high-kinematic or urban environments, this method may falter due to significant measurement anomalies. In summary, there are several limitations to anomaly detection algorithms for smartphones. Combination algorithms have high requirements for GNSS observation data quality and some methods are only suitable for post-processing, whereas smartphones are often used in real-time kinematic scenarios. TurboEdit-like algorithms require observation values from at least two frequencies, but the availability and reception of multi-frequency data for smartphones are often inadequate. Time-correlated algorithms may struggle to achieve precise repairs in high-kinematic or complex environments.

In response to the frequent occurrence of abnormal errors in smartphone observations and the limitations of currently proposed anomaly detection methods when applied to smartphones, this paper proposes a real-time detection and correction method in smartphone observations, thereby aiming at providing higher-quality GNSS observation data for high-precision smartphone positioning.

2. Methods

2.1. *Analysis of Abnormal Errors in Smartphone Pseudorange/Carrier Measurements*

Due to the significantly higher reliability of geodetic receivers compared to smartphones, their observations will be considered the true values for this analysis. This section will analyze the phenomenon of abnormal errors in GNSS observations gathered by smartphones in different environments. The data were concurrently obtained by two smartphones and a geodetic receiver, using the single-differenced method between epochs. The GNSS Receiver Chipsets and Supported Signal Frequencies for the three receivers are detailed in Table 1.

Table 1. GNSS Receiver Chipsets and Supported Signal Frequencies for Zhonghaida iRTK2, Xiaomi Mi 8, and Huawei Mate 40 Pro. The Zhonghaida iRTK2, with RTK version RS323.v2.1.2, is manufactured in Guangzhou, China. The Xiaomi Mi 8, version dipper, is produced in Beijing, China. The Huawei Mate 40 Pro, version NOH-AN00, is manufactured in Shenzhen, China.

Receiver Models	GNSS Chipset Model	Tracked Satellite Systems	Supported Signal Frequencies
Zhonghaida iRTK2	NovAtel OEM719 (2016)	GPS/GLONASS/Galileo/ BDS/QZSS/WAAS/MSAS/EAGAN	GPS (L1 C/A, L2C, L5) GLONASS (G1, G2) Galileo (E1, E5a, E5b) BDS (B1I, B1C, B2a)
Xiaomi Mi 8	Broadcom BCM47755 (2018)	GPS/GLONASS/Galileo/BDS/QZSS	GPS (L1 C/A, L5) GLONASS (G1) Galileo (E1, E5a) BDS (B1I)
Huawei Mate 40 Pro	Kirin 9000 integrated GNSS chipset (2020)	GPS/GLONASS/Galileo/BDS/QZSS/NavIC	GPS (L1 C/A, L5) GLONASS (G1) Galileo (E1, E5a) BDS (B1I, B1C, B2a, B2b)

2.1.1. Analysis of Abnormal Errors in Land Pseudorange/Carrier Measurements

To minimize factors like multipath effects, an open university playground field was chosen for the experiment. The three receivers were placed within 5 cm of each other on a single platform. A one-hour static observation with a 1-s sampling interval was conducted. See Figure 1 for details on the observation environment and receivers' placement.

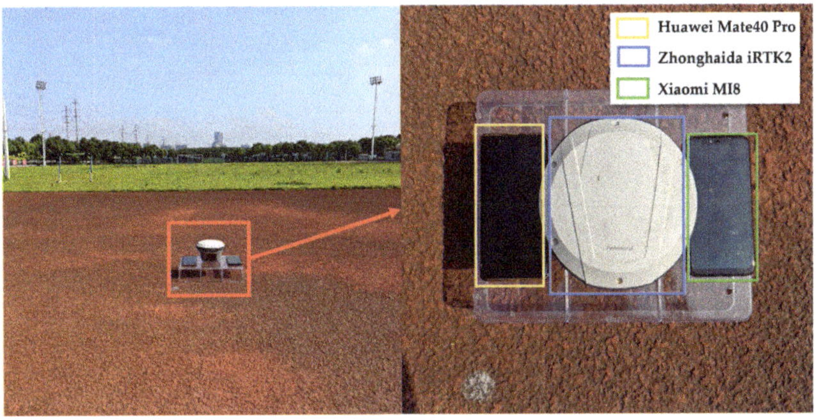

Figure 1. The layout of two smartphones and a geodetic receiver (**right**) at an open observation point on the university playground field (**left**).

This section focuses solely on the analysis of acquired observations from the first frequency of each GNSS system listed in the second column of Table 1 [35]. Following the acquisition of GNSS observations, single-differenced observations between epochs were analyzed in a static environment for the pseudorange and carrier phase observations from both smartphones and geodetic receivers. Figures 2–5 depict the single-differenced pseudorange and carrier phase observations from the geodetic receiver and the two smartphones.

A quantitative analysis was conducted to show the characteristics of abnormal errors in smartphone observations under various environmental conditions and frequencies. A third-order derivative method was used to calculate the noise values of pseudorange and carrier phase, followed by the determination of the standard deviation (STD). An observation epoch was considered to contain abnormal errors if it exceeded twice the standard deviation [36]. Finally, a statistical analysis of the occurrence of abnormal errors across observation epochs was performed. Detailed results are presented in Tables 2 and 3.

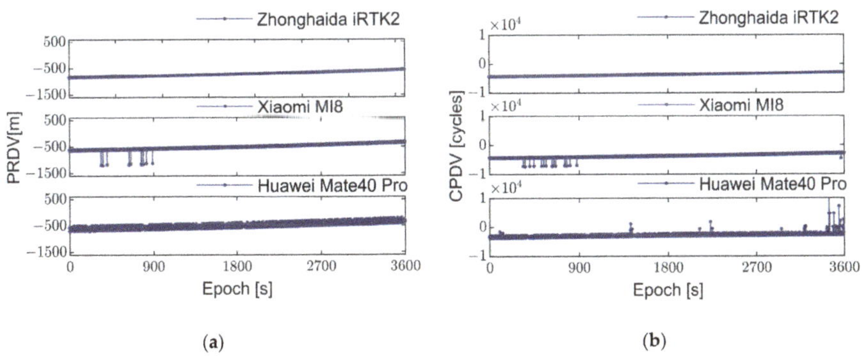

Figure 2. Observations of GPS satellite G03. The horizontal axis represents epochs, while PRDV (pseudorange single-differenced values between epochs) and CPDV (carrier phase single-differenced values between epochs) are depicted on the vertical axis. (**a**) Illustrates the single-differenced pseudorange between epochs for three receivers listed from top to bottom: Zhonghaida iRTK2, Xiaomi MI8 smartphone, and Huawei Mate40 Pro smartphone; (**b**) Displays the single-differenced carrier phase between epochs for the same receivers.

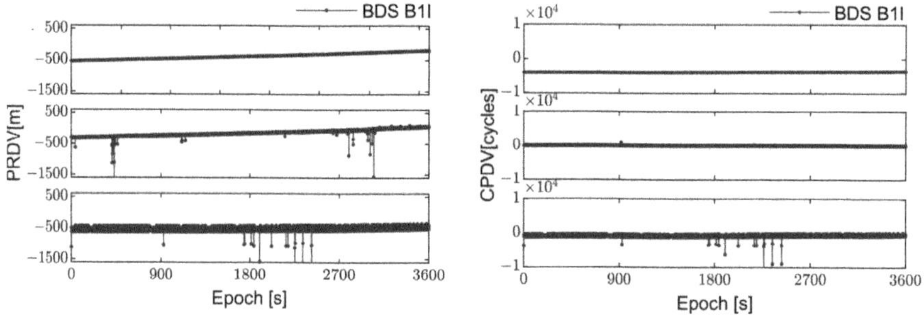

(a) (b)

Figure 3. Observations of BDS satellite C17. (**a**) Illustrates the single-differenced pseudorange between epochs for three receivers listed from top to bottom: Zhonghaida iRTK2, Xiaomi MI8 smartphone, and Huawei Mate40 Pro smartphone; (**b**) Displays the single-differenced carrier phase between epochs for the same receivers.

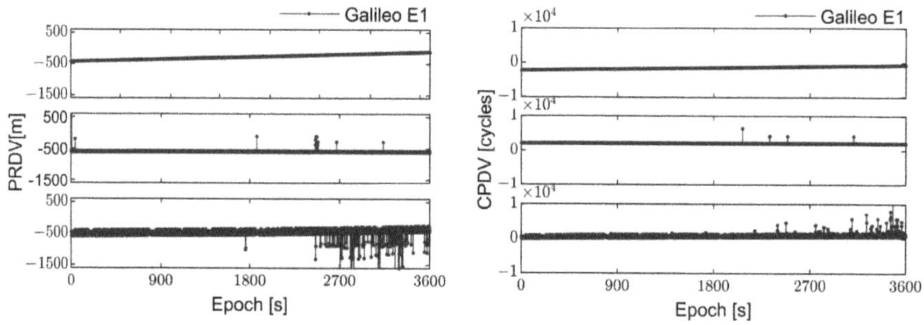

(a) (b)

Figure 4. Observations of Galileo satellite E05. (**a**) Illustrates the single-differenced pseudorange between epochs for three receivers listed from top to bottom: Zhonghaida iRTK2, Xiaomi MI8 smartphone, and Huawei Mate40 Pro smartphone; (**b**) Displays the single-differenced carrier phase between epochs for the same receivers.

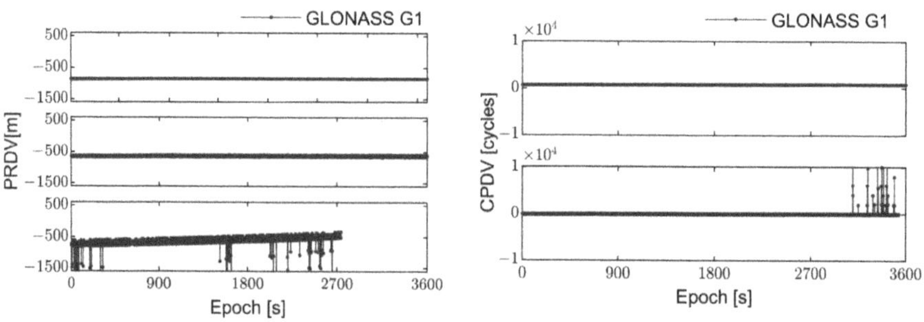

(a) (b)

Figure 5. Observations of GLONASS satellite R22. (**a**) Illustrates the single-differenced pseudorange between epochs for three receivers listed from top to bottom: Zhonghaida iRTK2, Xiaomi MI8 smartphone, and Huawei Mate40 Pro smartphone; (**b**) Displays the single-differenced carrier phase between epochs for the same receivers. The Huawei Mate40 Pro is unable to track GLONASS carrier phase observations.

Table 2. Presence of pseudorange and carrier abnormal errors in different GNSS during static land tests with the Xiaomi Mi8 and Huawei Mate40 Pro.

		Pseudorange Standard Deviation			Carrier Standard Deviation		
		Total Observation Epochs	Epochs with Abnormal Errors	Proportion	Total Observation Epochs	Epochs with Abnormal Errors	Proportion
Mi8	GPS L1	4048	31	0.76%	4032	57	1.41%
	BDS B1	4077	164	4.02%	3983	2	0.05%
	Galileo E1	4075	55	1.34%	3997	11	0.28%
	GLONASS G1	4080	0	0.00%	3865	82	2.12%
Mate40 Pro	GPS L1	3596	0	0.00%	3596	118	3.28%
	BDS B1	4027	45	1.12%	4027	27	0.67%
	Galileo E1	3838	297	7.74%	3796	203	5.34%
	GLONASS G1	2740	189	6.90%	/	/	/

Table 3. Presence of pseudorange and carrier abnormal errors at different GNSS frequencies during kinematic water tests with the Xiaomi Mi8 and Huawei Mate40 Pro.

		Pseudorange Standard Deviation			Carrier Standard Deviation		
		Total Observation Epochs	Epochs with Abnormal Errors	Proportion	Total Observation Epochs	Epochs with Abnormal Errors	Proportion
Mi8	GPS L1	2023	877	43.35%	1561	612	39.21%
	GPS L5	2052	22	1.02%	1360	494	36.32%
	BDS B1I	3013	221	7.33%	1657	328	19.79%
Mate40 Pro	GPS L1	2629	1021	38.84%	1463	710	48.53%
	GPS L5	2603	892	34.27%	1610	432	26.83%
	BDS B1I	2811	412	14.66%	1645	128	7.78%
	BDS B1C	2826	388	13.73%	1071	31	2.89%

As depicted in Figures 2–5, the pseudorange and carrier phase observation data from various GNSS systems, recorded by the geodetic receiver during a one-hour static observation experiment, display continuous and smooth trends. These observations show no significant outliers or systematic errors. Consequently, it can be inferred that under favorable observation conditions, the observation values from different GNSS systems should theoretically exhibit smoothness and are less likely to manifest in systematic errors or outliers.

However, the pseudorange and carrier phase observation values obtained from the two smartphones exhibit significant fluctuations and present frequent outliers. These outliers are prevalent across different systems and models of smartphones, indicating that the cause of such outliers is not attributed to satellite signals or propagation paths but rather to issues at the receiver end. It can also be seen from Table 2 that these outliers do not follow a consistent pattern and appear in both pseudorange and carrier phase observations. Therefore, they cannot be simply categorized as cycle slips [37].

Through comparative analysis of the pseudorange and carrier phase observation values from the Huawei Mate 40 Pro and Xiaomi Mi 8 smartphones, it was observed that the frequency of outliers in the Huawei Mate 40 Pro is significantly higher than that in the Xiaomi Mi 8. This suggests a notable difference in performance between the two devices. Particularly, as illustrated in Figure 5a, smartphones in the positioning process not only encounter issues with observation data quality but may also experience satellite signal loss. These outliers may stem from hardware limitations such as the linearly polarized antenna, low-precision clock, and low-cost GNSS chip adopted in smartphones. Given the considerable gap between smartphones and geodetic receivers in terms of satellite tracking count, observation frequency, and data quality, it is imperative to utilize existing observation values fully. This can be achieved by eliminating the outliers to enhance the overall quality of observation values, thereby further improving the positioning performance and accuracy of smartphones.

2.1.2. Analysis of Abnormal Errors in Water Pseudorange/Carrier Measurements

As shown in Section 2.1.1, smartphone pseudorange and carrier phase observations exhibit numerous abnormal errors in land environments. Navigation in water is also an important application of GNSS. Thus, to investigate the characteristics of abnormal errors in smartphone GNSS observations during water navigation, this study analyzes epoch-to-epoch differential values of GPS L1, L5, and BDS B1I B1C frequency pseudoranges and carrier phases by placing two smartphones on a platform mounted on a boat for kinematic testing. The experiment took place along the Yanshao Line route in Yancheng City, Jiangsu Province, China. Data were gathered during one hour with a sampling interval of 1 s. Figures 6 and 7 depict the pseudorange and carrier phase differential values of the two smartphones in a water environment.

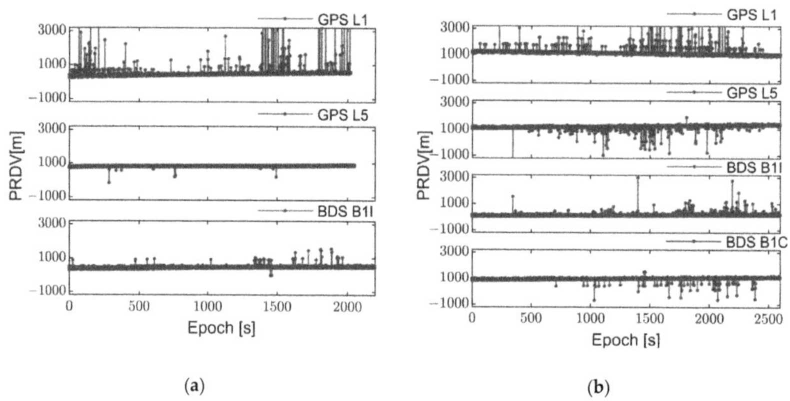

Figure 6. Observations of GPS satellite G05 and BDS satellite B10. Comparison of pseudorange differential values on water between two smartphones with different systems: (**a**) Xiaomi MI8 smartphone and (**b**) Huawei Mate40 Pro smartphone. From top to bottom, the figures display observations of GPS L1 and L5 frequencies and BDS B1I and/or BDS B1C frequencies, respectively. It should be noted that the Xiaomi MI8 cannot acquire BDS B1C data.

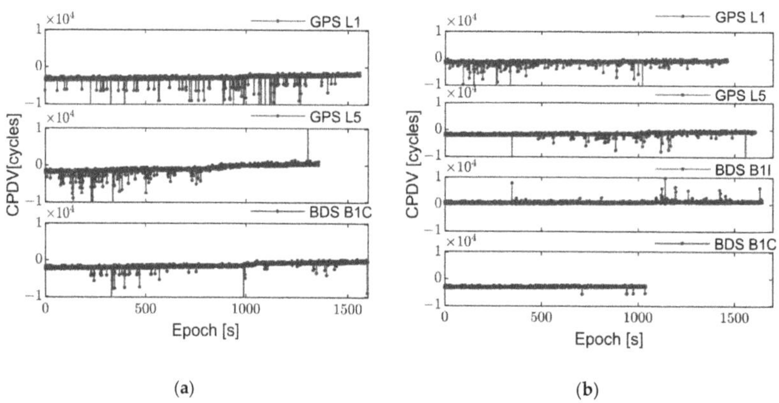

Figure 7. Observations of GPS satellite G20 and BDS satellite B05. Comparison of carrier phase differential values over water between two smartphones with different systems: (**a**) Xiaomi MI8 smartphone and (**b**) Huawei Mate40 Pro smartphone.

Figures 6 and 7 illustrate the abnormal errors in GNSS observations from smartphones during water navigation. The presence of multipath effects, caused by fluctuations and reflections on the water surface, substantially amplifies both the magnitude and frequency of these anomalies compared to terrestrial environments. As seen from Tables 2 and 3, the percentage of this type of abnormal error reaches a maximum of 6.90% in the experiments conducted on land. In contrast, in the kinematic experiments conducted on water, the rate peaks at 48.53%, occupying almost half of the observed epochs. The linearly polarized antennas typically employed in smartphones have limited capability to mitigate multipath effects, thereby exacerbating their influence on these observations.

Additionally, there are differences in the occurrence epochs and magnitude of abnormal errors in observation values across various GNSS systems and frequencies. In theory, observation values at the GPS L5 frequency should exhibit better quality than those at the L1 frequency. As shown in Table 3, the percentage of abnormal errors in the carriers of GPS L1 and L5 for the Xiaomi Mi8 smartphone is 39.21% and 36.32%, respectively. The difference between these two percentages is minimal. Similarly, the percentage of abnormal errors in the pseudorange of GPS L1 and L5 for the Huawei Mate 40 Pro smartphone is 38.84% and 34.27%, respectively. The advantage of L5 is not obvious in the smartphone. Moreover, as depicted in Figure 7b, the carrier data integrity of the Huawei Mate 40 Pro is notably lower, particularly with the BDS B1C frequency where the data discontinue at around 1000 epochs.

Considering the challenges related to the quality of acquired observables, available satellite count, and frequency of abnormal errors in smartphones, there is an urgent need to devise an error detection and elimination method tailored to these characteristic errors. This method should aim to enhance the quality of GNSS observation values acquired by smartphones.

2.2. Design of Detection and Repair Solutions

Based on the conditions mentioned above, real-time detection and reparation are performed on the pseudorange and carrier of each frequency obtained by the smartphone. The methodology initiates by using predicted values derived from Kalman filtering [38] to calculate the difference compared to the observed values. Subsequently, abnormal errors in the observations are identified in real time by evaluating whether these differences exceed predefined thresholds. Finally, depending on the presence of abnormal errors within the epoch, three distinct strategies are formulated to facilitate real-time repairs for these anomalies.

2.2.1. Establishment of State Equations

By establishing the satellite-to-ground distance state equation of the Kalman filter, the satellite-to-ground distance change can be predicted, and its predicted value can be used to correct abnormal errors. The satellite-to-ground distance ρ at the k epoch time is differentiated between epochs to obtain the differential component, as follows:

$$\Delta\rho_{k-1} = \rho_k - \rho_{k-1} \tag{1}$$

According to the motion characteristics of GNSS satellites, differential correction can be expressed as a second-order recursive equation

$$\begin{cases} \Delta\rho_k = \Delta\rho_{k-1} + \Delta\dot{\rho}_{k-1}t + \frac{\Delta\ddot{\rho}_{k-1}}{2}t^2 + \frac{w_{k-1}}{6}t^3 \\ \Delta\dot{\rho}_k = \Delta\dot{\rho}_{k-1} + \Delta\ddot{\rho}_{k-1}t + \frac{w_{k-1}}{2}t^2 \\ \Delta\ddot{\rho}_k = \Delta\ddot{\rho}_{k-1} + w_{k-1}t \end{cases} \tag{2}$$

where t represents the sampling interval, $\Delta\dot{\rho}_k$ denotes the rate of change in the satellite-to-ground distance, $\Delta\ddot{\rho}_k$ represents the acceleration of the satellite-to-ground distance change, and $w_{k-1} \sim N(0, D_{ww})$ stands for the system's kinematic noise.

Taking the state vector as

$$X_k = \left(\Delta\rho_k, \Delta\dot{\rho}_k, \Delta\ddot{\rho}_k\right)^T \tag{3}$$

The state equation can be represented as

$$X_k = FX_{k-1} + Gw_{k-1} \tag{4}$$

where F and G represent the system state transition matrix and the kinematic noise matrix, respectively.

$$F = \begin{pmatrix} 1 & t & t^2/2 \\ 0 & 1 & t \\ 0 & 0 & 1 \end{pmatrix} \tag{5}$$

$$G = \begin{pmatrix} t^3/6 \\ t^2/2 \\ t \end{pmatrix} \tag{6}$$

The state equation established through Equation (4) can effectively fit and predict the motion state of GNSS satellites.

2.2.2. Establishment of Observation Equations

By establishing the observation equation, it is possible to utilize the observed measurement information to update the state equation, thereby enhancing the accuracy of the state equation. Suppose $P_{k,i}$ represents the pseudorange of frequency i at time k and $\varphi_{k,i}$ represents the carrier of frequency i at time k. Utilizing Equation (1), the differential observation value can be obtained.

$$\begin{cases} \Delta P_{k,i} = P_{k+1,i} - P_{k,i} \\ \Delta\varphi_{k,i} = \varphi_{k+1,i} - \varphi_{k,i} \end{cases} \tag{7}$$

Furthermore, the variance can be calculated as

$$\begin{cases} D(\Delta P_{k,i}) = 2\delta_p^2 \\ D(\Delta\varphi_{k,i}) = 2\delta_\varphi^2 \end{cases} \tag{8}$$

where σ_P^2 and σ_φ^2 represent the variance of pseudorange and carrier observation values, respectively.

The observation equation can be formulated as follows:

$$\begin{cases} \delta\rho_k = \Delta P_{k,i} + \varepsilon_P \\ \frac{1}{\lambda_i}\delta\rho_k = \Delta\varphi_{k,i} + \varepsilon_\varphi \end{cases} \tag{9}$$

where ε_P and ε_φ represent the observation errors of pseudorange and carrier, respectively, and λ_i represents the wavelength of frequency i.

Let

$$L_{k,i} = \begin{pmatrix} \delta P_{k,i} \\ \delta\varphi_{k,i} \end{pmatrix} \sim \left(E(L_{k,i}), D_{ZZ}\right) \tag{10}$$

$$H = \begin{pmatrix} 1 & 0 & 0 \\ \frac{1}{\lambda_i} & 0 & 0 \end{pmatrix} \quad V = \begin{pmatrix} \varepsilon_P \\ \varepsilon_\varphi \end{pmatrix} \tag{11}$$

Then, the observation equation can be represented as

$$L_{k,i} = H_k X_k + V \tag{12}$$

where H is the observation matrix and V is the measurement noise matrix.

2.2.3. Abnormal Detection Solution Computation

Before using the Kalman filter for computation, it is necessary to provide appropriate initial values and variances for the state vector. Since the carrier observation values exhibit high precision, the differential satellite-to-ground distance $\Delta\rho_1$ between epochs at the initial time can be obtained through carrier phase measurements [39]. Given the stability of obtaining GPS L1 frequency carrier observations from smartphones, calculations of the satellite-to-ground distance are conducted using carrier measurements at the following frequency:

$$\Delta\rho_1 = \lambda_1 \cdot \Delta\varphi_1 \tag{13}$$

Subsequently, deduce the initial rate of change $\Delta\dot{\rho}_1$ and acceleration $\Delta\ddot{\rho}_1$ of the satellite-to-ground distance at the initial time.

$$\begin{cases} \Delta\dot{\rho}_1 = \frac{\Delta\rho_2 - \Delta\rho_1}{t} \\ \Delta\ddot{\rho}_1 = \frac{\Delta\dot{\rho}_2 - \Delta\dot{\rho}_1}{t} \end{cases} \tag{14}$$

The initial values of the state vector can be determined as follows:

$$X_1 = \left(\Delta\rho_1, \Delta\dot{\rho}_1, \Delta\ddot{\rho}_1\right)^T \tag{15}$$

Assuming that the initial state components are independent of each other, let $\sigma^2_{\varphi_1}$ be the variance of the original carrier of GPS L1 frequency at the initial epoch. Based on Equation (15), according to the law of error propagation, the initial variance can be obtained, as follows:

$$\begin{cases} D[X_1] = 2\sigma^2_{\varphi_1} \\ D[\dot{X}_1] = D[\frac{\delta\Delta\nabla\rho_2 - \delta\Delta\nabla\rho_1}{t^2}] = \frac{4\sigma^2_{\varphi_1}}{t^2} \\ D[\ddot{X}_1] = D[\frac{\delta\Delta\nabla\dot{\rho}_2 - \delta\Delta\nabla\dot{\rho}_1}{t^4}] = \frac{8\sigma^2_{\varphi_1}}{t^4} \end{cases} \tag{16}$$

The initial variance can be expressed as

$$D[X_1] = diag\left(2\sigma^2_{\varphi_1}, \frac{4\sigma^2_{\varphi_1}}{t^2}, \frac{8\sigma^2_{\varphi_1}}{t^4}\right) \tag{17}$$

Once the initial values and variances of the state vector are determined, the filtering recursion can be carried out.

Using the state equation, the predicted value of the satellite-to-ground distance can be recursively obtained. In theory, this value should be close to the observed value. Therefore, the difference between the predicted and observed values is calculated, and if it exceeds a predefined threshold, it indicates the presence of abnormal errors in the observations of that epoch.

Given the state vector at epoch $k-1$, it is possible to recursively derive the state at epoch k, as follows:

$$X'_k = FX_{k-1} \tag{18}$$

and the variance, as follows:

$$D(X'_k) = FD(X_{k-1})F^T + GD_{ww}G^T \tag{19}$$

The predicted value $\Delta\rho'_k$ of the satellite-to-ground distance can be obtained through X'_k.

By using Equations (7) and (8), we can obtain the pseudorange and carrier $\Delta P'_{k,i}$ and $\Delta\varphi'_{k,i}$, along with their variances. Moreover, we can then calculate the difference between the predicted and observed values, as follows:

$$\begin{cases} \Delta\nabla P_{k,i} = \Delta P'_{k,i} - \Delta\rho'_k \\ \Delta\nabla\varphi_{k,i} = \Delta\varphi'_{k,i} - \Delta\rho'_k \end{cases} \tag{20}$$

and the variance of the difference is as follows:

$$\begin{cases} D(\Delta\nabla P_{k,i}) = D(X_k') + D(\Delta P_{k,i}) \\ D(\Delta\nabla \varphi_{k,i}) = D(X_k') + D(\Delta\varphi_{k,i}) \end{cases} \quad (21)$$

When

$$\begin{cases} \Delta\nabla P_{k,i} > 2 \cdot \sqrt{D(\Delta\nabla P_{k,i})} \\ \Delta\nabla \varphi_{k,i} > 2 \cdot \sqrt{D(\Delta\nabla \varphi_{k,i})} \end{cases} \quad (22)$$

it is considered that the pseudorange or carrier of frequency i at epoch k contains abnormal errors.

When at time k, both $\Delta P_{k,i}$ and $\Delta\varphi_{k,i}$ satisfy

$$\begin{cases} \Delta P_{k,i} < 2 \cdot \sqrt{D(\Delta\nabla P_{k,i})} \\ \Delta \varphi_{k,i} < 2 \cdot \sqrt{D(\Delta\nabla \varphi_{k,i})} \end{cases} \quad (23)$$

the observations at k time are considered normal and do not need to be repaired.

2.2.4. Abnormal Repair Solution Computation

After completing the detection of abnormal errors, it is necessary to perform real-time repair on the observation values within that epoch containing abnormal errors. If some of the observation values within that epoch are abnormal data, the observation equation can be established using Equation (9) for the normal data in the observation values, yielding the Kalman gain matrix.

$$K_k = D(X_k')H_k^T\left(H_k D(X_k')H_k^T + D_{zz}\right)^{-1} \quad (24)$$

Subsequently, the optimal estimate of the state vector can be obtained

$$\widetilde{X}_k = X_k' + K_k(L_{k+1} - HX_k') \quad (25)$$

and the variance is as follows:

$$D(\widetilde{X}_k) = (I - K_k H_k)D(X_k')(I - K_k H_k)^T + K_k D_{ww} K_k^T \quad (26)$$

For the abnormal set of data, Equation (25) is utilized to obtain the state vector $\widetilde{X}_k$, which in turn leads to the optimal valuation of the star-earth distance $\Delta\widetilde{\rho}_k$. The abnormal data are replaced as follows:

$$\begin{cases} \Delta\hat{P}_{k,i} = \Delta\widetilde{\rho}_k \\ \Delta\hat{\varphi}_{k,i} = \frac{1}{\lambda_i}\Delta\widetilde{\rho}_k \end{cases} \quad (27)$$

where $\Delta\hat{P}_{k,i}$ and $\Delta\hat{\varphi}_{k,i}$ denote the amount of star-earth distance replaced by the optimal valuation.

The repair of the observation values is as follows:

$$\begin{cases} \overline{P}_{k,i} = \Delta\hat{P}_{k,i} + P_{k-1,i} \\ \overline{\varphi}_{k,i} = \Delta\hat{\varphi}_{k,i} + \varphi_{k-1,i} \end{cases} \quad (28)$$

where $\overline{P}_{k+1,i}$ and $\overline{\varphi}_{k+1,i}$ denote the restored pseudorange and carrier.

If all observation values within the epoch exhibit abnormal errors, then the predicted satellite-to-ground distance $\Delta\rho_k'$ calculated using Equation (18) is used to replace the abnormal data, as follows:

$$\begin{cases} \Delta\hat{P}_{k,i} = \Delta\rho_k' \\ \Delta\hat{\varphi}_{k,i} = \frac{1}{\lambda_i}\Delta\rho_k' \end{cases} \quad (29)$$

The repair of the observation values is as follows:

$$\begin{cases} \overline{P}_{k,i} = \Delta P'_{k,i} + P_{k-1,i} \\ \overline{\varphi}_{k,i} = \Delta \varphi'_{k,i} + \varphi_{k-1,i} \end{cases} \tag{30}$$

At this point, the repair of abnormal errors in the pseudorange and carrier has been completed.

As the number of epochs increases, the predicted values of the satellite-to-ground distance derived from the state equation become more accurate, providing more precise predictions and variances. For epochs with abnormal errors, especially those with significant errors, the repair is performed using the optimal estimates obtained from filtering predictions and observations from other frequencies without abnormal errors. Figure 8 presents the flowchart of the real-time detection and repair method for abnormal errors in smartphone GNSS.

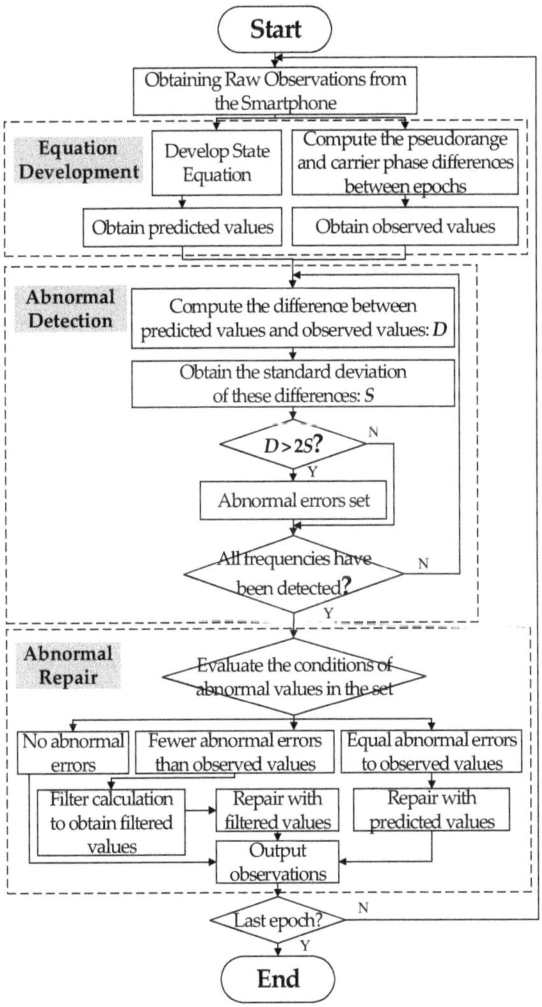

Figure 8. Flowchart of real-time detection and repair method of abnormal errors in smartphone GNSS observations.

As depicted in Figure 8, the key steps of the abnormal error detection and repair method based on Kalman filtering are outlined as follows:

1. Equation establishment: The smartphone observation values are used to compute epoch-to-epoch differences, resulting in single-differenced observation values. These are used to establish the observation equations. The state vector includes the satellite-to-ground distance, rate of change, and acceleration, forming the system state equation;
2. Anomaly detection computation: After determining the initial values and variances of the system state, Kalman filtering is employed to predict the satellite-to-ground distance. Differences between predicted values and pseudorange and carrier observation values for each epoch are calculated alongside their standard deviations. The standard deviation multiplied by two serves as the threshold to ascertain whether the differences exceed acceptable limits;
3. Anomaly repair computation: Based on the anomalies identified in the second step, three scenarios are considered. First, if no abnormalities are detected in the epoch, the original observation values are directly output without any alterations. Second, if abnormalities are present in some observations within the epoch, normal observations are used to establish observation equations for subsequent Kalman filtering. The resultant filtered values replace the abnormal observations to rectify errors. Third, if all observations within the epoch exhibit abnormalities, predicted values replace the observations to rectify errors.

3. Results

To validate the effectiveness of the proposed abnormal error detection and repair scheme, as well as its performance in land and water environments, multiple sets of experiments were conducted using Xiaomi MI 8 and Huawei Mate 40 Pro smartphones. These experiments included static tests on a university campus field and kinematic tests while boating on a waterway. Furthermore, the experimental results were analyzed.

3.1. Experimental Description

The experiment on land consisted of multiple static tests conducted on the playground of Southeast University in Jiangsu Province, China, from May to June 2023. Xiaomi MI 8 and Huawei Mate 40 Pro smartphones were placed on a ground platform with a distance of less than 15 cm between them. Each set of static tests lasted for 1 h with a sampling interval of 1 s. The water experiment occurred on 16 August 2023 aboard a boat in the waterways of Yancheng City, Jiangsu Province, China. The experimental setup mirrored that of the land experiment. Each session lasted approximately 1 h with a sampling interval of 1 s. The test route traversed urban areas, introducing a relatively complex environment, and the cargo vessel maintained a speed of approximately 10 km/h. The test trajectory is depicted in Figure 9. To validate the effectiveness of the method, the proposed approach was compared with a method that constructs state test statistics using the state-estimated value and the state-predicted value, hereafter denoted as the State-Based method [40].

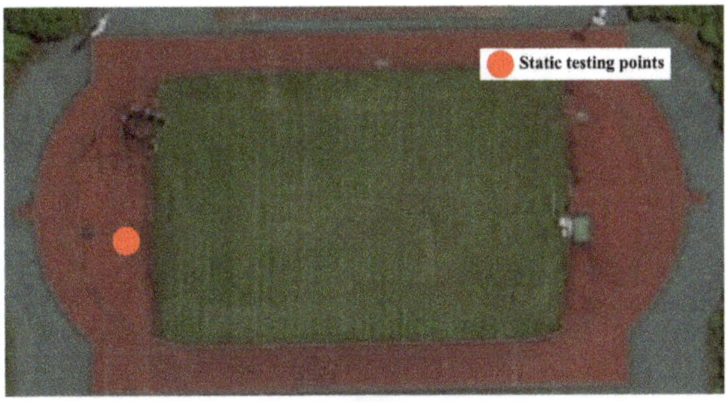

(**a**)

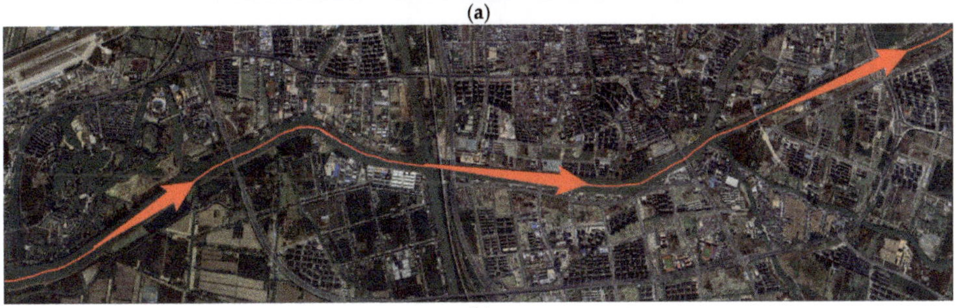

(**b**)

Figure 9. Smartphone testing in different environments. (**a**) Static test site on the land playground. (**b**) Waterway kinematic test track. The red line represents the navigation track and the red arrow indicates the navigation direction.

3.2. Performance Analysis of Detection and Repair Methods

To better showcase the effectiveness of the repair, the experiment evaluated the effectiveness of the repair method by comparing the pseudorange and carrier phase differentials before and after the repair. Due to the kinematic nature of real-time testing, conducting synchronous observations with a high-precision geodetic receiver and mobile phone for comparative analysis was not feasible. The observed values were fitted using the least squares method to generate a continuous and smooth curve. This curve was then utilized as a reference to quantitatively evaluate the effectiveness of the repair process.

The observations were processed using the real-time detection and repair methods proposed in this paper and the State-Based method to detect and repair abnormal errors. A comparison of the pseudorange and carrier before and after correction is presented, with the repair outcomes shown in Figures 10–13.

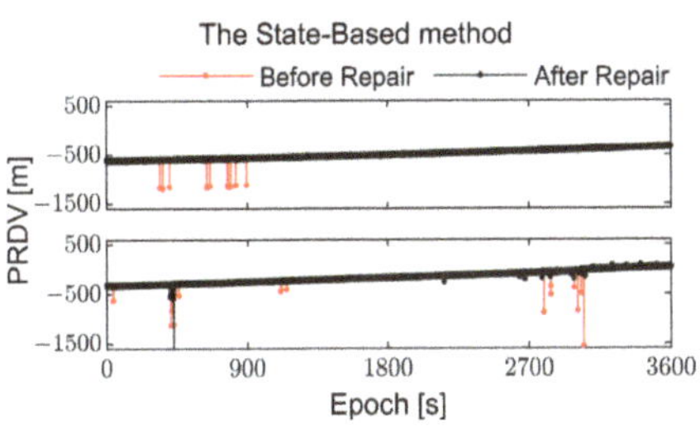

(**a**)

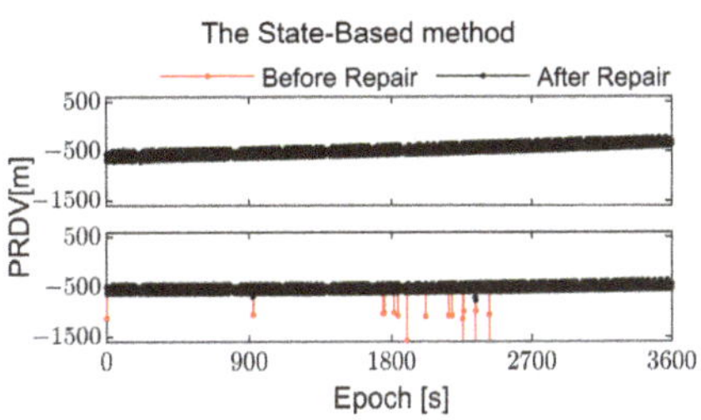

(**b**)

Figure 10. *Cont.*

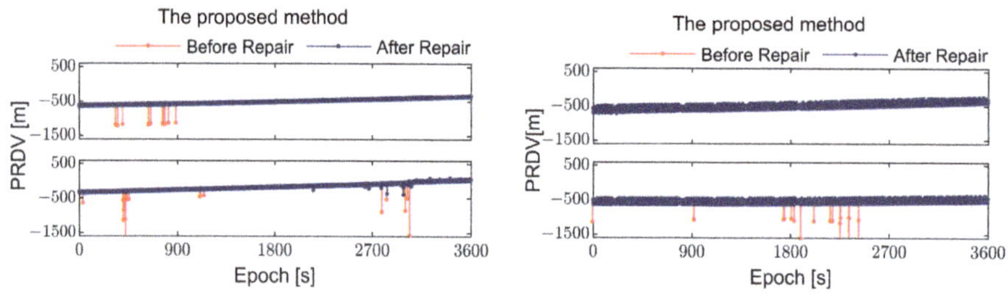

(c) (d)

Figure 10. Comparison of GNSS pseudorange differentials before and after repair using different methods during static tests on land for two smartphones. The red line represents the uncorrected observations, the green line indicates the results after repair using the State-Based method, and the blue line shows the results after repair using the method proposed in this paper. Subplots (**a**,**b**) show results obtained using the State-Based method, while subplots (**c**,**d**) display results from the method proposed in this paper. (**a**,**c**) show observed pseudorange differentials of GPS L1 frequency from satellite G02 and BDS B1I frequency from satellite C01 on the Xiaomi MI 8 smartphone, while (**b**,**d**) show corresponding observations on the Huawei Mate 40 Pro smartphone. This presentation format remains consistent across Figures 10–13.

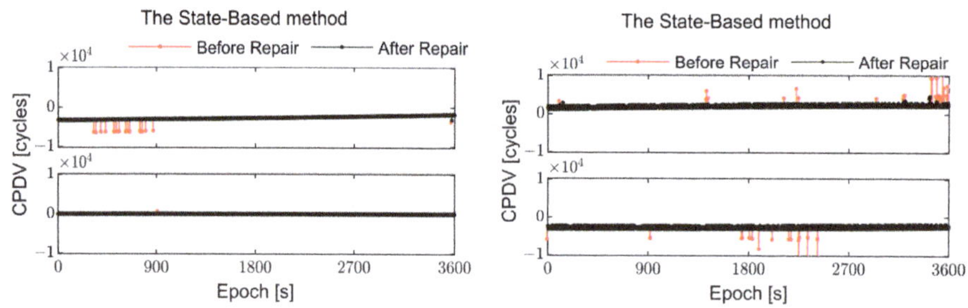

(a) (b)

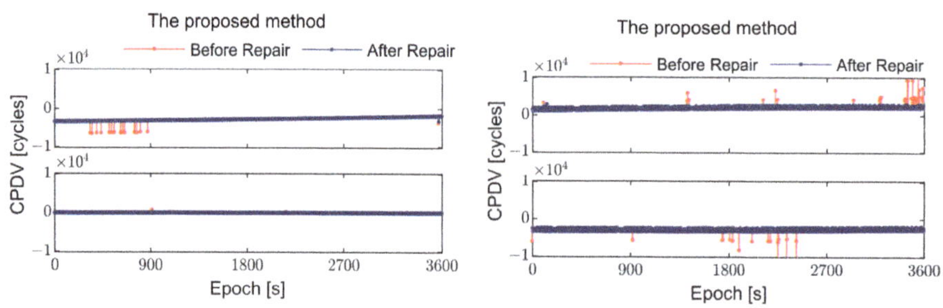

(c) (d)

Figure 11. Comparison of GNSS carrier phase differentials before and after repair using different methods during static tests on land for two smartphones. (**a**,**c**) show Xiaomi MI 8 smartphone, observed carrier phase differentials of GPS L1 frequency from satellite G02 and BDS B1I frequency from satellite C01. (**b**,**d**) show Huawei Mate 40 Pro smartphone, observed carrier phase differentials of GPS L1 frequency from satellite G02 and BDS B1I frequency from satellite C01.

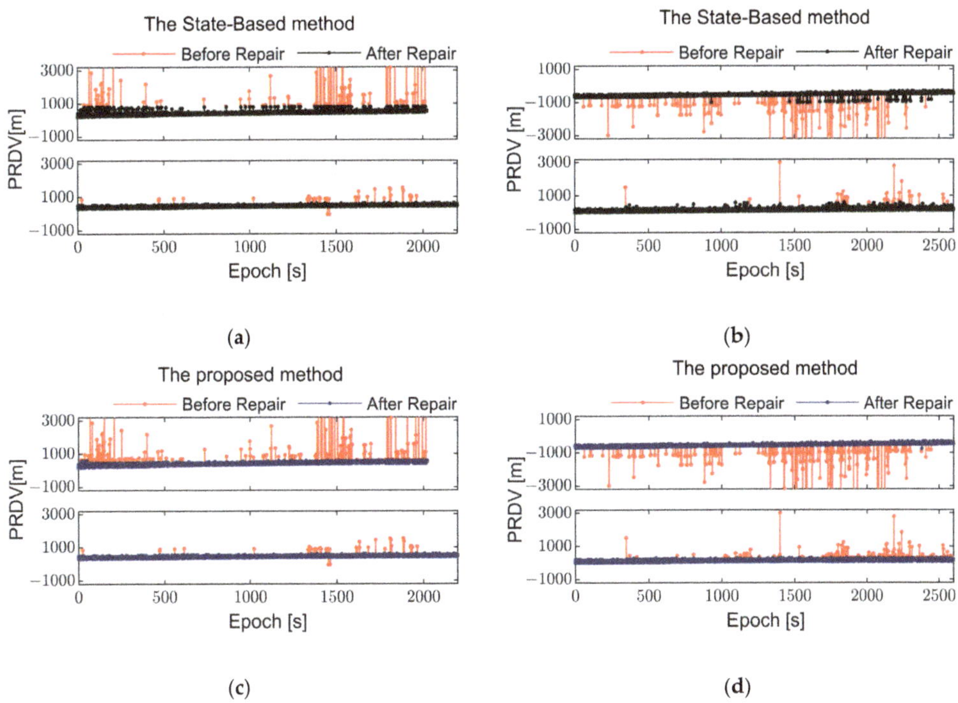

Figure 12. Comparison of GNSS pseudorange differentials before and after repair using different methods during kinematic tests on water for two smartphones. (**a**,**c**) show Xiaomi MI 8 smartphone, observed pseudorange differentials of GPS L1 frequency from satellite G04 and BDS B1I frequency from satellite C07. (**b**,**d**) show Huawei Mate 40 Pro smartphone, observed carrier phase differentials of GPS L1 frequency from satellite G04 and BDS B1I frequency from satellite C07. Similar to the land experiment, the green line represents the results repaired by the State-Based method, while the blue line represents the results repaired by the method proposed in this paper.

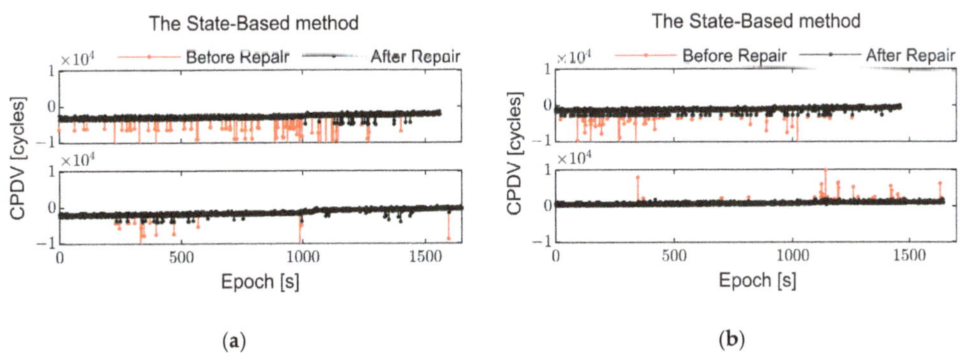

Figure 13. *Cont.*

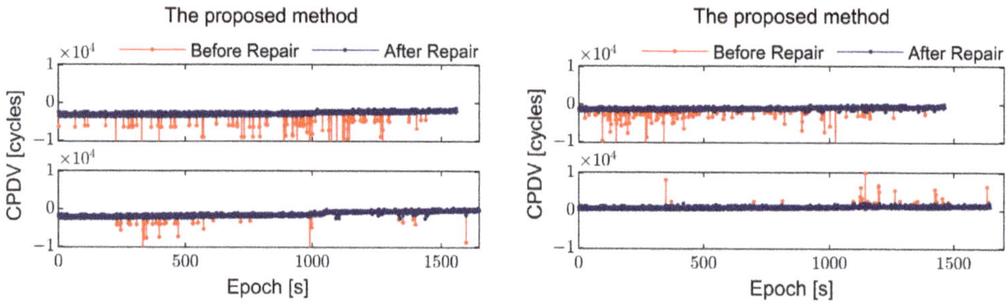

(c) (d)

Figure 13. Comparison of GNSS carrier phase differentials before and after repair using different methods during kinematic tests on water for two smartphones. (**a**,**c**) show Xiaomi MI 8 smartphone, observed carrier phase differentials of GPS L1 frequency from satellite G04 and BDS B1I frequency from satellite C07. (**b**,**d**) show Huawei Mate 40 Pro smartphone, observed carrier phase differentials of GPS L1 frequency from satellite G04, and BDS B1I frequency from satellite C07.

Figures 10 and 11 illustrate the comparison before and after repair of observation values for Xiaomi MI 8 and Huawei Mate 40 Pro during static land tests. In the open and spacious observation environment on land, the multipath effect is minimal, resulting in good quality observation values received by smartphones. However, during the observation process, significant fluctuations are observed in the values from most satellites, with only a few satellites showing nearly no abnormal errors. Compared to the Huawei Mate 40 Pro, the Xiaomi MI 8 experiences fewer abnormal errors. Although Huawei uses a different GNSS chip and antenna, this has not significantly improved the situation of abnormal errors in the observation values, necessitating the use of the algorithm proposed in this paper to enhance the quality of observations.

It can be seen that both methods are capable of repairing such abnormal errors. However, as shown in Figures 10c and 11b, when abnormal errors occur frequently, the State-Based method struggles to effectively repair these observations. In contrast, the method proposed in this paper shows a better capability to detect and repair abnormal errors even under such challenging conditions.

Tables 4 and 5 compare standard deviations of pseudorange and carrier phase observations for different satellites before and after repair during land positioning for the two smartphones used in this experiment. In most cases, the standard deviations decrease after repair, indicating that the repair method effectively improves the quality of observations. Specifically, for Xiaomi MI 8, satellites G02 and C01 exhibit significant reductions in pseudorange and carrier phase standard deviations after repair. Additionally, the proposed method in this paper outperforms the State-Based method by approximately 10% in the land experiment.

Table 4. Comparison of standard deviations of pseudorange and carrier observations from different satellites during land positioning for Xiaomi Mi 8 before and after repair.

		Pseudorange Standard Deviation (m)			Carrier Standard Deviation (m)		
Repair Method	Satellite Number	Before Repair	After Repair	Improvement	Before Repair	After Repair	Improvement
The proposed method	G02	34.157	17.679	48.29%	39.345	17.529	55.41%
	G09	46.742	35.352	24.36%	59.036	36.221	38.65%
	G12	27.321	25.894	5.22%	53.159	32.644	38.59%
	C01	54.792	20.581	62.52%	13.742	13.506	1.72%
	C04	39.235	30.231	22.94%	15.285	15.161	0.81%

Table 4. *Cont.*

		Pseudorange Standard Deviation (m)			Carrier Standard Deviation (m)		
Repair Method	Satellite Number	Before Repair	After Repair	Improvement	Before Repair	After Repair	Improvement
The proposed method	C15	64.238	27.346	57.46%	37.761	33.288	11.85%
	E11	365.829	58.164	46.78%	198.465	56.124	44.83%
	R05	168.432	50.387	46.19%	130.762	53.124	45.03%
The State-Based method	G02	34.157	18.972	44.46%	39.345	18.918	51.95%
	G09	46.742	35.944	23.12%	59.036	38.023	35.60%
	G12	27.321	26.035	4.74%	53.159	34.285	35.52%
	C01	54.792	22.263	59.38%	13.742	13.522	1.58%
	C04	39.235	30.996	21.02%	15.285	15.176	0.76%
	C15	64.238	29.672	53.81%	37.761	33.724	10.69%
	E11	365.829	209.547	42.72%	198.465	115.937	41.59%
	R05	168.432	98.272	41.66%	130.762	77.329	40.87%

Table 5. Comparison of standard deviations of pseudorange and carrier observations from different satellites during land positioning for Huawei Mate 40 Pro before and after repair.

		Pseudorange Standard Deviation (m)			Carrier Standard Deviation (m)		
Repair Method	Satellite Number	Before Repair	After Repair	Improvement	Before Repair	After Repair	Improvement
The proposed method	G02	50.353	50.353	0.00%	95.392	50.756	46.82%
	G09	47.742	31.263	34.55%	91.254	41.865	54.13%
	G12	87.323	45.401	48.02%	75.564	37.554	50.29%
	C01	92.090	50.205	45.50%	91.351	49.866	45.42%
	C04	107.340	52.412	51.17%	38.311	37.782	1.38%
	C15	91.234	40.413	55.72%	122.306	57.247	53.20%
	E11	397.812	59.214	40.92%	223.487	56.217	38.89%
	R05	154.892	47.238	33.98%	117.348	50.174	34.27%
The State-Based method	G02	50.353	50.353	0.00%	95.392	56.173	41.12%
	G09	47.742	33.041	30.79%	91.254	47.548	47.90%
	G12	87.323	50.615	42.05%	75.564	42.456	43.82%
	C01	92.090	54.885	40.41%	91.351	54.814	40.00%
	C04	107.340	59.363	44.70%	38.311	37.842	1.22%
	C15	91.234	46.372	49.18%	122.306	64.938	46.91%
	E11	397.812	255.614	35.75%	223.487	145.312	34.98%
	R05	154.892	108.626	29.88%	117.348	81.523	30.53%

However, in certain cases, the standard deviations after repair did not show significant improvement. For instance, the pseudorange observations of the G02 satellite for Huawei Mate40 Pro did not change in standard deviation before and after repair, as there were no abnormal errors present in its pseudorange observations during the observation process; thus, the method did not correct it. Additionally, there are variations in the standard deviation changes before and after repair among different smartphones and satellite numbers. Comparing the performance of the two smartphones, it is evident that in certain cases, the Huawei Mate40 Pro exhibits improved standard deviations post-repair compared to the Xiaomi Mi 8. This difference may stem from variations in the GNSS chips and antennas employed in these smartphones.

Figures 12 and 13 depict the observation repair results during kinematic testing on water for two smartphones. During water positioning, the frequency and magnitude of abnormal errors in observations are significantly higher than on land, indicating that smartphones equipped with linearly polarized antennas have almost no capability to resist multipath effects. In Figure 13, the GPS L1 frequency observations of the Huawei Mate 40 Pro exhibit numerous and continuous abnormal errors, whereas BDS B1I frequency observations show fewer abnormal errors.

In Figure 12, the BDS B1I frequency data of the Xiaomi MI 8 display abnormal errors around epoch 1500 that were not repaired in real time. This may be due to an inadequate threshold setting during filtering to detect abnormal errors or because other frequency data at that moment also contained abnormal errors, making it impossible to utilize them for repair. The State-Based method, when faced with a large number of abnormal errors during water testing, often fails to repair these errors adequately due to the lack of sufficient measurement information, leading to the replacement of some erroneous observations with predicted values, which is not sufficiently accurate. In contrast, the method proposed in this paper, by making full use of multi-frequency data, is able to better repair abnormal errors even under such conditions.

Tables 6 and 7 present the standard deviations of observations before and after repair for two smartphones across different systems. By applying the method proposed in this paper to correct abnormal errors in observations, the standard deviations of observations from various satellites have all been reduced, enhancing data quality. Specifically, observations of carrier quality were poorer without methodological repair, particularly for satellite G05, where the carrier standard deviations for the two smartphones were 377.612 m and 240.025 m, much lower than those of geodetic receivers. Such poor-quality observations are challenging to utilize in high-precision positioning. However, after repair using the method, the standard deviations of observations improved to 59.233 m and 40.598 m, respectively, representing a quality improvement of nearly 83%. The repair effect of the method proposed in this paper is 10–35% better than the State-Based method.

Table 6. Comparison of standard deviations of pseudorange and carrier observations from different satellites during water positioning for Xiaomi Mi 8 before and after repair.

		Pseudorange Standard Deviation (m)			Carrier Standard Deviation (m)		
Repair Method	**Satellite Number**	**Before Repair**	**After Repair**	**Improvement**	**Before Repair**	**After Repair**	**Improvement**
The proposed method	G04	512.000	87.895	82.81%	401.611	54.118	86.61%
	G05	173.341	51.287	70.42%	240.025	40.598	83.08%
	G17	266.932	37.281	86.04%	192.669	78.266	59.34%
	C07	106.402	44.599	58.11%	169.804	61.962	63.52%
	C12	537.453	81.235	84.87%	141.295	37.969	73.11%
	C22	311.510	67.359	78.39%	186.350	60.866	67.38%
	E17	450.187	64.239	72.34%	320.876	59.783	68.91%
	R22	251.476	52.831	61.27%	183.945	53.928	61.89%
The State-Based method	G04	512.000	132.132	74.19%	401.611	96.793	75.90%
	G05	173.341	83.811	51.65%	240.025	86.085	64.14%
	G17	266.932	95.970	64.05%	192.669	93.054	51.70%
	C07	106.402	61.515	42.19%	169.804	93.117	45.17%
	C12	537.453	216.217	59.77%	141.295	59.116	58.17%
	C22	311.510	130.773	58.02%	186.350	83.764	55.05%
	E17	405.872	187.154	58.43%	320.876	133.573	58.37%
	R22	487.234	136.869	45.58%	183.945	99.119	46.12%

The improvement in observation quality varies for different observations, primarily depending on the quantity and magnitude of abnormal errors present. Taking the pseudorange observations of satellites G05 and G17 with the Xiaomi MI 8 smartphone as an example, the standard deviation of G05 satellite pseudorange was 173.341 m before repair and improved to 51.287 m after repair, a quality improvement of 70.42%. For the G17 satellite, the improvement was 86.04%. Additionally, based on the original pseudorange observations of the two satellites, it is evident that the G17 satellite observations contained significantly higher quantities and magnitudes of abnormal errors compared to the G05 satellite.

Table 7. Comparison of standard deviations of pseudorange and carrier observations from different satellites during water positioning for Huawei Mate 40 Pro before and after repair.

		Pseudorange Standard Deviation (m)			Carrier Standard Deviation (m)		
Repair Method	Satellite Number	Before Repair	After Repair	Improvement	Before Repair	After Repair	Improvement
The proposed method	G04	418.108	50.213	88.04%	194.363	55.726	71.40%
	G05	511.642	70.885	86.11%	377.612	59.233	84.29%
	G17	488.341	49.631	89.84%	162.391	61.727	61.99%
	C07	131.039	59.004	55.00%	103.877	49.199	52.61%
	C12	179.353	41.962	76.65%	131.754	54.878	58.36%
	C22	611.274	39.741	93.47%	209.375	80.168	61.72%
	E17	450.187	64.239	72.34%	320.876	59.783	68.91%
	R22	251.476	52.831	61.27%	183.945	53.928	61.89%
The State-Based method	G04	418.108	116.083	72.24%	194.363	89.453	53.98%
	G05	511.642	160.752	68.58%	377.612	165.104	56.28%
	G17	488.341	100.521	79.42%	162.391	79.816	50.85%
	C07	131.039	78.654	39.98%	103.877	63.817	38.57%
	C12	179.353	75.106	58.13%	131.754	65.093	50.60%
	C22	611.274	134.967	77.92%	209.375	116.212	44.50%
	E17	450.187	199.693	55.64%	320.876	137.872	57.03%
	R11	251.476	116.422	53.70%	183.945	86.707	52.87%

4. Discussion

Comparison between the GNSS observations obtained from geodetic receivers and smartphones revealed distinct characteristics. Geodetic receiver data exhibited a consistent smooth trend with no abnormal or systematic errors. In contrast, smartphone observations displayed noticeable fluctuations and frequent outliers. Given the limited availability of dual-frequency data and fewer visible satellites for smartphones, optimizing the utilization of acquired observations is essential to enhance the quality of smartphone-derived observations [41,42].

An analysis of experimental results in both land and water environments reveals that the abnormal errors in smartphone GNSS observations do not exhibit a clear pattern. The frequency of outliers in the Huawei Mate 40 Pro smartphone is higher than in the Xiaomi Mi 8. While the Huawei Mate 40 Pro is more advanced than the Xiaomi Mi 8, the quality of its observational data does not match that of older models. This suggests that current smartphone hardware, such as GNSS chips, has not undergone significant improvements. The positioning performance of smartphones is severely constrained by hardware limitations. Abnormal errors vary across different systems, satellites, and smartphones, with pseudorange errors reaching up to 6000 m. Carrier phase errors can reach tens of thousands of cycles. kinematic and static tests indicate that these abnormal errors are closely correlated with the quality of the observation environment, with users often encountering severe multipath effects in practical usage scenarios.

The real-time detection and repair method proposed in this paper performed well in both environments, effectively reducing abnormal errors and improving data quality. In water environments, the frequency and magnitude of abnormal errors in smartphone observations increased by almost 40%, affecting positioning accuracy and stability. After repair using the proposed method, the standard deviation noticeably decreased, especially for lower-quality observations, with a more pronounced enhancement effect. For observations of lower quality, the standard deviation of the observations after repair increased by over 80%. In certain instances, observation data were unaffected by abnormal errors and thus did not necessitate correction.

In our experiments, especially during tests conducted on water, there was a high probability of abnormal errors in the smartphone observations. As depicted in Figure 12a, within 100 epochs, more than 50 epochs contained observations with abnormal errors. Our proposed method is constrained when all frequency observations simultaneously exhibit prolonged anomalies without recent updates using measurement values. Moreover, as evidenced in Section 2.1 experiments, different smartphones employ varying hardware devices, leading to differing occurrences of abnormal errors in their observations. Our method uses twice the standard deviation as the threshold for anomaly detection, which may hinder effective real-time detection and repair of abnormal errors for some smartphones. Adapting the threshold to accommodate different smartphone hardware is a key focus of our future research.

5. Conclusions

This paper proposes a real-time detection and repair method to mitigate frequent abnormal errors in GNSS observations during smartphone navigation. Based on the continuous smoothness of pseudorange and carrier observations, this method establishes state equations and observation equations separately. It utilizes the residuals between the predicted state vector from the Kalman filter and the observed values. Real-time detection of abnormal errors in observations is achieved by determining whether the residuals exceed the current epoch threshold. Subsequently, using the optimal estimation derived from normal data, the abnormal observations are replaced in real time, thereby realizing the real-time repair of smartphone observations. The method employs data synthesis from multiple frequencies and systems to repair continuous abnormal errors. Compared to filtering individual observations in isolation, using multi-frequency multi-system data for simultaneous filtering and replacement can prevent a decrease in prediction accuracy and also mitigate the risk of filter divergence or instability in cases where continuous abnormal errors occur in a single observation.

To validate the effectiveness of the proposed abnormal error detection and repair solution, experiments were conducted using Xiaomi MI 8 and Huawei Mate 40 Pro smartphones at the Southeast University campus in Jiangsu Province, China, and in the waterway of Yancheng City. Multiple sets of static/kinematic experiments were performed. The experimental results show that the method effectively repaired various types of abnormal errors, especially continuous ones, by integrating data from multiple frequencies and systems. This approach successfully prevented degradation in prediction accuracy or filter divergence. In scenarios with frequent occurrences of abnormal errors in the observations, the method demonstrated a maximum enhancement of 93.47% in pseudorange quality and 86.61% in carrier phase quality in the most favorable cases. The method proposed in this paper outperformed the State-Based method by approximately 10% on land and by 10–35% on water. However, during the repair process, there may be 2–3% of abnormal errors that are not detected and repaired in real time. This could be attributed to the failure to establish appropriate thresholds for detecting abnormal errors.

The method demonstrates high stability and reliability in repairing observations. By comparing the standard deviations of observations before and after repair, it is verified that the method significantly reduces the influence of abnormal errors on observations, thereby enhancing data quality. This provides crucial support for achieving high-precision positioning with smartphones in scenarios heavily affected by multipath, such as maritime navigation, and contributes to improving their positioning accuracy and reliability.

Author Contributions: Methodology, H.M., X.Y. and A.A.-A.; software, H.M., J.W. and Y.W.; validation, H.M. and Y.W.; data curation, Y.W.; writing—original draft preparation, H.M.; writing—review and editing, A.A.-A. and J.W.; supervision, X.Y.; project administration, X.Y.; funding acquisition, X.Y. All authors have read and agreed to the published version of the manuscript.

Funding: This research was funded by the National Key R&D Program of China, grant number 2023YFB3907103.

Data Availability Statement: The raw data supporting the conclusions of this article will be made available by the authors on request.

Conflicts of Interest: The authors declare no conflicts of interest.

References

1. Bahadur, B. A Study on the Real-Time Code-Based GNSS Positioning with Android Smartphones. *Measurement* **2022**, *194*, 111078. [CrossRef]
2. Li, B.; Miao, W.; Chen, G.; Li, Z. Ambiguity Resolution for Smartphone GNSS Precise Positioning: Effect Factors and Performance. *J. Geod.* **2022**, *96*, 63. [CrossRef]
3. Paziewski, J.; Fortunato, M.; Mazzoni, A.; Odolinski, R. An Analysis of Multi-GNSS Observations Tracked by Recent Android Smartphones and Smartphone-Only Relative Positioning Results. *Measurement* **2021**, *175*, 109162. [CrossRef]
4. Wang, L.; Li, Z.; Wang, N.; Wang, Z. Real-Time GNSS Precise Point Positioning for Low-Cost Smart Devices. *GPS Solut.* **2021**, *25*, 69. [CrossRef]
5. Yun, J.; Lim, C.; Park, B. Inherent Limitations of Smartphone GNSS Positioning and Effective Methods to Increase the Accuracy Utilizing Dual-Frequency Measurements. *Sensors* **2022**, *22*, 9879. [CrossRef]
6. Aggrey, J.; Bisnath, S.; Naciri, N.; Shinghal, G.; Yang, S. Multi-GNSS Precise Point Positioning with next-Generation Smartphone Measurements. *J. Spat. Sci.* **2020**, *65*, 79–98. [CrossRef]
7. Zhang, X.; Tao, X.; Zhu, F.; Shi, X.; Wang, F. Quality Assessment of GNSS Observations from an Android N Smartphone and Positioning Performance Analysis Using Time-Differenced Filtering Approach. *GPS Solut.* **2018**, *22*, 70. [CrossRef]
8. Shinghal, G.; Bisnath, S. Conditioning and PPP Processing of Smartphone GNSS Measurements in Realistic Environments. *Satell. Navig.* **2021**, *2*, 10. [CrossRef]
9. Weng, D.; Hou, Z.; Meng, Y.; Cai, M.; Chan, Y. Characterization and Mitigation of Urban GNSS Multipath Effects on Smartphones. *Measurement* **2023**, *223*, 113766. [CrossRef]
10. Peng, Z.; Gao, Y.; Gao, C.; Shang, R.; Gan, L. Improving Smartphone GNSS Positioning Accuracy Using Inequality Constraints. *Remote Sens.* **2023**, *15*, 2062. [CrossRef]
11. Liu, W.; Li, J.; Zeng, Q.; Guo, F.; Wu, R.; Zhang, X. An Improved Robust Kalman Filtering Strategy for GNSS Kinematic Positioning Considering Small Cycle Slips. *Adv. Space Res.* **2019**, *63*, 2724–2734. [CrossRef]
12. Li, Z.; Wang, L.; Wang, N.; Li, R.; Liu, A. Real-Time GNSS Precise Point Positioning with Smartphones for Vehicle Navigation. *Satell. Navig.* **2022**, *3*, 19. [CrossRef]
13. Lachapelle, G.; Gratton, P.; Horrelt, J.; Lemieux, E.; Broumandan, A. Evaluation of a Low Cost Hand Held Unit with GNSS Raw Data Capability and Comparison with an Android Smartphone. *Sensors* **2018**, *18*, 4185. [CrossRef] [PubMed]
14. Paziewski, J.; Sieradzki, R.; Baryla, R. Signal Characterization and Assessment of Code GNSS Positioning with Low-Power Consumption Smartphones. *GPS Solut.* **2019**, *23*, 98. [CrossRef]
15. Pesyna, K.M.; Heath, R.W.; Humphreys, T.E. Centimeter Positioning with a Smartphone-Quality GNSS Antenna. In Proceedings of the 27th International Technical Meeting of the Satellite Division of The Institute of Navigation (ION GNSS+ 2014), Tampa, FL, USA, 8–12 September 2014.
16. Geng, J.; Jiang, E.; Li, G.; Xin, S.; Wei, N. An Improved Hatch Filter Algorithm towards Sub-Meter Positioning Using Only Android Raw GNSS Measurements without External Augmentation Corrections. *Remote Sens.* **2019**, *11*, 1679. [CrossRef]
17. Kirkko-Jaakkola, M.; Söderholm, S.; Honkala, S.; Koivula, H.; Nyberg, S.; Kuusniemi, H. Low-Cost Precise Positioning Using a National GNSS Network. In Proceedings of the 28th International Technical Meeting of the Satellite Division of The Institute of Navigation (ION GNSS+ 2015), Tampa, FL, USA, 14–18 September 2015.
18. Zabala Haro, M.; Martín Furones, Á.; Anquela Julián, A.; Jiménez-Martínez, M.J. Comprehensive Analysis of Xiaomi Mi 8 GNSS Antenna Performance. *Sensors* **2024**, *24*, 2569. [CrossRef]
19. Tomaštík, J.; Chudá, J.; Tunák, D.; Chudý, F.; Kardoš, M. Advances in Smartphone Positioning in Forests: Dual-Frequency Receivers and Raw GNSS Data. *For. Int. J. For. Res.* **2021**, *94*, 292–310. [CrossRef]
20. Gao, H.; Groves, P.D. Environmental Context Detection for Adaptive Navigation Using GNSS Measurements from a Smartphone: Environment Detection Using GNSS Measurements. *J. Inst. Navig.* **2018**, *65*, 99–116. [CrossRef]
21. Tomaštík, J., Jr.; Tomaštík, J., Sr.; Saloň, Š.; Piroh, R. Horizontal Accuracy and Applicability of Smartphone GNSS Positioning in Forests. *For. Int. J. For. Res.* **2017**, *90*, 187–198. [CrossRef]
22. Angrisano, A.; Gaglione, S. Smartphone GNSS Performance in an Urban Scenario with RAIM Application. *Sensors* **2022**, *22*, 786. [CrossRef]
23. Ch, V.R.; Alekhya, N.L.; Rahaman, M.M.; Imran, M.; Pujasri, K.; Killamsetti, H.K. Analyzing Multipath Fading in Deep Water Through Signal to Noise Ratio. In Proceedings of the 2024 International Conference on Integrated Circuits and Communication Systems (ICICACS), Raichur, India, 23 February 2024; pp. 1–4.
24. Tranquilla, J.M.; Carr, J.P. GPS Multipath Field Observations at Land and Water Sites. *Navigation* **1990**, *37*, 393–414. [CrossRef]
25. Zangenehnejad, F.; Gao, Y. GNSS Smartphones Positioning: Advances, Challenges, Opportunities, and Future Perspectives. *Satell. Navig.* **2021**, *2*, 24. [CrossRef] [PubMed]

26. Paziewski, J. Recent Advances and Perspectives for Positioning and Applications with Smartphone GNSS Observations. *Meas. Sci. Technol.* **2020**, *31*, 091001. [CrossRef]
27. Specht, C.; Dabrowski, P.S.; Pawelski, J.; Specht, M.; Szot, T. Comparative Analysis of Positioning Accuracy of GNSS Receivers of Samsung Galaxy Smartphones in Marine Dynamic Measurements. *Adv. Space Res.* **2019**, *63*, 3018–3028. [CrossRef]
28. Bai, T.; Chai, H.; Wang, M.; Tian, X.; Du, Z.; Zhang, F. Analysis of GNSS Data Quality and Dynamic Positioning Assessment on Smartphones at Sea. *Hydrogr. Surv. Mapp.* **2022**, *42*, 42–47.
29. Robustelli, U.; Paziewski, J.; Pugliano, G. Observation Quality Assessment and Performance of GNSS Standalone Positioning with Code Pseudoranges of Dual-Frequency Android Smartphones. *Sensors* **2021**, *21*, 2125. [CrossRef]
30. Pan, C.; Li, Z.; Zhang, Q.; Soja, B.; Gao, J. Smartphone-Based Vision/MEMS-IMU/GNSS Tightly Coupled Seamless Positioning Using Factor Graph Optimization. *Measurement* **2024**, *229*, 114420. [CrossRef]
31. Blewitt, G. An Automatic Editing Algorithm for GPS Data. *Geophys. Res. Lett.* **1990**, *17*, 199–202. [CrossRef]
32. Zhang, X.; He, X.; Liu, W. Characteristics of Systematic Errors in the BDS Hatch–Melbourne–Wübbena Combination and Its Influence on Wide-Lane Ambiguity Resolution. *GPS Solut.* **2017**, *21*, 265–277. [CrossRef]
33. Li, Y.G. Zuofa Cycle Slip Detection and Ambiguity Resolution Algorithms for Dual-Frequency GPS Data Processing. *Mar. Geod.* **1999**, *22*, 169–181. [CrossRef]
34. Li, W.; Song, J.; Li, J.; Zhu, X. Improving GNSS Positioning Performance of Android Smart Devices by a Novel Pseudorange Correction Method. *Meas. Sci. Technol.* **2023**, *34*, 045010. [CrossRef]
35. Li, M.; Huang, T.; Li, W.; Zhao, Q.; Jiang, K. Precise Point Positioning with Mixed Single- and Dual-Frequency GNSS Observations from Android Smartphones Considering Code-Carrier Inconsistency. *Adv. Space Res.* **2023**, *74*, 2664–2679. [CrossRef]
36. Li, G.; Geng, J. Characteristics of Raw Multi-GNSS Measurement Error from Google Android Smart Devices. *GPS Solut.* **2019**, *23*, 90. [CrossRef]
37. Chen, K.; Chang, G.; Chen, C.; Zhu, T. An Improved TDCP-GNSS/INS Integration Scheme Considering Small Cycle Slip for Low-Cost Land Vehicular Applications. *Meas. Sci. Technol.* **2021**, *32*, 055006. [CrossRef]
38. Zhao, J.; Wang, J.; Han, H.; Jiang, T. A Study on the Model of Robust Fractional-Order Extended Kalman Filtering with Gross Error. *GPS Solut.* **2024**, *28*, 87. [CrossRef]
39. Xia, S.; Yu, X. A GNSS Triple-Frequency Combination Cycle Slip Detection and Repair Method without Blind Spots. *Sci. Surv. Mapp.* **2020**, *45*, 62–69. [CrossRef]
40. Sun, A.; Zhang, Q.; Yu, Z.; Meng, X.; Liu, X.; Zhang, Y.; Xie, Y. A Novel Slow-Growing Gross Error Detection Method for GNSS/Accelerometer Integrated Deformation Monitoring Based on State Domain Consistency Theory. *Remote Sens.* **2022**, *14*, 4758. [CrossRef]
41. Gao, R.; Xu, L.; Zhang, B.; Liu, T. Raw GNSS Observations from Android Smartphones: Characteristics and Short-Baseline RTK Positioning Performance. *Meas. Sci. Technol.* **2021**, *32*, 084012. [CrossRef]
42. Zangenehnejad, F.; Gao, Y. Quality Analysis of Smartphone Gnss Observations and Impact on Precise Positioning. *Int. Arch. Photogramm. Remote Sens. Spat. Inf. Sci.* **2023**, *48*, 1169–1177. [CrossRef]

remote sensing

MDPI

Article

An EOF-Based Global Plasmaspheric Electron Content Model and Its Potential Role in Vertical-Slant TEC Conversion

Fengyang Long, Chengfa Gao *, Yanfeng Dong and Zhenhao Xu

School of Transportation, Southeast University, Nanjing 211189, China; fylong1990@seu.edu.cn (F.L.); 230189842@seu.edu.cn (Y.D.); 22023035@seu.edu.cn (Z.X.)
* Correspondence: gaochfa@seu.edu.cn

Abstract: Topside total electron content (TEC) data measured by COSMIC/FORMAT-3 during 2008 and 2016 were used to analyze and model the global plasmaspheric electron content (PEC) above 800 km with the help of the empirical orthogonal function (EOF) analysis method, and the potential role of the proposed PEC model in helping Global Navigation Satellite System (GNSS) users derive accurate slant TEC (STEC) from existing high-precision vertical TEC (VTEC) products was validated. A uniform gridded PEC dataset was first obtained using the spherical harmonic regression method, and then, it was decomposed into EOF basis modes. The first four major EOF modes contributed more than 99% of the total variance. They captured the pronounced latitudinal gradient, longitudinal differences, hemispherical differences, diurnal and seasonal variations, and the solar activity dependency of global PEC. A second-layer EOF decomposition was conducted for the spatial pattern and amplitude coefficients of the first-layer EOF modes, and an empirical PEC model was constructed by fitting the second-layer basis functions related to latitude, longitude, local time, season, and solar flux. The PEC model was designed to be driven by whether solar proxy or parameters derived from the Klobuchar model meet the real-time requirements. The validation of the results demonstrated that the proposed PEC model could accurately simulate the major spatiotemporal patterns of global PEC, with a root-mean-square (RMS) error of 1.53 and 2.24 TECU, improvements of 40.70% and 51.71% compared with NeQuick2 model in 2009 and 2014, respectively. Finally, the proposed PEC model was applied to conduct a vertical-slant TEC conversion experiment with high-precision Global Ionospheric Maps (GIMs) and dual-frequency carrier phase observables of more than 400 globally distributed GNSS sites. The results of the differential STEC (dSTEC) analysis demonstrated the effectiveness of the proposed PEC model in aiding precise vertical-slant TEC conversion. It improved by 18.52% in dSTEC RMS on a global scale and performed better in 90.20% of the testing days compared with the commonly used single-layer mapping function.

Keywords: plasmaspheric electron content; COSMIC; GNSS; empirical orthogonal function; empirical model; total electron content; mapping function; vertical-slant TEC conversion

1. Introduction

The plasma environment of the Earth's atmosphere varies at different altitudes. The ionosphere, including the F-layer, contributes the majority of the total electron content (TEC) [1–6]. In the plasmasphere, however, the electron density is generally several orders of magnitude less than that in the ionosphere, but it can extend from the topside of the F-layer to an altitude of 3∼5 times the Earth's radius [7–9]. In ground-based GNSS applications, the length of the GNSS signals' propagation path in the plasmasphere usually accounts for more than 95% of the satellite-receiver distance, and in this case, the integral plasmaspheric electron content (PEC) along the ray path is still considerable, although with the quite low electron densities. Therefore, it is of great significance to research the plasmasphere in consideration of its potential role in the space weather forecasts and ionospheric delay error mitigation for GNSS users.

The PEC is a meaningful parameter for the physical description of the plasmasphere. In the early days, PEC data were usually derived from the residuals between the TEC retrieved from the Global Positioning System (GPS) observations and that measured by Ionosonde or incoherent scatter radar (ISR), etc., and some studies have demonstrated that the PEC can account for a considerable proportion of the GPS-TEC, particularly during nighttime [10–12], but they were only targeting a few specific regions because of the limited global distribution of Ionosonde or ISR. In the past two decades, with booming development of the low-Earth orbit (LEO) satellite, researchers have carried out many studies on spatiotemporal variations in the global PEC using PEC data derived from the residuals between the TECs retrieved from LEO platforms and the GNSS [13–18]. Yizengaw et al. studied the PEC above the orbit of JASON-1 satellite (1335 km) in 2008 using GNSS-derived and altimetry-derived TECs and found that the PEC may contribute 10% and 60% to the GNSS-TEC during daytime and nighttime, respectively [13]. Cherniak et al. proposed a technique for mapping the PEC above an altitude of 700 km with GNSS-derived TEC and EDP measurements from the Constellation Observing System for Meteorology, Ionosphere, and Climate/Formosa Satellite 3 (COSMIC/FORMAT-3) and studied the seasonal and diurnal variation of global PEC under the minimum solar activity condition. They pointed out that the plasmasphere may contribute 50∼60% and 25∼45% electron contents to the GNSS-TEC during daytime and nighttime, respectively [15]. Chen et al. utilized a similar method to study the seasonal and diurnal variations of the global PEC in 2011 and found significant latitudinal differences in the global PEC and its percentage contribution to the GNSS-TEC [16]. González-Casado et al. conducted a series of studies on the differences of fractional contributions between the ionosphere and plasmasphere to the GNSS-TEC using the COSMIC radio occultation data and GIM [18].

However, the PEC data used in the aforementioned studies all come from the residuals between GNSS-TEC and the TECs measured by LEO platforms or traditional ground-based observations. Significant uncertainty might be introduced because of the observation errors caused by the involvement of different observation equipment or methods. In comparison, the topside TECs measured directly with the precise orbit determination (POD) receiver of the LEO platform can provide more accurate PEC observations. Yue et al. systematically analyzed the main error sources of the LEO-based slant TEC derivation and found that the accuracy of the LEO slant TEC can be estimated at 1∼3 TECU for the COSMIC/FORMAT-3 mission [19]. Zhang et al. studied the PEC above around 800 km in altitude with the PEC data from the COSMIC and found significant longitudinal differences in seasonal variations of global PEC [20]. Jin et al. studied long-term variations in the global PEC using the same dataset and found that the PEC reaches its minimum between 3:00∼6:00 in magnetic local time and reaches its peak between 14:00∼17:00, and the PEC in the northern hemisphere is highest in spring and lowest in summer. The solar dependency was also found with a linear correlation coefficient of up to 0.93 with solar proxy [21]. Zhong et al. also found significant longitudinal differences in PEC and pointed out that the PEC over the Pacific Ocean is highest near the June solstice but extremely low near the winter solstice [22]. Chen et al. proposed a mapping technique for generating gridded PEC maps using topside measurements of the COSMIC and MetOp-A constellations at an orbital altitude of around 800 km [23]. Tao et al. analyzed the seasonal variations in the global PEC at LT 9:30 and 21:30 in 2020 using the EOF analysis method and dense PEC measurements from the MetOp-A, -B, and -C, and they further compared the differences of the major EOF modes between PEC and GNSS-TEC [24].

In addition to the spatiotemporal analysis of the global PEC, efforts have been conducted to simulate the PEC or EDP of the plasmasphere. Lunt et al. simulated the PEC above 1100 km with GPS/NIMS measurements and the Sheffield University plasmasphere ionosphere model (SUPIM) and found the contribution of the plasmasphere to the GPS-TEC may be over 50% under high solar activity condition [25]. Webb and Essex developed an ionosphere–plasmasphere global electron density model with the International Reference Ionosphere (IRI) [26]. However, a lot of assumptions are usually needed for the 3-D electron density models in dealing with the EDP of the plasmasphere because of the limited achievable data [27–30]. For example, the EDP of the plasmasphere is usually considered to be an extension of the ionospheric F-layer, and an artificially specified exponential function is used to extend the EDP to an altitude of thousands of kilometers [7,30]. On this condition, the EDP of the plasmasphere strongly relies on the shape of the EDP in the ionosphere, and some artificial mistakes may be introduced, such as the two-peak structure of the Equatorial Ionization Anomaly (EIA) that rarely occurs in plasmasphere [2,9]. So, it is still a challenge to accurately model the EDP of the plasmasphere with the limited reliable global topside electron density observations. In this case, an empirical PEC model can serve as an alternative to meet the increasing requirements in practical use. Zhong et al. developed a set of empirical models for the PEC with the topside TECs from the MetOp-A and TerraSAR-X satellites [9], but only PEC predictions above 830 km at 9:30/21:30 LT and PEC predictions above 520 km at 6:00/18:00 LT were considered. Chen et al. developed an empirical PEC model based on the uniform PEC maps with 4-h resolution [31], but the longitudinal differences were not discussed. Moreover, to meet real-time requirements, an operable empirical model should be easily accessed, and the driven parameters can be provided in real time. Therefore, there are still some limitations in the current PEC modeling research.

In addition, accurate plasmasphere information has a potential role in the ionospheric delay correction of precise GNSS positioning applications [32–34]. Wen et al. pointed out that the limitation of the commonly used single-layer mapping function can be attributed to its neglecting the electron content contributed by the plasmasphere [32]. Lyu et al. developed a two-layer mapping function, namely, BIMF, to achieve more accurate vertical-slant TEC conversion when using high-precision VTEC products, and the reason for the success of BIMF model is precisely because it takes the plasmasphere information into account [33]. Wu et al. investigated the horizontal gradients of the plasmasphere and proposed a mapping function for LEO satellites by parametrizing the time, location, elevation, and azimuth angle-based horizontal gradients using the global core plasma model (GCPM) [34]. However, there have been no reports related to the applications of an empirical PEC model in ionospheric delay error mitigation yet. The potential role of new development in PEC modeling research in helping improve the ionospheric delay correction for GNSS users remains to be discovered.

In this paper, we use the PEC data measured by COSMIC/FORMAT-3 from 2008 to 2016 to analyze and model the global PEC above around 800 km and use the rapid Global Ionospheric Map (GIM) products provided by the Universitat Politècnica de Catalunya (UPC) in 2014, 2016, and 2018 to explore the potential role of the proposed PEC model in vertical-slant TEC conversions. The innovations of this paper can be concluded as follows: (1) The EOF analysis method is used to analyze the major spatiotemporal patterns of the global PEC, and an empirical PEC model is constructed by fitting the major basis functions after a second-layer EOF decomposition. The users can compute the PEC with the model coefficients rather than interpolate the PEC with the dense matrix data of the EOF basis function. (2) The proposed PEC model is designed to be driven by whether solar proxy or parameters derived from the Klobuchar model meet the potential requirements of real-time applications. (3) A two-layer thin-shell mapping function based on the PEC model is proposed to explore the accurate plasmaspheric electron content information in aiding precise vertical-slant TEC conversion with the existing high-precision vertical VTEC products for global GNSS users.

2. PEC Dataset and Binning Procedure

2.1. PodTec Dataset

As the first satellite constellation system dedicated primarily for an RO observing the Earth's atmosphere and ionosphere, the COSMIC/FORMAT-3 was jointly launched by the United States and Taiwan of China on 15th April 2006. It comprises six identical micro-satellites running on the circular low-Earth orbit with an orbital period of 100 min and an orbital inclination of 72°. Each satellite carries a GPS RO receiver and a POD receiver, which can be used not only to obtain continuous TEC observations between COSMIC and GPS satellite pair during an occultation event and retrieve the ionospheric electron densities via Abel inversion [2,15,35] but also to receive signals from GPS satellites running on its line of sight directly and retrieve the topside TEC observations [7,9,22,31]. Various observation data from COSMIC/FORMAT-3 have provided great convenience for the study of the ionosphere and plasmasphere, and they are now available for free from the COSMIC Data Analysis and Archive Center (CDAAC).

The orbital altitude of COSMIC satellites is usually around 800 km, which is close to the lower boundary of the plasmasphere (1000 km), so the TEC within altitudes around 800∼20,000 km measured by COSMIC satellites can be approximated as the total electron content of the plasmasphere. We gathered the daily topside TEC dataset, namely, "PodTec", from 2006 to 2019 from the "level1b" catalog of CDAAC. However, just as shown in Figure 1a, we only chose the PodTec dataset for a period of 9 years from 2008 to 2016 to build and evaluate the PEC empirical model in this study because most LEO satellites had significantly lower orbital altitude than 800 km in 2006 and 2007, and the number of retrievable data decreased sharply in 2017∼2018.

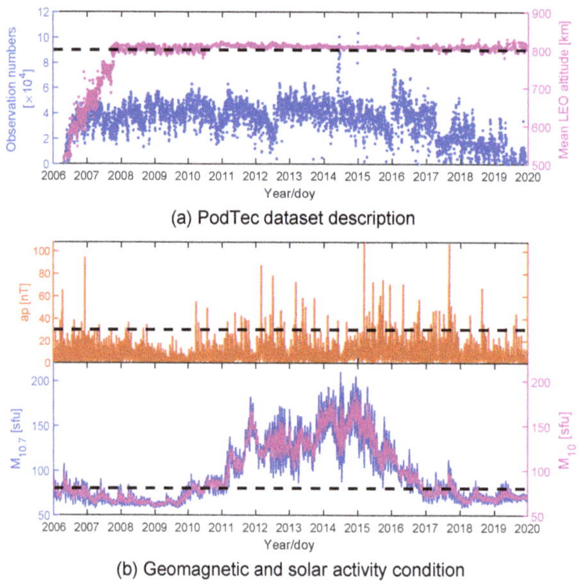

Figure 1. (**a**) Daily description of PodTec dataset measured by COSMIC/FORMAT-3 satellites; (**b**) daily mean geomagnetic activity indices and daily solar activity conditions from 2006 to 2019. The black dotted lines denote the altitude of 800 km, ap of 30 nT and $M_{10.7}$ of 80 sfu respectively.

The daily geomagnetic and solar activity conditions from 2006 to 2019 are shown in Figure 1b, where $M_{10.7}$ is a solar proxy derived from the MgII index ($M_{10.7} = 8083 \times MgII - 1154$) to replace the solar radio flux at 10.7 cm ($F_{10.7}$) since $F_{10.7}$ may not represent the solar extreme ultraviolet (EUV) radiation well during solar minimum [9,36]. In this work, the study period (2008∼2016) spans almost a complete solar cycle, including a solar minimum year (2009) and a solar maximum year (2014). In addition, we were only interested in the PEC analysis and model construction under quiet geomagnetic activity conditions, so the data with daily ap larger than 30 nT were excluded.

2.2. PEC Data Binning Procedure

The impact of the POD receiver and GPS satellite differential code bias in the PodTec dataset had been eliminated already [19]. We examined the variations in the RMS error of LEO DCB and found that they were generally between 0.5~2 TECU during the study period. To achieve a successful EOF analysis and model construction, we conducted a binning procedure for acquiring uniform gridded PEC data that included the following steps:

(1) We only considered the signal with an elevation angle larger than 40° to minimize the slant-vertical PEC conversion error. In this case, the numerical differences between different single-layer mapping functions were actually very small, so we used the simplest single-layer model (SLM) mapping function to convert the slant PEC (SPEC) into a vertical PEC (VPEC):$\mathrm{VPEC} = \frac{\mathrm{SPEC}}{\mathrm{MF}}$, and the MF can be expressed by

$$\mathrm{MF} = \frac{1}{\sqrt{1 - \left(\frac{R_e}{R_e+H}\right)^2 \sin^2(90 - E)}} \tag{1}$$

where $R_e = 6371.4$ km is the Earth' s average radius, E is the elevation angle of the signal, and H is the effective altitude of the SLM, which is generally set to hundreds to thousands of kilometers. We set $H = 1300$ km according to the suggestion of Zhong et al. [37]. It was also followed in the subsequent self-consistency assessment of GIMs. We declared that the subtle differences in VPEC values caused by different H would not affect the main conclusions of this study as the signals with low elevation were excluded.

(2) Similar to the binning strategy used by Li et al. in dealing with the EDPs of ionospheric F region [2], we binned the PodTec dataset from 2008 to 2016 by month (108 months in total) with a 40-day window and geographic longitude (every 15°, 25 longitudinal bands in total) with a 30° window, and we obtained 108 × 25 bins in total. In each bin, the spherical harmonic (SH) regression method was adopted to fit the VPEC in terms of geographic latitude, local time, and solar flux, which can be expressed by

$$\mathrm{VPEC}(\varphi, LT) = \sum_{n=0}^{n_{max}} \sum_{m=0}^{n} \tilde{P}_{n,m}(\sin\varphi)\left[A_{nm}\cos\left(2\pi m\frac{LT}{24}\right) + B_{nm}\sin\left(2\pi m\frac{LT}{24}\right)\right] \tag{2}$$

where n_{max} is the maximum degree of the SH function, which is set to 4 to avoid overfitting the VPEC because of the extremely uneven distribution of PodTec measurements; $\tilde{P}_{n,m}$ are the normalized associated polynomials of degree n and order m; φ is geographic latitude; and LT is local time. The SH coefficients A_{nm} and B_{nm} associated to solar flux are denoted as

$$\begin{cases} A_{nm} = a_{nm} + a'_{nm} \cdot M_{10} + a''_{nm} \cdot M_{10}^2 \\ B_{nm} = b_{nm} + b'_{nm} \cdot M_{10} + b''_{nm} \cdot M_{10}^2 \end{cases} \tag{3}$$

where M_{10} is the solar proxy derived from the daily $M_{10.7}$ and its average with a 27-day sliding window: $M_{10} = \frac{M_{10.7}+M_{10.7}^{27}}{2}$; A_{nm} and B_{nm} are denoted as quadratic regression of M_{10} indices here to weaken the impact of the 27-day solar rotation [38]; and a_{nm}, a'_{nm}, a''_{nm}, b_{nm}, b'_{nm}, and b''_{nm} are all the expansion coefficients to be estimated.

(3) The expansion coefficients in each bin were estimated by least squares, and then, the VPEC at any geographic latitude ($glat$), longitude ($glon$), and local time (LT) could be calculated by bringing the expansion coefficients into Equations (2) and (3). In this step, the 3-D matrix of monthly median VPEC on the 151 × 73 × 49 ($glat \times glon \times LT$) mesh grids of geographic latitude (75°, 74°, 73°, ..., −74°, −75°), longitude (−180°, −175°, −170°, ..., 175°, 180°), and local time (00:00, 00:30, 01:00, ..., 23:30, 24:00) were computed, and the bilinear interpolation method was used to enhance the continuity of PEC distribution between adjacent longitudes. We only considered the PEC from 75°S to 75°N because of the rarely available observables over the Arctic and Antarctic.

We computed the monthly averaged M_{10} from 2008 to 2016, and we finally obtained a 4-D (151 × 73 × 49 × 108, $glat \times glon \times LT \times month$) PEC space–time field. The yearly RMS error of the gridded PEC data compared with the raw PodTec dataset were found to be between 1 and 2 TECU, and the values increased slightly under high solar activity conditions. Figure 2 shows an example of binning results at 12:00 LT in March 2014. One can see that the electrons of the plasmasphere were generally distributed around the magnetic equator, and the PEC value decreased sharply with increasing latitude. We can also see that the uniform gridded data through our binning procedure matched well with the original PodTec dataset, though the distribution of PEC measurements was extremely uneven.

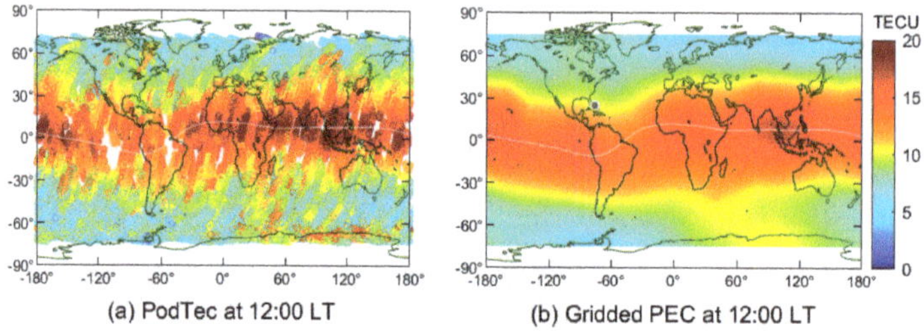

Figure 2. (**a**) Original PodTec dataset and (**b**) gridded PEC data at 12:00 LT in March 2014.

3. PEC EOF Analysis and Model Construction

3.1. PEC EOF Analysis

The EOF analysis is a data dimensionality reduction technique that can be used to analyze and reconstruct the original space–time field through a set of orthogonal basis functions or principal components [39]. For a 2-D matrix with sets of vector columns $\boldsymbol{X}_{m\times m} = \{\boldsymbol{x}_1, \boldsymbol{x}_2, \ldots, \boldsymbol{x}_n\}$, it can be decomposed into $\boldsymbol{V}_{m\times m}$ and $\boldsymbol{T}_{m\times n}$ via singular value decomposition, as follows:

$$\boldsymbol{X}_{m\times n} = \boldsymbol{V}_{m\times m} \times \boldsymbol{T}_{m\times n} \quad (4)$$

where $\boldsymbol{v}_{m\times m} = \{\boldsymbol{v}_1, \boldsymbol{v}_2, \ldots, \boldsymbol{v}_n\}$, $\boldsymbol{t}_{m\times n} = \{\boldsymbol{t}_1, \boldsymbol{t}_2, \ldots, \boldsymbol{t}_n\}^T$, and the combination of $\boldsymbol{v}_k$ and $\boldsymbol{t}_k$ are called the k-th EOF mode.

The physical meaning of different EOF modes can usually be explained by combining the original space–time field, and the importance of EOF modes can be ranked with the variance contribution ratios calculated by the eigenvalues:

$$r_i = \frac{\lambda_i}{\sum_{i=1}^{m} \lambda_i} \times 100\% \quad (5)$$

where λ_i is the *i*-th eigenvalue.

The first few major EOF modes can usually contribute the majority of the total variance and represent major spatiotemporal patterns of the original space–time field [2,9,31,40,41]. The procedure of the first-layer EOF decomposition is similar to that in the study of Zhong et al. [9], but the diurnal variations of the global PEC were also taken into account in this work. The gridded PEC dataset was regrouped into a 11,023 × 5292 matrix, and it could be decomposed into the combinations of spatial patterns E_k and time-associated amplitude coefficients P_k,

$$\mathrm{PEC}(glat, glon, LT, month, solar) = \sum_{k=1}^{m} E_k(glat, glon) \times P_k(LT, month, solar) \quad (6)$$

where $E_k(glat, glon)$ is the k-th EOF basis function associated with the geographic location and $P_k(LT, month, solar)$ is the k-th amplitude coefficient related to the local time, season, and solar activity condition.

We conducted the first-layer EOF decomposition using Equation (6), and Table 1 lists the variance contribution ratios and cumulative variance ratios of the six major EOF modes.

The Pearson correlation coefficients between the solar indices and the daily mean P_k are also shown in the table.

Table 1. Variance of the first 6 EOF modes through the first-layer EOF decomposition.

EOF Modes	Variances [%]	Cumulative Variances [%]	R between M_{10} and P_k
$E_1 \times P_1$	96.39	96.39	0.912
$E_2 \times P_2$	1.99	98.38	0.195
$E_3 \times P_3$	0.51	98.89	−0.695
$E_4 \times P_4$	0.26	99.15	−0.093
$E_5 \times P_5$	0.17	99.32	0.636
$E_6 \times P_6$	0.10	99.42	−0.03

The first-layer EOF decomposition can be used to separate the temporal and spatial variation of the original space–time field and provide physically meaningful spatiotemporal variation patterns of the global PEC. One can see that the first four major EOF modes contribute more than 99% of the total variance. We analyzed the first four EOF modes with emphasis due to the fast convergence of the eigen series, and the global distribution of the spatial patterns (E_1~E_4) and variations in amplitude coefficients (P_1~P_4) are shown in Figure 3a–h. The variations of monthly averaged M_{10} during 2008~2016 are shown in Figure 3e.

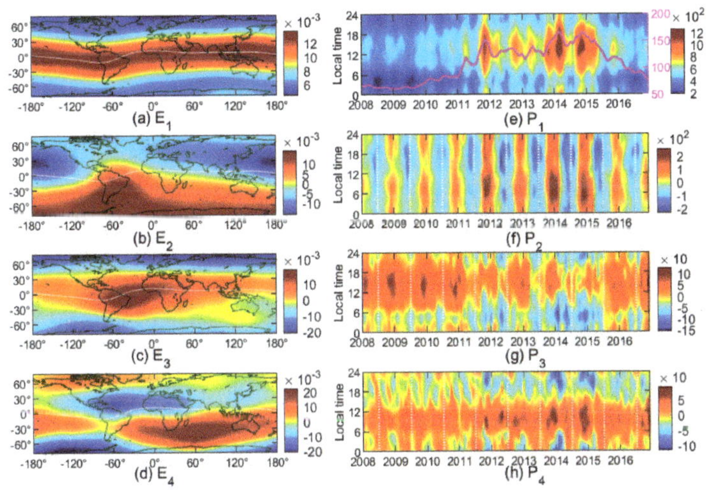

Figure 3. (**a–d**) Global distribution of E_1–E_4 and (**e–h**) variations of P_1–P_4 with month and local time in 2008~2016.

The variance contribution ratio of the first mode ($E_1 \times P_1$) is 96.39%, suggesting that it captures most of the variance of the original field. Only positive values can be found for both E_1 and P_1. In general, the first EOF mode can represent the predominant spatial and temporal pattern of the global PEC. Figure 3a shows that the E_1 exhibits a symmetrical distribution relative to the magnetic equator and a monotonically decreasing trend from low to high latitudes. This is different from the two-peak structure of the EIA in the global TEC or NmF2 [42]. The latitudinal gradient of E_1 may be associated with the shape of the plasmasphere, as it is usually closely constrained by the Earth's magnetic field lines, and the closer to the equator, the thinner the plasmasphere [9,13]. The time-associated amplitude coefficients P_1 are shown in Figure 3e, which shows a significant long-term variation from 2008~2016, with a clear ascending phase from 2009 to 2014 and a sharp reduction from 2015 to 2016, which is consistent with the variations in solar activity levels. The P_1 values are larger near the northern winter solstice than in June, and significantly larger values can also be found near equinox months, similar to the annual and semi-annual variation patterns in global NmF2 or TEC [41–44]. The linear correlation coefficient between M_{10} and P_1 is 0.912 (as shown in Table 1), which also indicates a strong solar activity dependence of the global PEC.

The variance ratio of the $E_2 \times P_2$ mode is only 1.99%, which is much smaller than that of the $E_1 \times P_1$ mode. Both negative and positive values can be found for both E_2 and P_2, as shown in Figure 3b,f. Opposite E_2 values can be observed for different hemispheres, but they are highly asymmetric. There are two distinct peak regions in both hemispheres; one is located over the North Pacific, and the other one is over South America and extends toward the Weddell Sea. The hemisphere boundary line is also irregular: it deviates significantly from both the magnetic equator and the geographical equator. P_2 shows significant annual variation. It generally reaches its minimum in June and maximum in December and approaches 0 near the equinox months. The amplitude variation of P_2 also suggests its strong solar activity dependence, though the linear correlation coefficient shown in Table 1 is rather small. For the northern hemisphere, the seasonal component represented by the $E_2 \times P_2$ mode is positive in June and negative in December, and the opposite is true for the southern hemisphere. In general, the $E_2 \times P_2$ mode mainly represents the hemispherical differences in the summer-to-winter variation pattern of the global PEC, which is similar to that of global NmF2 and TEC [22,41,44], but significant hemispherical asymmetry different from the NmF2 and TEC can also be found in the spatial pattern.

As for the third EOF mode ($E_3 \times P_3$) and the fourth EOF mode ($E_4 \times P_4$), they only contribute 0.51% and 0.26% to the total variance, respectively. The $E_3 \times P_3$ mode can present latitudinal differences in seasonal and diurnal variations of the global PEC, and two boundary lines can be found near 40°N and 40°S. The PEC components over the Weddell Sea region between 120°W∼40°W are usually positive during nighttime, and it is opposite during daytime, which is similar to the Weddell Sea anomalies in TEC and NmF2 [41,42,44]. The significantly larger E_3 values near the equator, particularly over South America indicate that the amplitude of seasonal variations in low to mid-latitudes is much larger than that in higher latitudes. The spatiotemporal patterns represented by the $E_3 \times P_3$ mode also indicate the differences in the PEC between different latitudes in solar activity dependence. The $E_4 \times P_4$ mode shows an opposite spatial pattern compared to the $E_2 \times P_2$ mode, and its amplitude coefficients (P_2) show a minor correlation to the solar activity. As the $E_3 \times P_3$ and $E_4 \times P_4$ modes contribute only 0.77% of the total variance, we did not find out more about the physical meaning of their spatiotemporal patterns in this work.

For the ease of model construction, we suggest performing a second-layer decomposition on the major EOF modes after the first-layer EOF decomposition. In this case, the spatial patterns E_k and amplitude coefficients P_k can be decomposed into

$$\begin{cases} E_k(glat, glon) = \sum_{i=1}^{m} e_i^k(glat) \times u_i^k(glon) \\ P_k(LT, month, solar) = \sum_{i=1}^{n} p_i^k(LT) \times v_i^k(month, solar) \end{cases} \tag{7}$$

where $e_i^k(glat)$, $u_i^k(glon)$, and $p_i^k(LT)$ are the second-layer EOF basis functions associated with geographic latitude, longitude, and local time and $v_i^k(month, solar)$ is the time-associated basis function related to the seasonal variation and solar activity dependence of the global PEC.

As the $E_1 \times P_1$ mode contributes 96.39% of the total variance, the second-layer EOF analysis is focused on the E_1 and P_1 pattern. The variance ratio of the $e_1 \times u_1$ mode is 99.44%, which thus represents the major feature of E_1, and the shape of the latitude-associated basis function (e_1) and longitude-associated function (p_1) are shown in Figure 4.

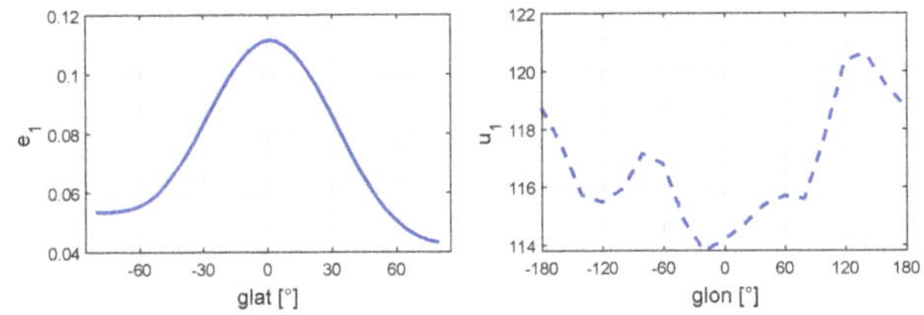

Figure 4. Normalized e_1 (**left**) and u_1 (**right**) after a second-layer EOF decomposition for E_1.

We can find in Figure 4 that the distribution of the PEC is not strictly symmetrical to the equator according to the shape of e_1, and there are also pronounced longitudinal differences that are not significantly reflected in Figure 3a.

As for the amplitude coefficient P_1, the $p_1 \times v_1$ mode contributes 99.72% of the total variance, which thus can represent the major pattern of the P_1 series, as shown in Figure 5.

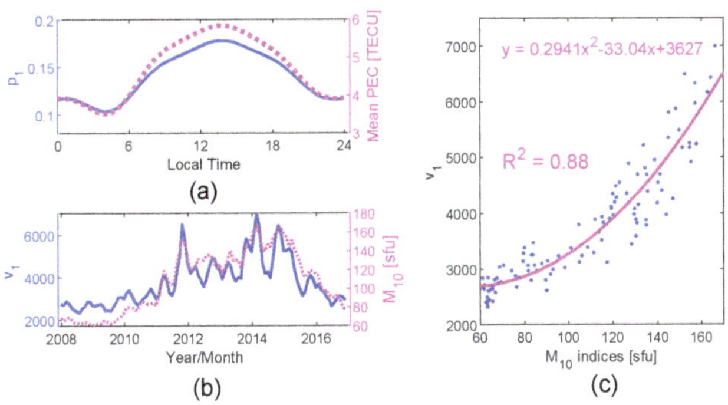

Figure 5. (**a**) p_1 and (**b**) v_1 after a second-layer decomposition from P_1 and (**c**) scatter plot and fitting curve for M_{10} against v_1.

We can see that p_1 mainly represents the diurnal variation of the global PEC, and the similar variation pattern of v_1 and M_{10} illustrates that the global average PEC is strongly controlled by the solar activity condition in the long term. However, the fitting curve shown in Figure 5c indicates that the solar activity dependence of the global PEC seems not to be linear, so it is best to consider a quadratic regression of the solar indices in model construction.

We conducted a second-layer decomposition for the remaining EOF modes according to Equation (7), and the variance contribution ratios of the second-layer basis EOF modes are shown in Figure 6. The cumulative variances of the first three major second-layer modes for E_1–E_4 and P_1–P_4 are also indicated at the bottom.

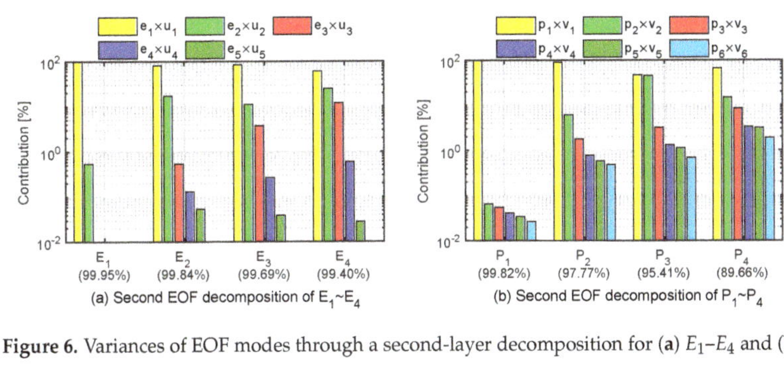

Figure 6. Variances of EOF modes through a second-layer decomposition for (**a**) E_1–E_4 and (**b**) P_1–P_4.

One can see in Figure 6a that the first three major second-layer basis modes contain 99.95%, 99.84%, 99.69%, and 99.40% of the total variances for E_1–E_4 pattern, respectively. The amplitude coefficients P_1–P_4 can be decomposed into basis functions response to the diurnal/semi-diurnal variation (p_i^k) and long-term variations (v_i^k) of the global PEC. As shown in Figure 6b, the three major second-layer modes related to local time, season, and solar activity contribute 99.82%, 97.77%, 95.41%, and 89.66% of the total variance for P_1–P_4, respectively. Similar analyses like those shown in Figures 4 and 5 have also been applied to the second-layer major EOF modes for $E_2 \sim E_4$ and $P_2 \sim P_4$. The second-layer EOF analysis demonstrates that we can reconstruct the global PEC by fitting the several major modes after a second-layer decomposition because of the fast eigenseries convergence for E_k and P_k.

3.2. Model Construction

We conducted a successful PEC model reconstruction by fitting the second-layer basis functions, and the PEC model can be described by

$$PEC^{mod} = \sum_{k=1}^{M} \left\{ \left[\sum_{i=1}^{m_k} e_i^k(glat) \times u_i^k(glon) \right] \times \left[\sum_{i=1}^{n_k} p_i^k(LT) \times v_i^k(month, solar) \right] \right\} \tag{8}$$

where m_k and n_k are the numbers of second-layer basis modes used in the PEC modeling; we set them to 3 according to the second-layer EOF analysis result shown in Figure 6.

The latitude and longitude associated functions e_i^k and u_i^k can be fitted by

$$\begin{cases} e_i^k(glat) = a_0^{glat} + \sum_{j=1}^{3} a_j^{glat,k,i} \cos\left(j \cdot glat \cdot \frac{2\pi}{360}\right) + b_j^{glat,k,i} \sin\left(j \cdot glat \cdot \frac{2\pi}{360}\right) \\ u_i^k(glon) = a_0^{glon} + \sum_{j=1}^{3} a_j^{glon,k,i} \cos\left(j \cdot glon \cdot \frac{2\pi}{360}\right) + b_j^{glon,k,i} \sin\left(j \cdot glon \cdot \frac{2\pi}{360}\right) \end{cases} \tag{9}$$

where $a_j^{glat,k,i}$, $b_j^{glat,k,i}$, $a_j^{glon,k,i}$, and $b_j^{glon,k,i}$ are the coefficients to be estimated.

We take the fitting results for E_1–E_4 shown in Figure 7 as examples.

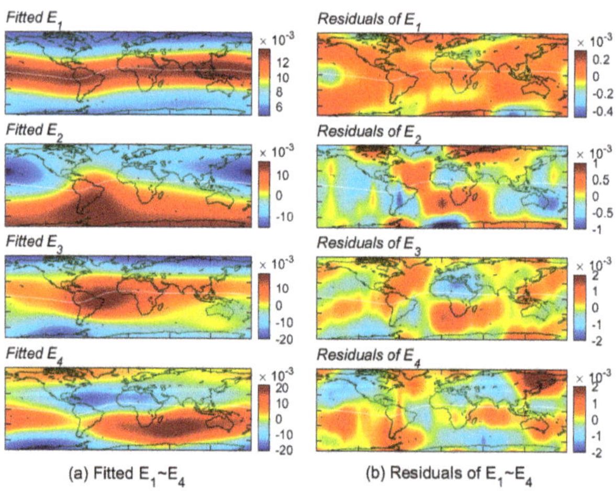

Figure 7. (**a**) Fitting results and (**b**) residuals of E_1–E_4 respectively.

Figure 7 indicates that the predominant spatial patterns, such as latitudinal gradient, longitudinal difference, hemispherical asymmetry, etc., in E_1–E_4 can be reconstructed with fidelity using Equation (9).

The local time-associated functions p_i^k can be fitted by

$$p_i^k(LT) = a_0^{LT} + \sum_{j=1}^{2} a_j^{LT,k,i} \cos\left(j \cdot LT \cdot \frac{2\pi}{24}\right) + b_j^{LT,k,i} \sin\left(j \cdot LT \cdot \frac{2\pi}{24}\right) \quad (10)$$

And, the v_i^k response to the seasonal variation and solar activity dependence of global PEC can be fitted by

$$\begin{aligned} v_i^k(month, M_{10}) = {} & a_0^{seas,k,i} + a_1^{seas,k,i} \cdot M_{10} + a_2^{seas,k,i} \cdot M_{10}^2 \\ & + \sum_{j=1}^{2} \left[b_{1,2j-1}^{seas,k,i} \cos\left(j \cdot month \cdot \frac{2\pi}{12}\right) + b_{1,2j}^{seas,k,i} \sin\left(j \cdot month \cdot \frac{2\pi}{12}\right) \right] \\ & + \sum_{j=1}^{2} \left[b_{2,2j-1}^{seas,k,i} \cos\left(j \cdot month \cdot \frac{2\pi}{12}\right) + b_{2,2j}^{seas,k,i} \sin\left(j \cdot month \cdot \frac{2\pi}{12}\right) \right] \cdot M_{10} \end{aligned} \quad (11)$$

A quadratic regression term of solar proxy, a seasonal term containing annual/semi-annual variation, and a cross-term modulation of solar proxy and seasonal terms are included. The fitting results for P_1–P_4 are shown in Figure 8.

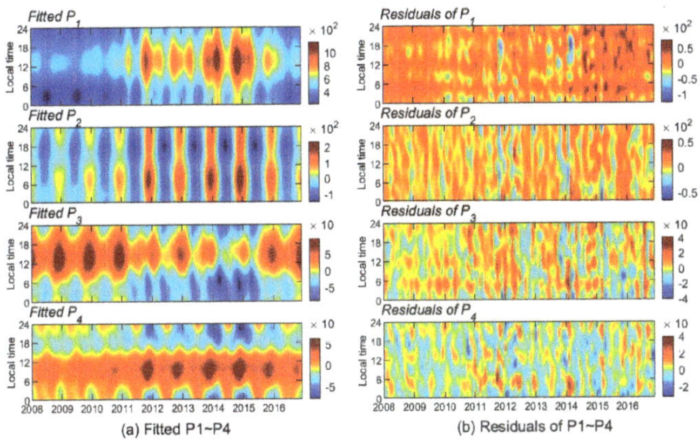

Figure 8. (**a**) Fitting results and (**b**) residuals of P_1–P_4, respectively.

We can also see good fitting results except for some outliers under extremely strong solar activity conditions in Figure 8.

To investigate the impact of the first-layer major EOF mode numbers on model accuracy, we conducted a second-layer decomposition for E_5–E_{10} and P_5–P_{10} and built the PEC model with different numbers (one to ten) of first-layer major EOF modes. The results of the correlation coefficient and root-mean-square (RMS) error of the PEC model compared to the gridded PEC dataset are shown in Figure 9.

We find in Figure 9 that the accuracy of the PEC model can reach high fidelity with an RMS error of 0.74 TECU when only three first-layer major EOF modes are considered, and the PEC model accuracy seems to not be significantly improved when more first-layer EOF modes are used. It also can be seen in Figure 9c that with three first-layer major EOF modes, the residuals of the PEC model compared to gridded PEC data are mainly concentrated between −1 to 1 TECU. Therefore, we set $M = 3$ in Equation (8).

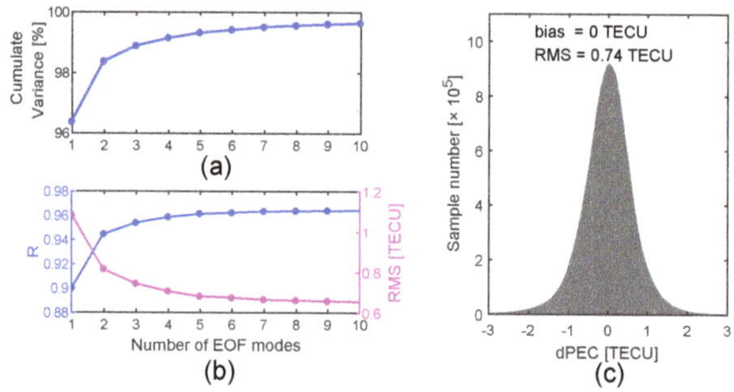

Figure 9. (**a**) Cumulate variance for different numbers of major EOF modes; (**b**) the correlation coefficients and RMS errors of the PEC model compared to those of the gridded PEC dataset when using different numbers of major EOF modes; (**c**) histograms of model residuals when three first-layer major EOF modes are considered.

The PEC model can be driven by solar proxies, but sometimes, for real-time operability, such as the ionospheric delay error mitigation in precisely GNSS positioning application, a driven parameter that can be obtained in real-time becomes more important. In this work, we try to replace the M_{10} indices with the Klobuchar model parameters, hereinafter $Kpar$, which can be derived from the Klobuchar model with the daily broadcasted 8 coefficients according to the research of Hoque et al. [45]:

$$Kpar = \text{VTEC}_\text{A} + \text{VTEC}_\text{B} = \left(T_{ion}^{A}(\varphi_A, \lambda_A) + T_{ion}^{B}(\varphi_B, \lambda_B)\right) \cdot c/\gamma \tag{12}$$

where c = 299,792,458 m/s is the light speed, $\gamma = 6.1587$ is the factor converting T_{ion} to TEC, and T_{ion}^{A} and T_{ion}^{B} are ionospheric delay values over A (10°N, 90°W) and B (10°S, 90°W), respectively, computed with Klobuchar model [46]:

$$T_{ion} = \begin{cases} A_0 + A_1 \cos\left(2\pi \cdot \frac{t-t_0}{A_2}\right) & , \left|\frac{t-t_0}{A_2}\right| \le \frac{1}{4} \\ A_0 & , \left|\frac{t-t_0}{A_2}\right| > \frac{1}{4} \end{cases} \tag{13}$$

where $A_0 = 5$ ns is the constant of the ionospheric delay during night time, $t_0 = 50{,}040$ s denotes the local time at 14:00, t is the epoch to compute the ionospheric delay error, and A_1 and A_2 are the amplitude and period, respectively, associated with geomagnetic latitudes:

$$A_1 = \begin{cases} \sum_{i=0}^{3} \alpha_i |\phi_m|^i & , A_1 \ge 0 \\ 0 & , A_1 < 0 \end{cases} \tag{14}$$

$$A_2 = \begin{cases} \sum_{i=0}^{3} \beta_i |\phi_m|^i & , A_2 \ge 72{,}000 \\ 0 & , A_2 < 72{,}000 \end{cases} \tag{15}$$

where $\alpha_i, \beta_i (i = 0, 1, 2, 3)$ are the Klobuchar model coefficients, which are with the broadcast ephemeris in the navigation message of the GPS; they are generally updated every 1~7 days.

Figure 10 shows the M_{10} and $Kpar$ indices variations during the nearly 20 years and the fitting curve of $Kpar$ against M_{10}.

It can be seen that the long-term variation pattern of $Kpar$ is similar to that of M_{10}. We conducted a cubic regression from $Kpar$ to M_{10} and obtained the following equation to derive M_{10} from $Kpar$:

$$M_{10} = 0.0003043 \cdot Kpar^3 - 0.07179 \cdot Kpar^2 + 5.917 \cdot Kpar - 15.03 \tag{16}$$

For ease of description, we give the empirical PEC model constructed in this work the name Plasmaspheric Electron Content Model (PECM) and the PECM model driven by M_{10} and $Kpar$ indices the name PECM-M10 and PECM-Kpar, respectively.

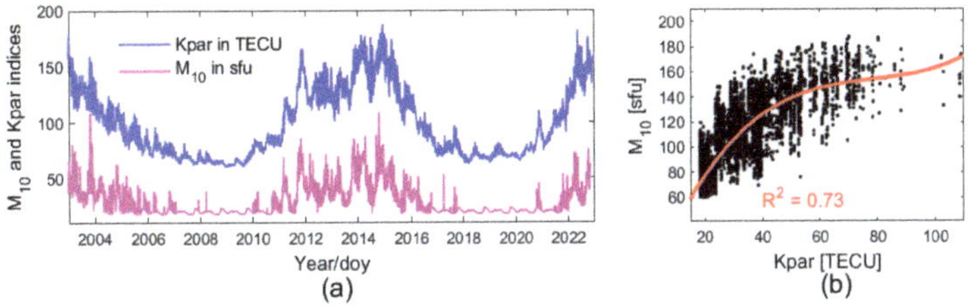

Figure 10. (**a**) Daily M_{10} and $Kpar$ indices in 2003~2022 and (**b**) scatter plot and fit curve of $Kpar$ against M_{10}.

4. Model Validation

4.1. Model Errors

We reconstructed the PEC model with the gridded PEC dataset that excluded the data from 2009 and 2014. In this section, the model error analysis for both PECM-M10 and PECM-Kpar is conducted with the raw PodTec dataset in 2009 and 2014 and with the topside TECs derived from the NeQuick2 model as a comparison. The histograms of residuals and mean/daily bias/RMS error for different competing models in 2009 and 2014 are shown in Figure 11.

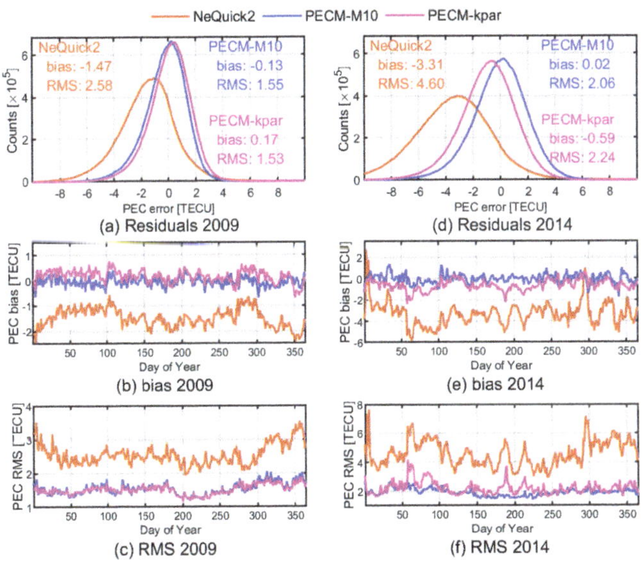

Figure 11. (**a**,**d**) Histograms of model residuals; (**b**,**e**) daily bias; and (**c**,**f**) daily RMS errors of different competing models in 2009 and 2014, respectively.

One can see that the NeQuick2 model underestimates global PEC consistently, whether in 2009 or 2014, with a mean bias of −1.47 TECU and −3.31 TECU, respectively. By contrast, deviations close to zero can be observed for both PECM-M10 and PECM-Kpar in 2009, and their daily deviations also fluctuate around zero, but in 2014, the PECM-Kpar slightly underestimates the global PEC with a mean bias of −0.59 TECU and the underestimations mainly occur in March and June. The PECM-M10, however, still performs the best with a mean bias of only 0.02 TECU. In terms of RMS error, we can also observe that both PECM-M10 and PECM-Kpar perform better than NeQuick2 model, with improvements of 39.92% and 40.70% in 2009 and 55.22% and 51.74% in 2014, respectively. In 2009, the PECM-M10 and PECM-Kpar perform similarly, and minor seasonal differences between equinox and solstice can be found. In 2014, however, remarkably larger daily RMS errors for PECM-Kpar can be found in several days near the March equinox and June solstice, which suggest significant differences sometimes between the true solar activity level and that derived from *Kpar* indices. The above analysis preliminarily indicates that the proposed empirical PEC models can more accurately describe the global PEC, and the PEC model driven by *Kpar* has only a slight decrease in accuracy compared to that driven by M_{10}.

4.2. Validity of Climatological Description

The validity of the climatological description of an empirical model is also very important and sometimes even more important than the mathematical errors [2]. We first checked the validity of different competing models in describing the latitudinal gradient of global PEC, and the results are shown in Figures 12 and 13.

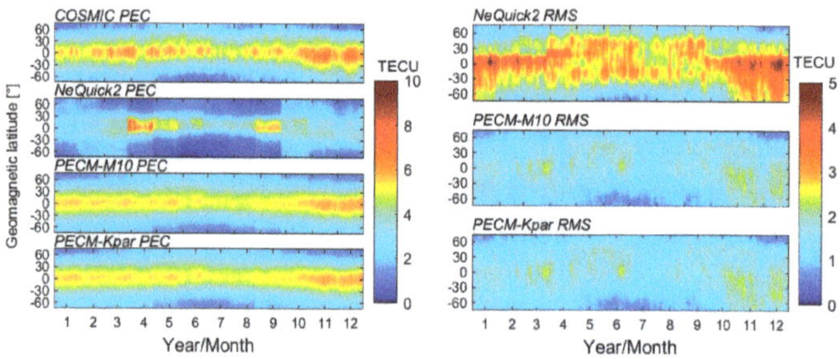

Figure 12. Daily average PEC distributed with geomagnetic latitude of PodTec dataset, NeQuick2 model, PECM-M10 model, and PECM-Kpar model (**left**) and daily RMS error at different geomagnetic latitudes for the three competing models (**right**) in 2009.

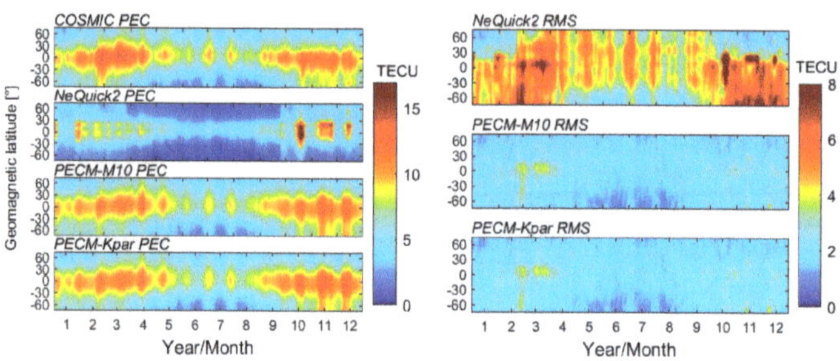

Figure 13. Similar to Figure 12 but for 2014.

In 2009, the PEC values are generally small, and the electrons in the plasmasphere are mainly distributed near the magnetic equator. The PEC exhibits a significant latitude gradient. The most prominent feature in terms of seasonal variation is the annual anomaly, where the PEC is significantly higher in northern winter than in summer, but the semi-annual anomaly is unclear. Both PECM-M10 and PECM-Kpar can capture the seasonal variations in the latitudinal gradient of the global PEC, with much smaller RMS errors than that of NeQuick2, and there seems to be no differences between PECM-M10 and PECM-Kpar. However, slight PEC underestimations can be found for both PECM-M10 and PECM-Kpar, particularly in December, and larger RMS errors can also be found in low latitudes of the southern hemisphere. In 2014, both annual and semi-annual anomalies can be found for PEC in low latitudes, and the proposed PEC models capture the seasonal variations of global PEC well but tend to slightly overestimate PEC in northern winter months and underestimate PEC near the March equinox. In terms of RMS error, the PECM-M10 and PECM-Kpar models have relatively uniform and better accuracy than NeQuick2 in different months except for March. The significant accuracy decrease in March 2014 for the proposed PEC models can be attributed to the abnormally active geomagnetic conditions, which indicates that the current empirical PEC models cannot accurately capture the PEC variations under complex space weather yet.

In addition to the latitudinal gradient, the longitudinal differences in seasonal variations of the global PEC are also prominent, especially for daytime PEC at low latitudes. We further checked the consistency of different competing models compared to the PodTec dataset by taking the daytime (LT10:00∼16:00) global PEC in July and December under different solar activity conditions as examples, and the results are shown in Figures 14 and 15.

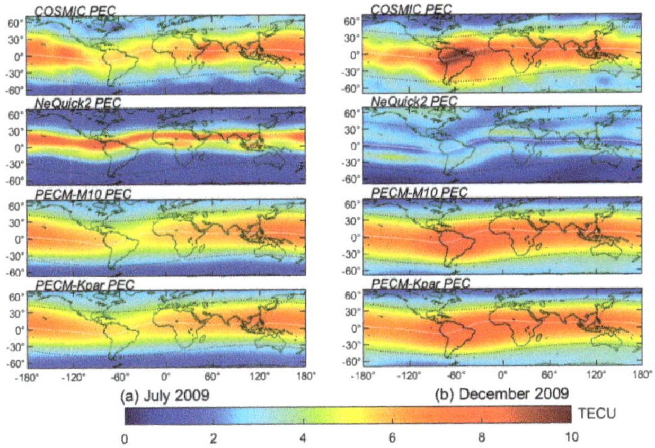

Figure 14. The monthly average PEC computed with PodTec dataset, NeQuick2 model, PECM-M10 model, and PECM-Kpar model in (**a**) July and (**b**) December 2009.

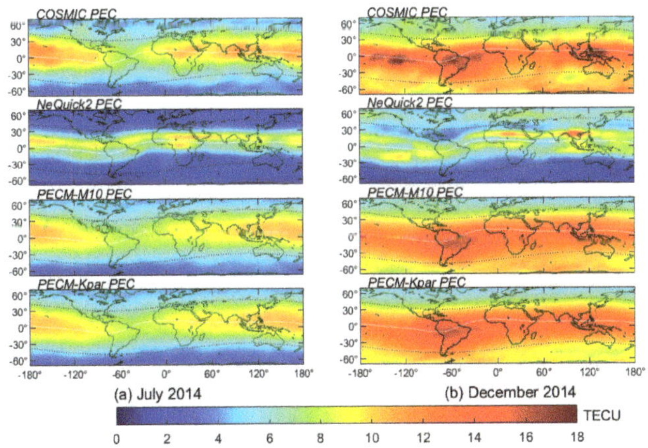

Figure 15. Similar to Figure 14 but for 2014.

The average $F_{10.7}$ in July 2009 and 2014 is 68 sfu and 137 sfu, respectively, and we can observe significant longitudinal differences manifested as a higher PEC over the Pacific Ocean at mid to low latitudes than that over the Atlantic Ocean under both low and high solar activity levels, which can correspond to the shape of spatial pattern of the $E_1 \times P_1$ mode shown in Figure 3. Both PECM-M10 and PECM-Kpar models capture the spatial pattern of the global PEC in July 2009 and 2014 well. The NeQuick2 model usually underestimates the PEC in mid to high latitudes and shows a sharper latitudinal gradient than those of PECM-M10 and PECM-Kpar models under different solar activity conditions.

In low latitudes, a prominent EIA structure can also be observed in the PEC computed with the NeQuick2 model, which is not consistent with the climatological behavior of electrons in the plasmasphere.

The average $F_{10.7}$ in December 2009 and 2014 is 74 sfu and 159 sfu, respectively. It can be observed that the global PEC in December exhibits completely opposite longitudinal differences to that in June. The PEC over South America and the Southern Atlantic is generally higher than that over the Pacific region, which is even more pronounced under low solar activity conditions. The different spatial patterns in Figures 14 and 15 can be attributed to the differences over different longitudinal zones in the summer-to-winter variation, and the PEC over South America generally exhibits more significant annual variations than other longitudinal zones [9,22]. Both PECM-M10 and PECM-Kpar can capture these changes well, but the shape of the modeled PEC distribution is generally smoother, which indicates that the monthly median model still struggles to capture the details of local PEC variations accurately. The NeQuick2 model, however, still shows overall underestimation, and significant EIA structures can still be observed in both 2009 and 2014. As we know, two Ne peaks at about ±15° magnetic latitude and a Ne trough near the magnetic equator are usually caused by the fountain effect, but they usually merge into one peak with increasing height, according to the study of Li et al. [2], so the global PEC should typically exhibit a more monotonic latitudinal gradient compared with the ionospheric F-layer. The erroneous description of the PEC distribution in the NeQuick2 model suggests that the altitude extension strategy of the current 3-D electron density empirical model in dealing with the EDP of the top ionosphere and plasmasphere may not be suitable.

We further check the long-term trend of the proposed PEC model's mean deviation on a global scale, and the results are shown in Figure 16, where some available PodTec data in 2007 and 2017~2018 are also included.

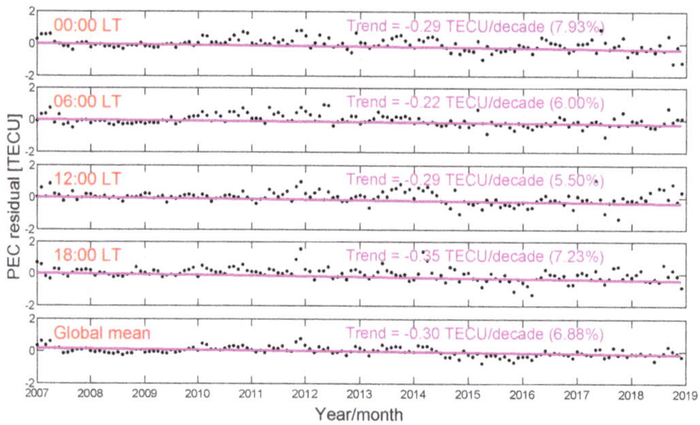

Figure 16. Monthly mean deviations of PEC models compared to the PodTec dataset. The red lines represent the linear fitting of the residuals, and the linear trends per decade for different local times are also shown in each subgraph.

We can find in Figure 16 that the global mean residuals of the proposed PEC models usually randomly fluctuate near the zero line, with positive monthly mean deviations in 2010~2013 and negative values in 2014~2016. The linear fitting results of the global mean PEC residuals demonstrate that the long-term trends of the PEC are −0.29, −0.22, −0.29, and −0.35 TECU per decade (7.93%, 6.00%, 5.50%, and 7.23%) at LT00:00, 06:00, 12:00, and 18:00, respectively, and with −0.30 TECU (6.88%) on a global scale. The results at different local times show a consistent decreasing trend in 2007~2018, which is different from the study of Zhong et al., who found that the long-term trend of the mean PEC above the orbit of MetOp-A (830 km) behaves completely opposite at LT09:30 and 21:30, with −0.061

and 0.023 TECU per decade, respectively [9]. Lean et al. analyzed the long-term trend of the global mean TEC and found that it has a long-term decrease trend with −0.15 per decade [43]. It was even smaller than our result. More available data for a longer period are needed to further validate the truth of the consistent decline trend found in the current work, and if so, an empirical global PEC model that corrects the long-term trend of the global PEC can be further developed.

Overall, the verification work above demonstrates that the proposed empirical PEC model, whether driven by M_{10} or *Kpar* indices, can accurately capture the major spatial and temporal patterns of the global PEC. They have uniform and better accuracy compared with the NeQuick2 model, although the PECM-Kpar model sometimes slightly underperforms compared to the PECM-M10 model under complex space weather.

5. Model Application in Vertical-Slant TEC Conversion

5.1. Two-Layer Mapping Function Based on PECM Model

The ionospheric delay information is crucial for GNSS users for precise positioning, and various VTEC products, such as the GIMs of different Ionosphere Associate Analysis Centers (IAACs), can provide high-precision ionospheric delay error correction information [33,47–49], but they are all usually two-dimensional, and a mapping function is necessary to achieve vertical-slant TEC conversion in practical use. The commonly used single-layer mapping functions are simple enough, but they usually do not consider the electrons contributed by the plasmasphere, and larger errors may be introduced into the vertical-slant TEC conversion, particularly for the signals with low elevation angles [32].

In this work, a two-layer mapping function based on the PECM-Kpar model, namely, PECM-2L, is further proposed to investigate whether introducing plasmasphere information can help with vertical-slant TEC conversion, of which the basic principle is to divide the STEC of a given GNSS ray into two parts with an altitude of 800 km as the boundary artificially and then conduct a vertical-slant TEC or PEC conversion, respectively, at two assumed thin shells; the diagrammatic sketch is shown as Figure 17.

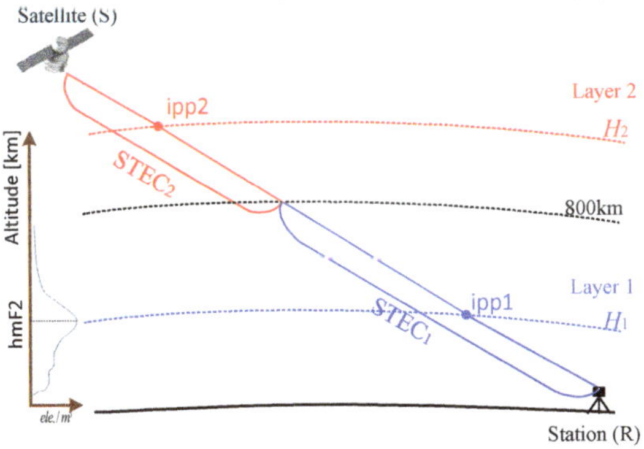

Figure 17. Diagrammatic sketch of the PECM-2L mapping function.

Under the two-layer thin-shell assumption, the STEC for the given GNSS ray can be expressed by

$$\text{STEC} = \text{STEC}_1 + \text{STEC}_2 = \int_{Rec}^{800} Ne(l_1)dl + \int_{800}^{Sat} Ne(l_2)dl \approx \bar{N}_1 \cdot l_1 + \bar{N}_2 \cdot l_2 \quad (17)$$

where l_1 and l_2 are the lengths of the ray paths included in the first and second layers, respectively, and $\bar{N}_1$, $\bar{N}_2$ are the average electron density near the two assumed thin shells, respectively.

We assume that both the VTEC products and PEC model are precise enough and neglect the horizontal gradient of electrons within each layer. Then, the STEC can be computed with the VPEC and VTEC values over the two ionospheric pierce points (IPP) of the given GNSS ray:

$$\mathrm{STEC} = \mathrm{STEC}_1 + \mathrm{STEC}_2 = \left(\mathrm{VTEC}_{ipp1} - \mathrm{VPEC}_{ipp1}\right) \cdot \mathrm{MF}_{ipp1} + \mathrm{VPEC}_{ipp2} \cdot \mathrm{MF}_{ipp2} \quad (18)$$

where MF_{ipp1} and MF_{ipp2} are the mapping function values at the first and second assumed thin shell of a given GNSS ray, respectively; VPEC_{ipp1} and VPEC_{ipp2} are the vertical PEC values over ipp1 and ipp2, respectively, computed with PEC model; and VTEC_{ipp1} is the GNSS-VTEC over ipp1, which can be computed with high-precision VTEC products, such as GIM.

The effective thin-shell height for the single-layer model is generally set to 350~450 km, which is usually appropriate when neglecting the electrons contributed by the plasmasphere [32,33]. This setting is also followed for the first layer of PECM-2L since the PEC over ipp1 was excluded from the VTEC. As for the second layer, we follow the setting of the PEC model where $H = 1300$ km and use the SLM mapping function to achieve vertical-slant PEC conversion. It is worth mentioning that the proposed PECM-2L mapping function is user-orientated, which can serve as an augmentation tool for helping GNSS users derive more accurate STEC from the existing high-precision VTEC products rather than generate any VTEC models.

5.2. Assessment and Discussion

One of the most accurate methods to evaluate the quality of VTEC products is the direct comparison to the GNSS observations, hereinafter dSTEC, which is defined as the differential STEC between a specific epoch and the epoch with the maximum elevation angle on a continuous arc [48]:

$$\mathrm{dSTEC}_t = \mathrm{STEC}_t - \mathrm{STEC}_{E_{max}} = \frac{L_{GF}^t - L_{GF}^{E_{max}}}{\xi_{GF}} = \frac{L1^t - L2^t - \left(L1^{E_{max}} - L2^{E_{max}}\right)}{\xi_{GF}} \quad (19)$$

where E_{max} is the epoch with the maximum elevation angle; $L_{GF}^t = L1^t - L2^t$ and $L_{GF}^{E_{max}} = L1^{E_{max}} - L2^{E_{max}}$ are the geometry-free (GF) linear combination of $L1$ and $L2$ carrier phase observables at epoch t and E_{max}, respectively; $\xi_{GF} = 40.3 \cdot 10^{16} \cdot \left(1/f_2^2 - 1/f_1^2\right) \approx 0.105$ is the factor to convert L_{GF} to TEC; and f_1 and f_2 are the $L1$ and $L2$ GPS signal frequencies, respectively.

The accuracy of the dSTEC observation is typically much below 0.1 TECU. It thus can serve as a reference truth independent of any ionospheric model. The dSTEC value is very sensitive to the relative changes in the STEC with time and elevation angle. For the same VTEC product, the dSTEC analysis results can also reflect the role of different mapping functions in vertical-slant TEC conversion.

We chose the UPC's rapid GIM product, UQRG, to assess the proposed VTEC mapping function, as it has been proven to be better-quality than GIMs from other IAACs when compared to the altimetry-derived VTEC from JASON satellites or the dSTEC from dual-frequency GPS observables [49]. UQRG GIMs are generated with kriging interpolation and a two-layer voxel model [47,48], but they are still two-dimensional, and a mapping function is still necessary to conduct vertical-slant TEC conversion in practical use. Moreover, all IAACs, including UPC, define $H = 450$ km in the IONEX headers. In our opinion, it is the suggested effective height for the interpolation procedure and vertical-slant TEC conversion with a single-layer mapping function. In this work, the SLM mapping function with $H = 450$ km serves as a comparison, and we give it the name SLM450. As for the proposed PECM-2L mapping function, we set $H_1 = 450$ km and $H_2 = 1300$ km, use the SLM mapping function at each layer, and compute the PEC value of the second layer with the PECM-Kpar model.

We gathered the GPS dual-frequency observables in 2014, 2016, and 2018 of more than 400 International GNSS Service (IGS) sites distributed in the range of 70°S and 70°N to perform the dSTEC analysis. In this work, only the continuous arc with a maximum elevation angle larger than 70° was considered, and the dSTEC analyses were conducted for (a) signals with elevation angles between 15° and 45° (low-elevation signals) and (b) all the signals with elevation angle larger than 15° (full data). Several statistical parameters are used to perform the comparison:

(1) dSTEC RMS error. The RMS error between the dSTEC computed by GNSS dual-frequency carrier phase observables and that by GIMs:

$$\mathrm{RMS} = \sqrt{\frac{1}{N}\sum_{i=1}^{N}\left(\mathrm{dSTEC}_{GIM}^{i} - \mathrm{dSTEC}_{OBS}^{i}\right)} \quad (20)$$

where N is the number of available dSTEC observations and dSTEC_{GIM} denotes the relative changes in the STEC computed with the GIM from UPC: $\mathrm{dSTEC}_{GIM} = \mathrm{STEC}_{GIM}^{t} - \mathrm{STEC}_{GIM}^{E_{max}}$. The SLM and the proposed PECM-2L mapping function are used here. dSTEC_{OBS} is the relative change in the STEC computed by Equation (19) from the dual-frequency carrier phase observables.

(2) dRMS. The percentage when PECM-2L performs better than SLM450 in RMS error:

$$\mathrm{dRMS} = \frac{\mathrm{RMS}_{\mathrm{SLM450}} - \mathrm{RMS}_{\mathrm{PECM-2L}}}{\mathrm{RMS}_{\mathrm{SLM450}}} \times 100\% \quad (21)$$

where $\mathrm{RMS}_{\mathrm{SLM450}}$ and $\mathrm{RMS}_{\mathrm{PECM-2L}}$ are the dSTEC RMS errors of the SLM450 and PECM-2L mapping functions, respectively.

(3) $\mathrm{dRMS}_{\mathrm{better}}$. The percentage of testing days when PECM-2L performs better than the SLM450 mapping function.

We first take the dSTEC analysis for the station ORID (39.1°N, 20.8°E) and ALIC (23.7°S, 133.9°E) in 2014 as examples, and the results are shown in Figure 18.

Figure 18 clearly demonstrates that the daily dSTEC RMS errors of PECM-2L are smaller than those of the standard single-layer mapping function for both the ORID and ALIC stations when applying to the UQRG GIMs, and larger RMS error reductions can be found particularly for the low-elevation signals. The RMS error of PECM-2L is over 25% and 30% better than that of SLM450, and it performs better in over 90% of the testing days throughout 2014 compared with SLM450, particularly in the ALIC station, at nearly 100%. An interesting aspect is the seasonal performance of different mapping functions, especially for the signals with low elevation angles. We can see that the dSTEC RMS errors of SLM450 are much larger in March than those in June, which can be attributed to the seasonal differences in the quality of GIMs, as it is well known that the ionosphere is much more active in the equinox than in solstice months. However, we can obtain much better dSTEC analysis results with PECM-2L than SLM450 in different months, and the dSTEC RMS errors of PECM-2L in equinox months do not show significant increases compared with those in other months. The above analysis preliminary demonstrates that the quality of GIMs in practical use can be further improved when considering the impact of the plasmasphere, and the seasonal variations of PEC should be accurately modeled when applying the PECM model to the two-layer thin-shell assumption mapping function.

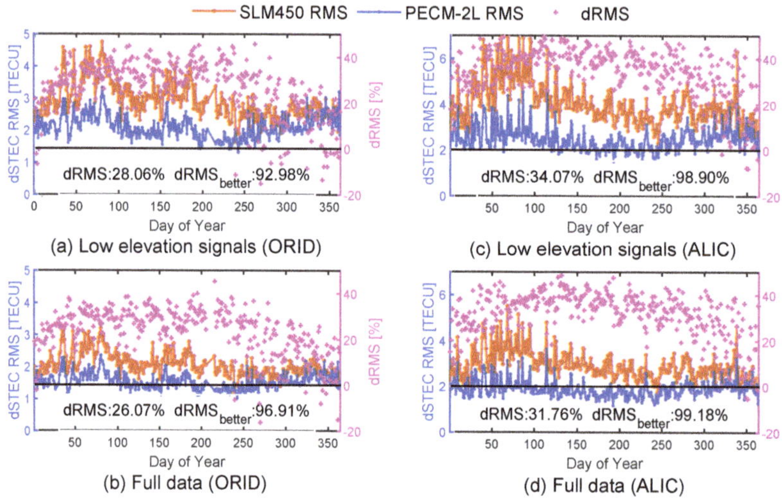

Figure 18. Daily RMS errors and dRMS for (**a**,**c**) low-elevation signals and (**b**,**d**) full data in ORID and ALIC station when using SLM450 and PECM-2L, respectively. The $dRMS_{better}$ and mean dRMS throughout 2014 are also shown at the bottom of each subgraph.

We further computed the dRMS and $dRMS_{better}$ parameters for more than 400 IGS sites distributed globally to explore the role of PECM-2L in aiding vertical-slant TEC conversion on a global scale; the results are shown in Figures 19 and 20.

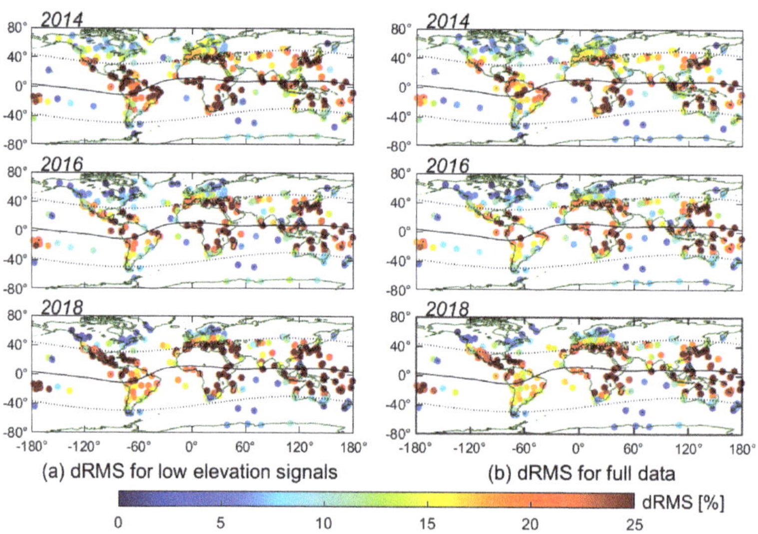

Figure 19. The dRMS distribution of global GNSS sites for (**a**) low-elevation signals and (**b**) full data in 2014, 2016, and 2018 when using SLM450 and PECM-2L, respectively.

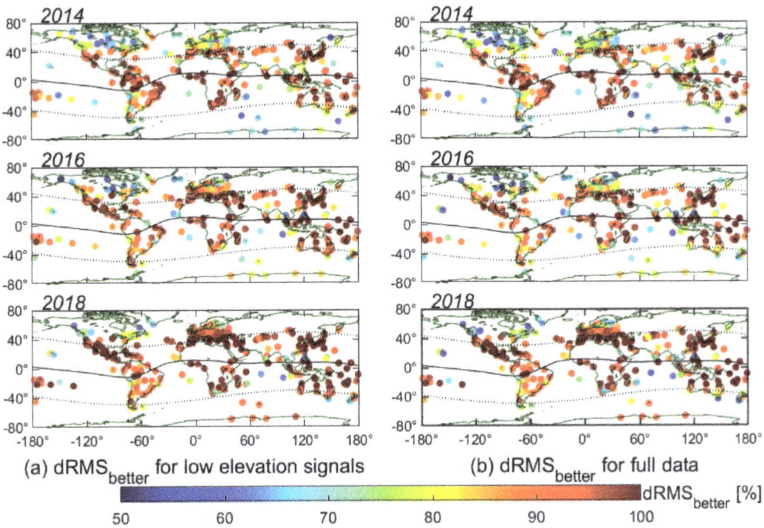

Figure 20. Similar to Figure 19 but for $dRMS_{better}$ parameter.

Figures 19 and 20 clearly demonstrate that the dSTEC analysis results with PECM-2L are better than those with SLM450 for both low-elevation-angle signals and full data in the majority of the global testing IGS sites. The improvement of PECM-2L compared with SLM450 shows a significant latitude effect manifested by the larger dRMS and $dRMS_{better}$ values for testing IGS sites located in low to mid latitudes than those in higher latitudes under different solar activity conditions. The PECM-2L improves dSTEC RMS error by over 25% for the IGS sites located in low to mid latitudes and outperforms the SLM450 in more than 90% of the testing days, but in high latitudes, such as North America, Northern Europe, and Antarctica, the advantages of PECM-2L are not significant any more, and it even performs poorer than SLM450 in some GNSS sites with a negative dRMS value and a $dRMS_{better}$ value less than 50%. Another interesting aspect is that the improvement of PECM-2L is more pronounced in the regions with denser IGS stations, such as Europe, Australia, Japan, South Africa, and the near-equatorial regions of North America. In our opinion, the regional differences in the performance of the PECM-2L model on a global scale can be explained in terms of both the quality of the PECM model and GIMs themselves. On the one hand, rarely available PodTec data in high latitudes result in a lower accuracy of the PECM model and thus affect the performance of PECM-2L in vertical-slant TEC conversion. On the other hand, the quality of GIM is closely related to the denseness of GNSS stations used for modeling, and its accuracy is generally poorer over ocean and continental margins. In this case, there would not be much distinction in the accuracy of a vertical-slant TEC conversion when using different mapping functions.

The average dSTEC RMS error, dRMS, and $dRMS_{better}$ parameters for all the testing IGS sites in different solar activity years were further computed, and the results are shown in Table 2.

As for all the testing IGS sites for computing statistics, the dSTEC RMS errors of both PECM-2L and SLM450 applied to UQRG GIMs are much larger in 2014 than those in 2016 and 2018, particularly for the low-elevation-angle signals, but the PECM-2L always performs better than SLM450 under different solar activity conditions. Overall, PECM-2L applied to UQRG GIMs performs better than SLM450 in 86.92% (90.20%) of the testing days and has an improvement of 20.20% (18.52%) in dSTEC RMS error when validating with low-elevation signals (full data).

The dSTEC analysis results demonstrate that the proposed PECM-2L mapping function can help improve the quality of high-precision VTEC products in practical use, particularly for the signals with low elevation angles.

Table 2. Statistical results of global IGS sites during different testing years when using PECM-2L and SLM450.

Dateset Group	Testing Year	dSTEC RMS [TECU]		PECM-2L with Respect to SLM450	
		SLM450	PECM-2L	dRMS [%]	$dRMS_{better}$ [%]
Low-elevation signals	2014	3.81	3.03	20.51	83.89
	2016	2.68	2.18	18.70	85.94
	2018	2.42	1.92	20.56	89.13
	Mean	2.97	2.37	20.20	86.92
Full data	2014	2.71	2.23	17.64	86.51
	2016	1.96	1.63	17.07	88.89
	2018	1.78	1.44	18.85	92.11
	Mean	2.16	1.76	18.52	90.20

6. Conclusions

In this work, a study on the EOF analysis and model construction of the global PEC above 800 km in altitude is conducted with the topside TECs measured by COSMIC/FORMAT-3 during 2008 through 2016, and the effectiveness of the proposed PEC model in aiding GNSS users to achieve accurate vertical-slant TEC conversion is validated. The main findings and conclusions are as follows:

(1) The first four EOF modes contain more than 99% of the total variance of the PEC space–time field, and the pronounced patterns of latitudinal gradient, longitudinal differences, hemispherical differences, diurnal and seasonal variations, as well as the solar activity dependence, etc., of global PEC were captured. A second-layer decomposition was conducted, and the first three major modes related to geographic latitude and longitude contain over 99% variance of the four major space basis functions (E_1–E_4). The three major modes related to local time, season, and solar activity contribute 99.82%, 97.77%, 95.41%, and 89.66% variance for P_1–P_4, respectively;

(2) An empirical PEC model is constructed with the major EOF modes after a second-layer decomposition, and only three first-layer major modes are utilized. The PEC model is designed to be driven by whether the solar proxy (PECM-M10) or the parameters derived from the Klobuchar model (PECM-Kpar), and there is no significant distinction in model accuracy between them. The PECM-Kpar (PECM-M10) model has accuracy with RMS errors of 1.53 (1.55) TECU and 2.24 (2.06) TECU (see Figure 11a,d), improvements of over 40% and 50% compared to the NeQuick2 model validated with the original topside TEC measurements in 2009 and 2014, respectively. The proposed PEC model can also accurately capture the major spatiotemporal patterns of global PEC;

(3) The proposed PEC model is applied to conduct a vertical-slant TEC conversion experiment based on a two-layer thin-shell assumption. The dSTEC observations of more than 400 globally distributed IGS sites and the high-precision UPC's UQRG GIMs are used for validation. The results show that the proposed PEC model can help improve the consistency between UQRG GIMs and dSTEC observables by 20.20% and 18.52% in dSTEC RMS errors and performs better in 86.92% and 90.20% of the testing days for the low-elevation signals and full data, respectively, compared with the commonly used single-layer mapping function.

This work preliminarily demonstrates the potential role of precise plasmasphere information in helping GNSS users obtain a more accurate slant TEC from the existing VTEC maps. In the following work, the proposed method can be further validated with other high-precision VTEC products and extended to the positioning domain as well.

Author Contributions: Conceptualization, F.L. and C.G.; methodology, F.L.; software, F.L.; validation, F.L. and Y.D.; formal analysis, F.L. and C.G.; investigation, F.L. and Z.X.; resources, F.L.; data curation, F.L.; writing—original draft preparation, F.L.; writing—review and editing, F.L., Y.D. and Z.X.; visualization, F.L.; supervision, C.G.; project administration, F.L. and C.G. All authors have read and agreed to the published version of the manuscript.

Funding: This work was funded by the Postgraduate Research & Practice Innovation Program of Jiangsu Province (grant number KYCX17_0150) and the "Research on Key Technologies for High-precision Navigation and Positioning through GNSS/INS/SLAM Multi-source Fusion" project granted by Nanjing Insitute of Measurement and Testing Technology (grant number KJ2023034).

Data Availability Statement: The PodTec dataset of COSMIC/FORMAT-3 is available at https://cdaac-www.cosmic.ucar.edu/ (accessed on 25 March 2024), the RINEX format GNSS data and IONEX GIMs are available at https://cddis.nasa.gov/archive/gnss (accessed on 18 April 2024), and the long-term series of composite Mg II index are available at https://www.iup.uni-bremen.de/gome/gomemgii.html (accessed on 1 March 2024).

Acknowledgments: The authors would like to thank the COSMIC Data Analysis and Archive Center (CDAAC) for providing PodTec dataset, the NASA's Crustal Dynamics Data Information System (CDDIS) for providing GIMs in the IONEX format and GNSS observables in the RINEX format, and the Global Ozone Monitoring Experiment (GONE) of Universität Bremen for providing long-term series of composite Mg II index. We also thank Angela Aragón-Ángel and Jiafu Wang for their help in manuscript submission.

Conflicts of Interest: The authors declare no conflicts of interest.

References

1. Bilitza, D.; Altadill, D.; Truhlik, V.; Shubin, V.; Galkin, I.; Reinisch, B.; Huang, X. International Reference Ionosphere 2016: From ionospheric climate to real-time weather predictions. *Space Weather* **2017**, *15*, 418–429. [CrossRef]
2. Li, Q.; Liu, L.; He, M.; Huang, H.; Zhong, J.; Yang, N.; Cui, J. A Global Empirical Model of Electron Density Profile in the F Region Ionosphere Basing on COSMIC Measurements. *Space Weather* **2021**, *19*, e2020SW002642. [CrossRef]
3. Huang, H.; Liu, L.; Le, H.; Wan, W. An empirical model of the topside plasma density around 600km based on ROCSAT-1 and Hinotori observations. *J. Geophys. Res. Space Phys.* **2015**, *120*, 4052–4063. [CrossRef]
4. Gowtam, V.; Ram, S. An Artificial Neural Network-Based Ionospheric Model to Predict NmF2 and hmF2 Using Long-Term Data Set of FORMOSAT-3/COSMIC Radio Occultation Observations:Preliminary Results. *J. Geophys. Res. Space Phys.* **2017**, *122*, 11743–11755. [CrossRef]
5. Ram, S.; Gowtam, V.; Mitra, A.; Reinisch, B. The Improved Two-Dimensional Artificial Neural Network-Based Ionospheric Model (ANNIM). *J. Geophys. Res. Space Phys.* **2018**, *123*, 5807–5820.
6. Gowtam, V.; Ram, S.; Truhlik, V.; Reinisch, B.; Prajapati, A. A New Artificial Neural Network-Based Global Three-Dimensional Ionospheric Model (ANNIM-3D) Using Long-Term Ionospheric Observations: Preliminary Results. *J. Geophys. Res. Space Phys.* **2019**, *124*, 4639–4657. [CrossRef]
7. Prol, F.S.; Hoque, M.M. Topside Ionosphere and Plasmasphere Modelling Using GNSS Radio Occultation and POD Data. *Remote Sens.* **2021**, *13*, 1559. [CrossRef]
8. Jakowski, N.; Hoque, M.M. A new electron density model of the plasmasphere for operational applications and services. *J. Space Weather Space Clim.* **2018**, *8*, A16. [CrossRef]
9. Zhong, J.; Lei, J.; Yue, X.; Wang, W.; Burns, A.G.; Luan, X.; Dou, X. Empirical Orthogonal Function Analysis and Modeling of the Topside Ionospheric and Plasmaspheric TECs. *J. Geophys. Res. Space Phys.* **2019**, *124*, 3681–3698. [CrossRef]
10. Chong, X.; Zhang, M.; Zhang, S.; Wen, J.; Liu, L.; Ning, B.; Wan, W. An investigation on plasmaspheric electron content derived from ISR and GPS observations at Millstone Hill. *Chin. J. Geophys.* **2013**, *56*, 738–745.
11. Belehaki, A.; Jakowski, N.; Reinisch, B. Comparison of ionospheric ionization measurements over Athens using ground ionosonde and GPS-derived TEC values. *Radio Sci.* **2003**, *38*, 1105. [CrossRef]
12. Belehaki, A.; Jakowski, N.; Reinisch, B. Plasmaspheric electron content derived from GPS TEC and digisonde ionograms. *Adv. Space Res.* **2004**, *33*, 833–837. [CrossRef]
13. Yizengaw, E.; Moldwin, M.; Galvan, D.; Iijima, B.; Komjathy, A.; Mannucci, A. Global plasmaspheric TEC and its relative contribution to GPS TEC. *J. Atmos. Sol.-Terr. Phys.* **2008**, *70*, 1541–1548. [CrossRef]
14. Liu, L.; Yao, Y.; Kong, J.; Shan, L. Plasmaspheric Electron Content Inferred from Residuals between GNSS-Derived and TOPEX/JASON Vertical TEC Data. *Remote Sens.* **2018**, *10*, 621. [CrossRef]
15. Cherniak, I.; Zakharenkova, I.; Krankowski, A.; Shagimuratov, I. Plasmaspheric electron content derived from GPS TEC and FORMOSAT-3/COSMIC measurements: Solar minimum condition. *Adv. Space Res.* **2012**, *50*, 427–440. [CrossRef]
16. Chen, P.; Yao, Y. Research on global plasmaspheric electron content by using LEO occultation and GPS data. *Adv. Space Res.* **2015**, *55*, 2248–2255. [CrossRef]

17. Angel, A.; Sanz, J.; Juan, J.; Pajares, M.; Altadill, D. Plasmaspheric Electron Content contribution inferred from ground and radio occultation derived Total Electron Content. In Proceedings of the 2012 Fourth International Conference on Communications and Electronics, Hue, Vietnam, 1–3 August 2012; pp. 257–262.
18. González-Casado, G.; Juan, J.; Sanz, J.; Rovira-Garcia, A.; Aragon-Angel, A. Ionospheric and plasmaspheric electron contents inferred from radio occultations and global ionospheric maps. *J. Geophys. Res. Space Phys.* **2015**, *120*, 5983–5997. [CrossRef]
19. Yue, X.; Schreiner, W.; Hunt, D.; Rocken, C.; Kuo, T. Quantitative evaluation of the low Earth orbit satellite based slant total electron content determination. *Space Weather* **2011**, *9*, S09001. [CrossRef]
20. Zhang, M.; Liu, L.; Wan, W.; Ning, B. Variation of the plasmaspheric electron content derived from the podTEC observations of COSMIC LEO satellites to GPS signals. *Chin. J. Geophys.* **2016**, *59*, 1–7.
21. Jin, S.; Gao, C.; Yuan, L.; Guo, P.; Luo, P. Long-Term Variations of Plasmaspheric Total Electron Content from Topside GPS Observations on LEO Satellites. *Remote Sens.* **2021**, *13*, 545. [CrossRef]
22. Zhong, J.; Lei, J.; Wang, W.; Burns, A.; Yue, X.; Dou, X. Longitudinal variations of topside ionospheric and plasmaspheric TEC. *J. Geophys. Res. Space Phys.* **2017**, *122*, 6737–6760. [CrossRef]
23. Chen, P.; Yao, Y.; Li, Q.; Yao, W. Modeling the plasmasphere based on LEO satellites onboard GPS measurements. *J. Geophys. Res. Space Phys.* **2017**, *122*, 1221–1233. [CrossRef]
24. Tao, X.; Li, S.; Wu, Z.; Ji, M.; Mao, P.; Xiong, S. Modeling and empirical orthogonal function analysis of plasmaspheric electron content based on MetOp satellites. *Astrophys. Space Sci.* **2023**, *368*, 22. [CrossRef]
25. Lunt, N.; Kersley, L.; Bishop, G.; Mazzella, A.; Bailey, G. The effect of the protonosphere on the estimation of GPS total electron content: Validation using model simulations. *Radio Sci.* **1999**, *34*, 1261–1271. [CrossRef]
26. Webb, P.; Essex, E. An ionosphere-plasmasphere global electron density model. *Phys. Chem. Earth* **2000**, *25*, 301–306. [CrossRef]
27. Nava, B.; Coisson, P.; Radicella, S. A new version of the NeQuick ionosphere electron density model. *J. Atmos. Sol.-Terr. Phys.* **2008**, *70*, 1856–1862. [CrossRef]
28. Gallagher, D.; Craven, P.; Comfort, H. Global core plasma model. *J. Geophys. Res. Space Phys.* **2000**, *105*, 18819–18833. [CrossRef]
29. Zhang, M.; Radicella, S.; Shi, J.; Wang, X.; Wu, S. Comparison among IRI, GPS-IGS and ionogram-derived total electron contents. *Adv. Space Res.* **2006**, *37*, 972–977. [CrossRef]
30. Cherniak, I.; Zakharenkova, I. NeQuick and IRI-Plas model performance on topside electron content representation: Spaceborne GPS measurements. *Radio Sci.* **2016**, *51*, 752–766. [CrossRef]
31. Chen, P.; Liu, L.; Yao, Y.; Yao, W. A global empirical orthogonal function model of plasmaspheric electron content. *Adv. Space Res.* **2020**, *65*, 138–151. [CrossRef]
32. Wen, J.; Wan, W.; Ding, F.; Yue, X.; She, C.; Liu, L. Experimental observation and statistical analysis of the vertical TEC mapping function. *Chin. J. Geophys.* **2010**, *53*, 22–29.
33. Lyu, H.; Hernandez-Pajares, M.; Nohutcu, M.; Garcia-Rigo, A.; Zhang, H.; Liu, J. The Barcelona ionospheric mapping function (BIMF) and its application to northern mid-latitudes. *GPS Solut.* **2018**, *22*, 67. [CrossRef]
34. Wu, M.; Guo, P.; Xue, J.; Wang, Y.; Tian, Y.; Wu, J. Analysis and Empirical Modeling of Ionospheric Horizontal Gradients in the TEC Mapping Onboard LEO Satellites. *IEEE T. Geosci. Remote* **2022**, *60*, 5917514. [CrossRef]
35. Anthes, R.; Bernhardt, P.; Chen, Y.; Cucurull, L.; Dymond, K.; Ector, D.; Zeng, Z. The COSMIC/FORMOSAT-3 Mission: Early Results. *Bull. Am. Meteorol. Soc.* **2008**, *89*, 313–333. [CrossRef]
36. Ruan, H.; Lei, J.; Dou, X.; Liu, S.; Aa, E. An Exospheric Temperature Model Based On CHAMP Observations and TIEGCM Simulations. *Space Weather* **2018**, *16*, 147–156. [CrossRef]
37. Zhong, J.; Lei, J.; Dou, X.; Yue, X. Assessment of vertical TEC mapping functions for space-based GNSS observations. *GPS Solut.* **2016**, *20*, 353–362. [CrossRef]
38. de Haro Barbas, B.; Elias, A.; Venchiarutti, J.; Fagre, M.; Zossi, B.; Tan Jun, G.; Medina, F. MgII as a Solar Proxy to Filter F2-Region Ionospheric Parameters. *Pure Appl. Geophys.* **2021**, *178*, 4605–4618. [CrossRef]
39. Pearson, K. On lines and planes of closest fit to systems of points in space. *Philos. Mag.* **1901**, *2*, 559–572. [CrossRef]
40. Ercha, A.; Zhang, D.; Ridley, A.; Xiao, Z.; Hao, Y. A global model: Empirical orthogonal function analysis of total electron content 1999–2009 data. *J. Geophys. Res. Space Phys.* **2012**, *117*, 328.
41. Zhao, B.; Wan, W.; Liu, L.; Yue, X.; Venkatraman, S. Statistical characteristics of the total ion density in the topside ionosphere during the period 1996–2004 using empirical orthogonal function (EOF) analysis. *Ann. Geophys.* **2005**, *123*, 3615–3631. [CrossRef]
42. Vaishnav, R.; Jacobi, C.; Berdermann, J. Long-term trends in the ionospheric response to solar extreme-ultraviolet variations. *Ann. Geophys.* **2019**, *37*, 1141–1159. [CrossRef]
43. Lean, J.; Meier, R.; Picone, J.; Sassi, F.; Emmert, J.; Richards, P. Ionospheric total electron content: Spatial patterns of variability. *J. Geophys. Res. Space Phys.* **2016**, *121*, 10367–10402. [CrossRef]
44. Rishbeth, H.; Müller-Wodarg, I.; Zou, L.; Fuller-Rowell, T.; Millward, G.; Moffett, R.; Aylward, A. Annual and semiannual variations in the ionospheric F2-layer: II. Physical discussion. *Ann. Geophys.* **2000**, *18*, 945–956. [CrossRef]
45. Hoque, M.; Jakowski, N.; Berdermann, J. Ionospheric correction using NTCM driven by GPS Klobuchar coefficients for GNSS applications. *GPS Solut.* **2017**, *21*, 1563–1572. [CrossRef]
46. John A. Ionospheric Time-Delay Algorithm for Single-Frequency GPS Users. *IEEE Trans. Aerosp. Electron. Syst.* **1987**, *AES-23*, 325–331.

47. Hernández-Pajares, M.; Juan, J.; Sanz, J.; Orus, R.; Garcia-Rigo, A.; Feltens, J.; Krankowski, A. The IGS VTEC maps: A reliable source of ionospheric information since 1998. *J. Geod.* **2009**, *83*, 263–275. [CrossRef]
48. Hernández-Pajares, M.; Roma-Dollase, D.; Krankowski, A.; García-Rigo, A.; Orús-Pérez, R. Methodology and consistency of slant and vertical assessments for ionospheric electron content models. *J. Geod.* **2017**, *91*, 1405–1414. [CrossRef]
49. Wielgosz, P.; Milanowska, B.; Krypiak-Gregorczyk, A.; Jarmoowski, W. Validation of GNSS-derived global ionosphere maps for different solar activity levels: Case studies for years 2014 and 2018. *GPS Solut.* **2021**, *25*, 103. [CrossRef]

remote sensing

MDPI

Article

Effects of Strong Geomagnetic Storms on the Ionosphere and Degradation of Precise Point Positioning Accuracy during the 25th Solar Cycle Rising Phase: A Case Study

Yifan Wang [1], Yunbin Yuan [2], Min Li [2], Ting Zhang [2,3,*], Hao Geng [1], Guofang Wang [1] and Gang Wen [1]

1 Joint Laboratory of Power Remote Sensing Technology, Electric Power Research Institute, Yunnan Power Grid Company Ltd., Kunming 650217, China; wangyifan@yn.csg.cn (Y.W.); genghao@dlyjy.yn.csg.cn (H.G.); wangguofang@yn.csg.cn (G.W.); wengang01@im.yn.csg.cn (G.W.)
2 State Key Laboratory of Geodesy and Earth's Dynamics, Innovation Academy for Precision Measurement Science and Technology, Chinese Academy of Sciences, Wuhan 430077, China; yybgps@asch.whigg.ac.cn (Y.Y.); limin@whigg.ac.cn (M.L.)
3 University of Chinese Academy of Sciences, Beijing 100049, China
* Correspondence: zhangting@apm.ac.cn

Abstract: Approaching the peak year of the 25th solar activity cycle, the frequency of strong geomagnetic storms is gradually increasing, which seriously affects the navigation and positioning performance of GNSS. Based on the globally distributed GNSS station data and FORMOSAT-7/COSMIC-2 occultation data, this paper explores for the first time the effects of the G4-class geomagnetic storm that occurred on 23–24 April 2023 on the global ionosphere, especially the ionospheric equatorial anomalies and F-layer perturbations. It reveals the precise point positioning (PPP) accuracy degradation during a geomagnetic storm. The results show that the ionospheric rate of total electron content index (ROTI) and near high latitude GNSS phase scintillations index have varying levels of perturbation during geomagnetic storms, with the maximum ROTI and phase scintillations index exceeding 0.5 TECU/min and 0.8, respectively. The equatorial ionization anomaly (EIA) shows an enhanced state (positive ionospheric storms) during geomagnetic storms, and the cause of this phenomenon is most likely the equatorward neutral wind. The variation of the S4 index of the FORMOSAT-7/COSMIC-2 satellite reveals the uplift of the F-layer during geomagnetic storms. During geomagnetic storms, the PPP accuracy degrades most seriously at high latitudes, the maximum MAE exceeds 2.3 m, and the RMS in the three-dimensional (3D) direction exceeds 2.0 m. These investigations can provide case support for space weather and GNSS studies of the impact of geomagnetic storms during peak solar activity years.

Keywords: geomagnetic storms; EIA; ionospheric disturbances; ROTI; precise point positioning

1. Introduction

The ionosphere is an important part of the solar–terrestrial space environment and is a highly variable ionized region. For the global navigation satellite system (GNSS), ionospheric delays caused by somewhat regular spatial and temporal variations in the ionosphere can be better solved by numerical simulations or ionosphere-free combination methods [1–4]. However, the ionosphere does not exist alone in the upper atmosphere; for example, solar wind–magnetosphere–ionosphere coupling and lithosphere–atmosphere–ionosphere coupling increase the difficulty of the investigation of the ionosphere. Among these challenges, geomagnetic storms triggered by strong solar activity are a global magnetospheric disturbance phenomenon, which is the most important link in the solar–terrestrial energy coupling chain [5]. Along with the injection of magnetospheric energy, the ionosphere is significantly perturbed, forming ionospheric storms [6]. Ionospheric delay errors caused by ionospheric storms are very difficult to correct through modeling, and the resulting degradation of the positioning performance of the GNSS is exceptionally strong [7–9].

Investigations of the ionospheric response to strong geomagnetic storms are mostly developed with case studies [10–14]. In particular, the two geomagnetic storms of the 24th solar active cycle (on March 2015 and September 2017) are the most typical. The geomagnetic storm of March 2015, also known as the St. Patrick's Day Storm, was classified as G4 by the National Oceanic and Atmospheric Administration's (NOAA's) space weather classification system. Different scholars have studied the relationship between this geomagnetic storm and the ionosphere or its influences on GNSS positioning from different perspectives. Cherniak and Zakharenkova [15] found a high correlation between high-latitude ionospheric irregularities caused by geomagnetic storms and auroral activity based on data from more than 2500 GPS stations. Nava et al. [16] investigated the status of ionospheric variations in four sectors (Asia, Africa, the Americas, and the Pacific) at low and middle latitudes during the storms and provided explanations for various types of phenomena. Jacobsen and Andalsvik [17] observed ionospheric disturbances in northern Norway (based on the national real-time kinematic (RTK) positioning network) that may be related to ionospheric polar cap plasma patches and particle precipitation caused by magnetic reconnection. They also analyzed the performance metrics of RTK and precise point positioning (PPP) during the storm. Combining GPS and Swarm satellite data, Zakharenkova et al. [18] observed large-scale ionospheric plasma irregularities in Russia and highly structured auroral ionospheric regions extending over 850 km.

The September 2017 coronal mass ejection (CME) and high-speed flow reached near-Earth space and triggered a strong geomagnetic storm with successive violent disturbances in the ionosphere [19–21]. Li et al. [22] observed that prompt penetration undershielding electric fields (PPEF) caused by the southward shift of the interplanetary magnetic field (IMF) Bz during the storm was responsible for the significant enhancement of an equatorial plasma bubble (EPB) in the 120°E/60°W longitudinal sector. Zhang et al. [23] reported simultaneous large- and medium-scale traveling ionospheric disturbances (TIDs) at dusk during geomagnetic storms at subaurora and mid-latitudes. Based on observations from the U.S. Continuously Operating Reference Station (CORS) network, Liu et al. [24] explored the total electron content (TEC) of the ionosphere in response to this geomagnetic storm and found medium TIDs with intensities higher than 2.5 TECU. Wan et al. [25] discussed nighttime ionospheric behavior during the recovery phase of the geomagnetic storm. They found that the nighttime ionosphere in the American sector was largely depressed, while in the Asian sector, only a small TEC increase was observed at nighttime, which was much weaker than the prominent daytime TEC enhancement in this longitude sector. Zakharenkova and Cherniak [26] considered the effect of EPB generated by geomagnetic storms on the PPP in mid-latitudes, especially at 25°~35°N, where the three-dimensional (3D) error of GPS stations rises to the meter scale.

Strong geomagnetic storms occurred infrequently during the early part of the 25th solar cycle. However, the occurrence of medium geomagnetic storms also had a severe impact on navigation, communications, and aerospace [27]. For example, the SpaceX storm event in February 2022 resulted in 38 SpaceX Starlink satellites not being able to ascend to their assigned orbits and having to re-enter the atmosphere to be destroyed [28]. With the arrival of the peak year of the 25th solar activity cycle, there has been a gradual increase in intense geomagnetic storms. Based on the 11-year cycle of solar activity [29], October 2023 was in a phase in the 25th solar activity cycle of increasing intensity of solar activity (sunspot number and F10.7) that, as of the time of this writing, had already exceeded the peak of the 24th cycle. Even though many studies have been conducted to analyze the effects of geomagnetic storms occurring at different times on the ionosphere, the geomagnetic storm events under this condition still need further investigation. In this paper, based on the globally distributed GNSS station data and the FORMOSAT-7/COSMIC-2 occultation data, the geomagnetic storm on 23–24 April 2023 is taken as an example to investigate the ionospheric disturbances and the PPP accuracy of global GNSS stations during this period.

2. Data and Methods

2.1. Ionospheric Data

The International GNSS Service (IGS) has established 516 GNSS stations around the world (as of 25 October 2023), providing a large amount of raw observational data for geophysical and space physics investigations. Based on dual-frequency observations of GNSS stations worldwide (see Figure 1) on 23 and 24 April 2023 at a sampling rate of 30 s, the ionospheric TEC of the GNSS stations is solved using a carrier phase smoothing pseudorange:

$$< P_{i,j} + L_{i,j} >= -\left[(\lambda_i N_i - \lambda_j N_j) + 2c \cdot (DCB_{r,i,j} + DCB^{s,i,j})\right] \tag{1}$$

$$TEC = \frac{L_{i,j} - < P_{i,j} + L_{i,j} > -c \cdot (DCB_{r,i,j} + DCB^{s,i,j})}{40.3 \cdot (f_i^{-2} - f_j^{-2})} \tag{2}$$

where r and s denote receiver and satellite, respectively; i and j denote signals at different frequencies (f_i and f_j); $P_{i,j}$ is a combination of pseudorange observations; $L_{i,j}$ is a combination of carrier-phase observations; $< \cdot >$ denotes the average value of a continuous arc segment; and $\lambda_i N_i - \lambda_j N_j$ is a combination of integer ambiguity. $DCB_{r,i,j}$ and $DCB^{s,i,j}$ are the differential code biases (*DCB*) of the receivers and satellites, respectively, both from the CODE's (Center for Orbit Determination in Europe) *DCB* products. The GPS L1 and L2 measurements are used in this paper.

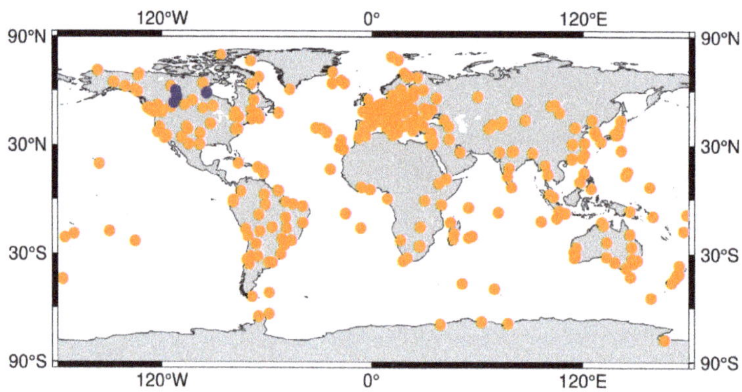

Figure 1. Global distribution of GNSS stations (dark orange dots) and CHAIN stations (dark blue dots) used in this paper.

The rate of total electron content index (*ROTI*) is defined as the standard deviation of the rate of *TEC*, which can indicate the extent to which the GNSS measurements are disturbed by the ionosphere, and is calculated as follows [30]:

$$ROT = \frac{TEC(t + \Delta t) - TEC(t)}{\Delta t} \tag{3}$$

$$ROTI = \sqrt{< ROT^2 > - < ROT >^2} \tag{4}$$

where *ROT* is the rate of *TEC* and $< \cdot >$ denotes the average value of a continuous time period. In this paper, we used the observation data with a sampling interval of 30 s, so the time period was set to 5 min. The unit of both *ROT* and *ROTI* is TECU/min.

The Massachusetts Institute of Technology (MIT) Haystack Observatory automatically processes original observations from about 6000 GPS receivers around the world every day, generating global ionospheric TEC products with a spatial resolution of $1^\circ \times 1^\circ$ (geographic latitude × geographic longitude) every 5 min [31]. Meanwhile, the global ionospheric maps (GIM) from CODE are used in this paper. The product is calculated from a spherical harmonic function of 15 orders and 15 degrees, taking 5° × 2.5° along the latitude and longitude, with a total of 5183 grid points.

The Canadian High Arctic ionospheric network (CHAIN) is a distributed terrestrial radio instrument array located in the Canadian Arctic. It includes 25 GNSS ionospheric scintillation and total electron content monitors (GISTMs) with high time resolution and 6 advanced digital Canadian ionosondes. Each GISTM provides GPS amplitude scintillation index S4, phase scintillation index σ_ϕ, and ionospheric TEC data. The CHAIN provides powerful data support for understanding solar wind–magnetosphere–ionosphere coupling [32].

The FORMOSAT-7/COSMIC-2 radio occultation mission satellite, jointly developed by the United States and the Taiwan Province of China, carries an advanced Tri-GNSS radio occultation system instrument (TGRS), which is capable of receiving GPS and GLONASS signals, as well as Galileo and BDS signals, and provides measurements of about 5000 occultation events per day [33,34]. The ionospheric amplitude scintillation *S4* index is calculated as follows, where I is the intensity of the signal and $< \cdot >$ denotes the average value over 1 min:

$$S_4 = \sqrt{\frac{< I^2 > - < I >^2}{< I >^2}} \quad (5)$$

2.2. Space Weather Data

Space weather data are widely used in various types of geophysical and space physics investigations. The space weather data used in this paper include the IMF Bx/By/Bz components, solar wind plasma speed, plasma flow pressure, Kp index, Dst index, and F10.7 index, where the Kp index and the Dst index characterize the degree of geomagnetic disturbance. According to the NOAA space weather scale, a Kp index from 5 to 9 represents minor, moderate, strong, severe, and extreme geomagnetic storms, respectively. The negative Dst index indicates the occurrence of geomagnetic storms, and the larger its absolute value, the greater the intensity of the geomagnetic storm. Its unit is nT. The Space Physics Data Facility of the National Aeronautics and Space Administration provided the space weather data for this paper.

2.3. Precision Point Positioning

PPP is typically based on dual-frequency GNSS observations, and it uses ionosphere-free (IF) combinations for positioning [35,36]:

$$P^s_{r,IF} = \rho^s_r + c(dt_r - dt^s) + T^s_r + \varepsilon_P \quad (6)$$

$$L^s_{r,IF} = \rho^s_r + c(dt_r - dt^s) + T^s_r + \lambda_{IF} A_{IF} + \varepsilon_L \quad (7)$$

where $P^s_{r,IF}$ is the ionosphere-free combination $(f_1^2 P_1 - f_2^2 P_2)/(f_1^2 - f_2^2)$ of pseudorange observations P_1 and P_2 for signal frequencies f_1 and f_2, $L^s_{r,IF}$ is the ionosphere-free combination $(f_1^2 L_1 - f_2^2 L_2)/(f_1^2 - f_2^2)$ of the corresponding carrier phases L_1 and L_2, ρ^s_r is the geometric range between the satellite and receiver antennas, c is the speed of light, dt_r is the receiver's clock offset (including the receiver code bias and delay), dt^s is the clock offset of the satellite (including the satellite code bias and delay), T^s_r is the signal path delay, λ_{IF} is the ionosphere-free combination of carrier phase wavelengths λ_{L1} and λ_{L2} of signals L_1 and L_2, A_{IF} is the nonperiodic ambiguity, and ε_P and ε_L are the noises in the measurements.

The RTKLIB software (version 2.4.3) developed by Takasu and Yasuda [37] is widely used in GNSS data processing. In this paper, RTKLIB software was used to solve the PPP Kinematic and the solution strategy is shown in Table 1.

Table 1. PPP Kinematic Processing Techniques.

Items	Strategy
Sampling interval	30 s
Frequencies	GPS L1 and L2
Filter type	Smoother combined solution with forward and backward filter solutions
Elevation mask angle	15°
Satellite orbit and clock	CODE precise ephemeris and precise clock files
Earth tides	Solid earth tides, ocean tide loading, and pole tide corrections
Ionospheric delay	Ionosphere-free linear combination with dual-frequency measurements is used for ionospheric correction
Tropospheric delay	Estimate ZTD + grad
Receiver reference coordinates	IGS SINEX files
Satellite antenna phase center correction	IGS20.atx
Differential code bias	CODE products (P1C1, P1P2, P2C2 files)

3. Results

3.1. Analysis of Space Weather Disturbances

Figure 2 shows the variations of the IMF Bx/By/Bz components, solar wind plasma speed, plasma flow pressure, Kp index, Dst index, and F10.7 index for 23–25 April 2023. As far as the IMF was concerned, the Bx, By, and Bz components were in a calm state with no large fluctuations until 08:00 UT on 23 April. The solar wind plasma speed was below 400 km/s, the plasma flow pressure was also at a low pressure, and the values of the Kp index and the Dst index showed no geomagnetic disturbance. The plasma flow pressure showed a small increase at 08:00 UT on 23 April, except for the solar wind plasma speed and IMF Bx component, which did not change significantly. The IMF By component gradually increased from −4.2 nT to 9.5 nT. The IMF Bz component turned southward, but not by much. The Kp index gradually increased and exceeded four at 12:00 UT, and the Dst index also showed a decreasing trend, falling to −35 nT at 12:00 UT.

Unlike a case of typical geomagnetic storms [6], there was no apparent storm surge commencement (SSC) in this geomagnetic storm and the geomagnetic storm process appeared to be a two-step process. From 12:00 to 16:00 UT, the Dst index continued to fall, the IMF By component decreased from 9.5 nT to −3.3 nT, and the Kp index increased to 5.3, indicating a small geomagnetic storm. At this time, the solar wind plasma speed and plasma flow pressure still did not show any significant variation. Starting from 16:00 UT, the solar wind plasma speed started to increase rapidly from 359 km/s to 711 km/s at 21:00 UT, the plasma flow pressure likewise started to increase from 1.15 to 11.72 at 20:00 UT, and the IMF Bx component also showed an increasing trend. The IMF Bz component turned southward from 17:00 UT, reaching −21 nT at 19:00 UT, and then quickly turned northward, reaching 11.8 nT at 21:00 nT. During this process, influenced by IMF and plasma, the Kp index increased to 8.3, indicating a severe geomagnetic storm, and the Dst index decreased from −56 nT to −165 nT. The IMF Bx component showed an increasing–decreasing–increasing trend, and the IMF By showed a fluctuating state of decreasing–increasing. The IMF Bz component quickly turned southward from 21:00 UT and decreased to −32.4 nT at 02:00 UT on 24 April, then turned northward and slowly increased to 22.5 nT at 16:00 UT before finally returning to a calm state. The solar wind plasma speed slowly decreased to about 500 km/s at 21:00 UT on 23 April. The plasma flow pressure first decreased to 8.21 at 22:00 UT on 23 April after reaching its maximum value, and then it increased to 11.33 at 23:00 UT followed by a rapid decrease and then gradually returned to a low-pressure state. After the occurrence of the severe geomagnetic storm, the Kp index decreased from 8.3 to 6.7, then increased to 8, and finally slowly decreased to a calm state. The Dst index did not continue to decrease after −165 nT. It increased to −146 nT at 01:00 UT on 24 April, decreased to −212 nT at 05:00 UT, and subsequently entered the recovery phase of the geomagnetic storm, returning to a state of geomagnetic calm. Due to this being a rising year of the 25th solar activity cycle, the solar activity parameter F10.7 has been maintained at a level of about 135 sfu during this period.

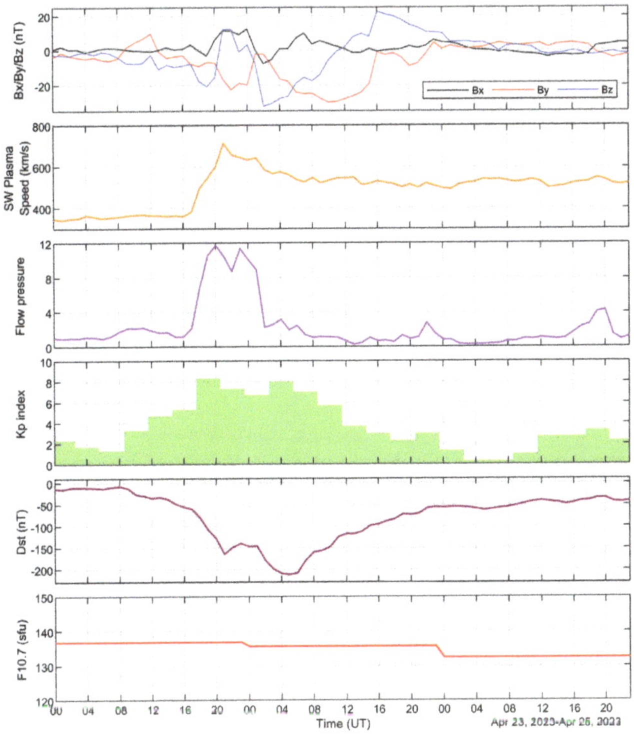

Figure 2. Space weather conditions from 23 to 25 April 2023.

3.2. Ionospheric Disturbances

In order to investigate the impact of this geomagnetic storm on the global ionosphere, we analyzed the TEC response to the geomagnetic storm using ionospheric TEC data provided by the GNSS receiver network from the Madrigal database. The time resolution of this product is 5 min, from which we calculated the hours-averaged ionospheric TEC values over the period of 0 to 23 h. The global distribution of ionospheric TEC hours-averaged values for 23 and 24 April 2023 is given in Figures 3 and 4, respectively. In order to visualize the effect of geomagnetic storms on the ionospheric TEC, we processed the global ionospheric maps of calm days based on the sliding interquartile range method to generate the global ionospheric TEC anomaly distribution, as shown in Figures 5 and 6. The calculation method of the sliding interquartile range is from Feng et al. [38].

Combining Figures 3 and 5, it can be found that on 23 April, the ionospheric TEC was in a relatively normal state before 20:00 UT on 23 April, i.e., the equatorial region had a significant double crest phenomenon, and there was little variation in the TEC in other regions. Starting from 20:00 UT, the ionospheric anomalies began to appear, as shown in Figure 5, which were manifested by the weakening of the TEC near the geomagnetic equator and the obvious enhancement of the TEC in the EIA double crest and the high latitude region of the Antarctic, where the TEC value of the EIA exceeded 100 TECU. The global ionospheric TEC anomaly expanded from 20:00 to 22:00 UT. At 22:00 UT, the intensity of the EIA reached the maximum value of the day, exceeding that of a quiet day by about 30 TECU, and the range of the EIA gradually spread to the poles. The variations of the EIA on 23 April show a highly temporal consistency with the first step of the geomagnetic storm process, indicating that this geomagnetic storm also affected the equatorial low-latitude region.

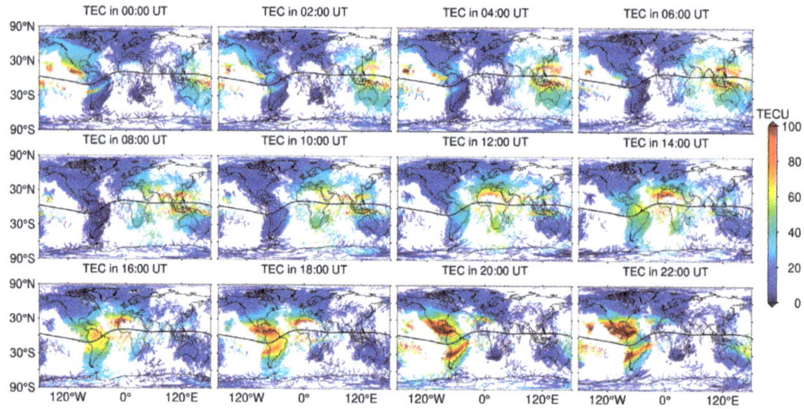

Figure 3. Distribution of global ionospheric TEC hours-averaged values on 23 April 2023. The solid black line indicates the geomagnetic equator (the same as below).

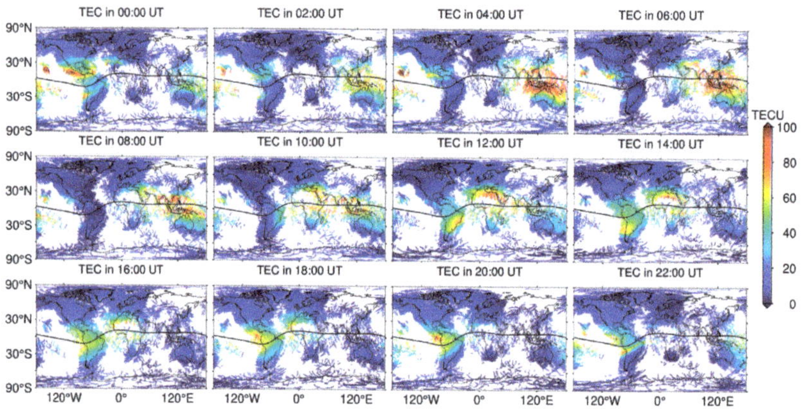

Figure 4. Distribution of global ionospheric TEC hours-averaged values on 24 April 2023.

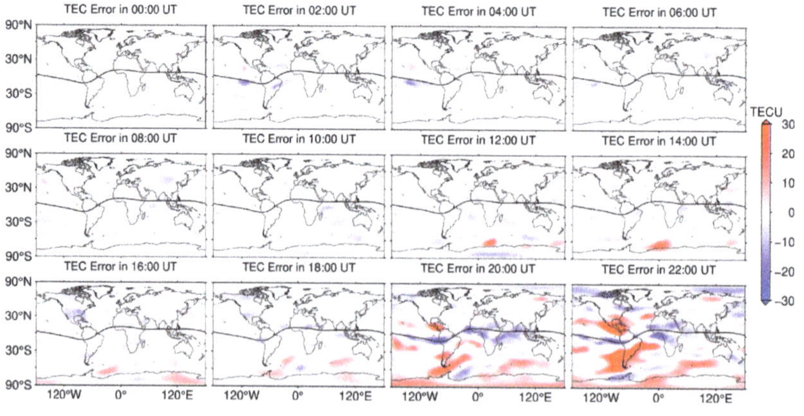

Figure 5. Distribution of CODE TEC error on 23 April 2023.

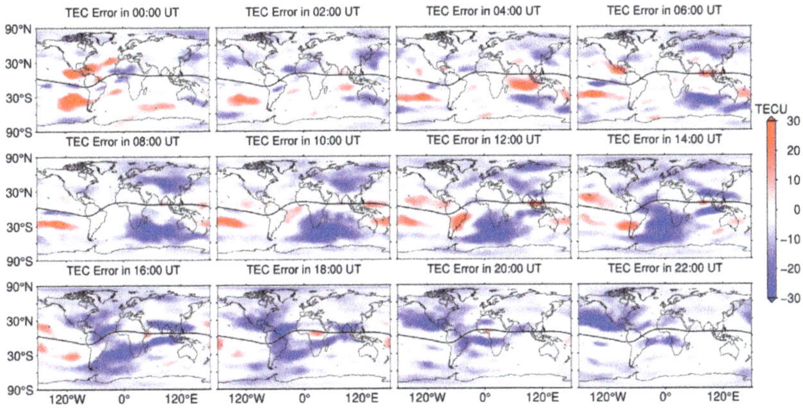

Figure 6. Distribution of CODE TEC error on 24 April 2023.

The global ionospheric TEC distribution and anomalies for 24 April are shown in Figures 4 and 6, further illustrating the effect of geomagnetic storms on the ionosphere and the second step in the process of geomagnetic storms. Starting from 22:00 UT on 23 April, the EIA crest and latitudinal range gradually decreased, the positive anomaly of the northern peak of the EIA disappeared, and the positive anomaly of the southern crest became weaker until 02:00 UT on 24 April. According to the status of the geomagnetic activity in Figure 1, this was the beginning time of the second step of the change process of the geomagnetic storm. Starting from 02:00 UT on 24 April, the crest and range of the EIA began to gradually increase again, and the double crest gradually changed to a single crest on the south side of the equator. The single crest of the EIA was close to the maximum at 04:00 UT, and this continued until 06:00 UT. At 08:00 UT, only a small range of positive anomalies of the south crest of EIA remained. Starting from 06:00 UT, the middle and high latitude regions of the longitude sector in which the EIA was located exhibited negative anomalies in the TEC, which continued until 22:00 UT. A phenomenon of weakening TEC intensity occurs mainly during the recovery phase of geomagnetic storms.

Above, we found the response of the ionospheric TEC to geomagnetic storms in different regions. Then, we calculated the ROTI for each station using observations distributed in more than 400 GNSS stations around the world, as shown in Figures 7 and 8. As can be seen in Figure 7, the number of GNSS stations with an increased ROTI gradually increased from 18:00 UT on 23 April, and the ROTI of all GNSS stations in the northern part of North America and the northern part of Europe increased to more than 0.2 TECU/min at 18:00 UT. From 18:00 UT to 20:00 UT, the number of GNSS stations affected by geomagnetic storms increased further, and the geographic latitude of GNSS stations responding to geomagnetic storms gradually decreased. At 20:00 UT, the ROTI increased to over 0.5 TECU/min at some GNSS stations in central North America and northern Africa. At 22:00 UT, GNSS stations with a ROTI exceeding 0.3 TECU/min were mainly located in central North America, northern Africa, and the polar regions. However, unlike at 20:00 UT, some GNSS stations had an increased ROTI near the geomagnetic equator in South America. On 24 April, the ROTI variations of the GNSS stations shown in Figure 8 continued to be affected by geomagnetic storms, but not as significantly as on 23 April. Starting from 00:00 UT to 06:00 UT, the ROTIs of the perturbed GNSS stations were higher than the 0.2 TECU/min, and they were mostly located in the central and northern parts of North America. After 12:00 UT, the ROTI of the global GNSS stations tended to calm down.

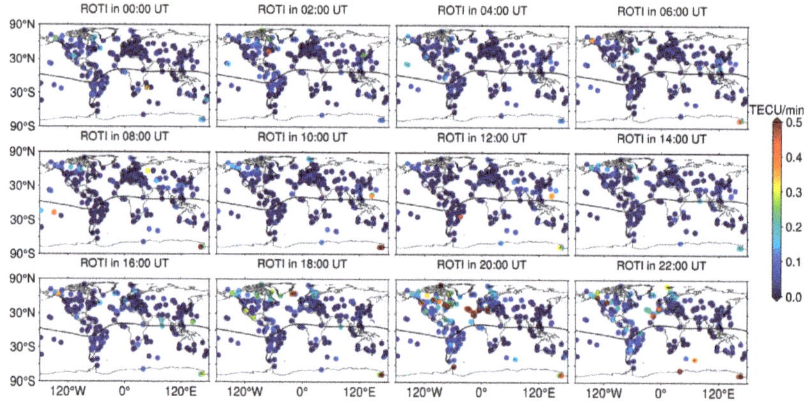

Figure 7. ROTI calculated from more than 400 GNSS stations worldwide on 23 April 2023.

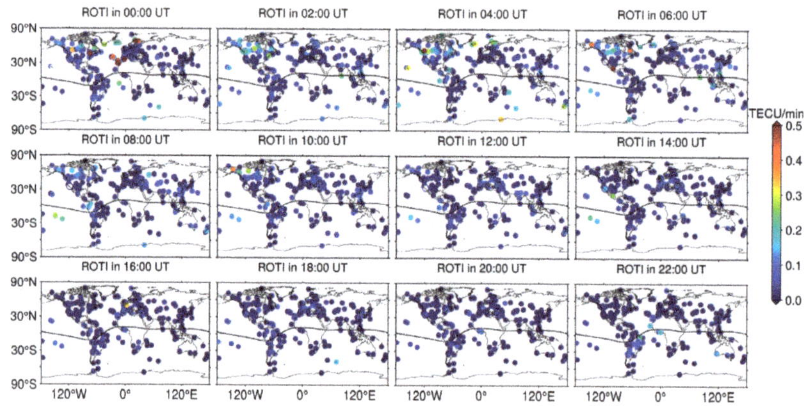

Figure 8. ROTI calculated from more than 400 GNSS stations worldwide on 24 April 2023.

Ionospheric irregularities in the polar regions make ionospheric amplitude scintillations in the polar regions weak [39]. Therefore, we selected ionospheric phase scintillation index (σ_{ϕ}) data from four GNSS stations in CHAIN to analyze the effect of geomagnetic storms on the ionosphere in the near high-latitude region. Figure 9 shows the sequence of mean phase flashes per minute for four high-rate GNSS stations (fsmc, chuc, mcmc, and edmc) on 23 and 24 April 2023, and also gives the IMF Bz and Dst indices for these time periods. As can be seen in Figure 9, there was a small fluctuation in the IMF Bz component from 10:00 UT to 14:00 UT on 23 April, during which all four stations showed different levels of ionospheric scintillation. Among them, the phase scintillation index of the mcmc station showed a peak of more than 0.4 at 11:00 UT. From 18:00 UT to 22:00 UT, the IMF Bz turned northward and the Dst decreased rapidly. The chuc station phase scintillation index showed an extreme value of more than 0.3 at 18:00 UT. An extreme value close to 0.3 and 0.4 was observed at 21:00 UT for the fcmc and mcmc stations, respectively. The edmc station did not have any obvious peaks appear, but the intensity of its fluctuations were higher than that of a quiet time. On 24 April, ionospheric scintillation was observed at the fsmc, chuc, and mcmc stations from 01:00 UT, which is when the second step of the geomagnetic storm started. The phase scintillation index at the mcmc station reached 0.4 at 02:00 UT and it even exceeded 0.8 at the chuc station close to 03:00 UT. During 04:00–06:00 UT, the four GNSS stations again showed various levels of ionospheric scintillations, and this is also the period when Dst reached its minimum. During the recovery phase of the geomagnetic storm, different levels of ionospheric disturbances continued to occur at the four GNSS stations, but there were basically no ionospheric disturbances after 12:00 UT.

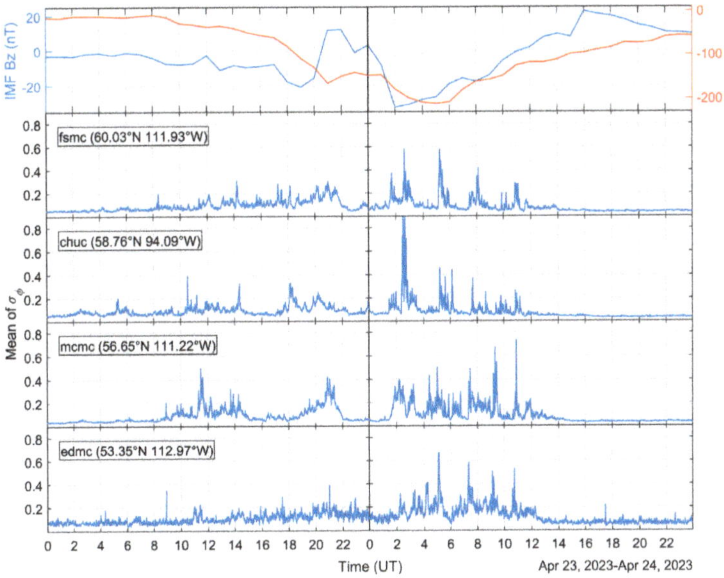

Figure 9. The sequence of mean phase scintillation index (σ_{ϕ}) per minute for 4 GNSS stations in CHAIN on 23 and 24 April 2023.

To explore the response of the ionospheric F-layer to geomagnetic storms in the low to middle latitude region, we used data from the FORMOSAT-7/COSMIC-2 mission's podTc2 product to calculate the position of the substar point at which the occultation event occurred. Since the F-layer of the ionosphere is the region with the highest electron density, the variation of parameters in the F-layer ultimately forms the variation of the ionosphere, but due to the presence of other layers of the ionosphere, some irregular structures increase the complexity of the ionosphere, especially the phenomenon of the episodic E-layer. To avoid such anomalies, we selected the S4 data in the L1 band of the GPS satellites from 200 to 600 km (i.e., the F-layer altitude) for analysis, and the results are shown in Figure 10.

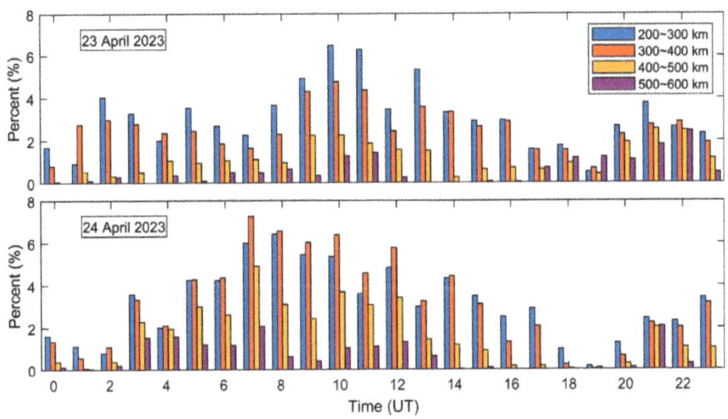

Figure 10. The percent of S4 index greater than 0.3 for 23 April 2023 (**top**) and 24 April 2023 (**bottom**) given by the FORMOSAT-7/COSMIC-2 products.

Figure 10 gives the percent of occultation points with S4 indices for the greater than 0.3 of all occultation points for different altitudes from 23 April 2023 to 24 April 2023. We mainly consider the variations of S4 under the time period of the occurrence of geomagnetic storms. As can be seen in Figure 11, on 23 April 2023, the number of S4 > 0.3 at different altitudes started to increase from 08:00 UT. The number of S4 > 0.3 at 200–300 km altitude was more than 50% higher at 10:00 UT than in the quiet time (02:00 UT). From 16:00 UT to 22:00 UT, as the Dst index reached the first minima, the number of S4 > 0.3 at all heights also tended to increase, but in the case of 200~400 km heights, the increase was not significant. On the contrary, for heights of 400~600 km, the increase in their number was more than that in the early stage of the geomagnetic storm. During the second phase of the geomagnetic storm on 24 April 2023, the number of S4 > 0.3 at different heights varied with the geomagnetic index. At 21:00 UT, the number of S4 > 0.3 increased again, and in combination with Figure 2, the increase in the number may be due to a sudden increase in the flow pressure index.

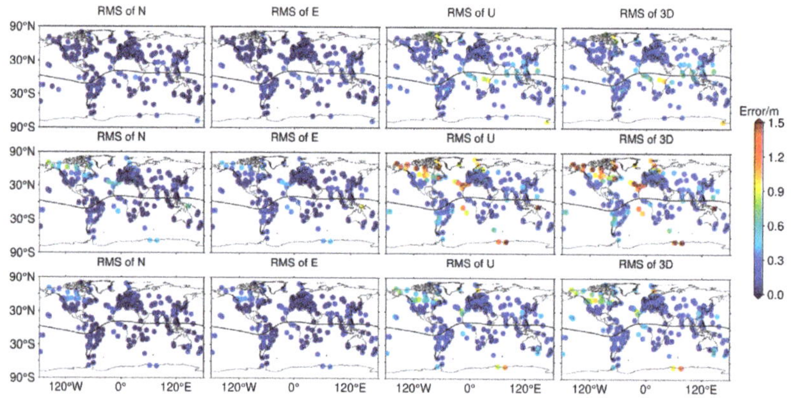

Figure 11. RMS for PPP errors in the N, E, U, and 3D directions on 20 April 2023 (**top**), 23 April 2023 (**middle**), and 24 April 2023 (**bottom**).

3.3. Degradation of PPP Accuracy

After a detailed analysis of the ionospheric response during geomagnetic storms, our focus turned to the investigation of the performance of PPP during geomagnetic storms. Based on the PPP solution strategy given in the Section 2, we processed more than 400 GNSS station observations worldwide. The errors in the north (N), east (E), vertical (U), and 3D directions corresponding to the coordinates of the real GNSS stations were obtained, and the root mean square (RMS) of the errors was calculated. In this process, we excluded some of the GNSS stations with positioning anomalies based on the data from a geomagnetic quiet day (20 April 2023) to visualize the effect of geomagnetic storms on PPP.

Figure 11 illustrates the RMS of the PPP errors on 20 April, 23 April, and 24 April 2023 in the N, E, U, and 3D directions. On geomagnetically quiet days, the PPP in the N and E directions has an RMS of about 0.3 m only near the magnetic equator, and the RMS in other regions is less than 0.1 m. The RMS of PPP in the U direction is larger, close to 1.0 m, and is mainly distributed in the magnetic equator and pole regions, while the RMS is less than 0.3 m in other regions. The RMS of global GNSS stations during geomagnetic storms in the N and E directions is small, the RMS of GNSS stations in the middle and low latitudes is below 0.2 m, and the PPP accuracy is mostly in the centimeter level, while the RMS of GNSS stations in high latitudes is below 1.0 m. However, in the U direction, the RMS of both middle and high latitude GNSS stations exceeded 1.0 m on 23 April, and the RMS of the GNSS stations in the polar region even exceeded 2.0 m. The RMS of both middle and high latitude GNSS stations on 24 April was close to 1.0 m. The degradation of the positioning accuracy in the vertical direction leads to the degradation of the accuracy in the 3D direction. The RMS of the 3D errors of the GNSS stations in the high latitude region all exceeded 1.2 m on 23 April. The RMS of the 3D errors of the GNSS stations in the high latitude region on 24 April were lower than that on 23 April, and the PPP accuracy degraded to the meter level.

Figure 12 shows the mean absolute errors (MAE) of the PPP on 20 April, 23 April, and 24 April 2023 in the N, E, and U directions for the high, middle, and low latitude GNSS stations. It is obvious from Figure 12 that during the development of geomagnetic storms, the positioning errors of GNSS stations at low, middle, and high latitudes fluctuate to different levels. The GNSS station positioning errors in the low latitude region occurred between 20:00 UT on 23 April and 02:00 UT on 24 April, with maximum MAEs exceeding 0.2 m, 0.2 m, and 0.5 m in the N, E, and U directions, respectively. The GNSS station positioning errors in the middle latitude region occurred between 18:00 UT on 23 April and 08:00 UT on 24 April. The MAE maximum in the N direction occurred at 20:00 UT on April 23, with a maximum value exceeding 0.4 m. The U direction showed a maximum value close to 1.0 m at the same time. However, the PPP MAE at the middle latitude was smaller than that at the low latitude in the E, N, and U directions on geomagnetically quiet days. The most obvious fluctuations in positioning errors were observed at the high latitude GNSS stations, and the significant degradation of PPP accuracy caused by the geomagnetic storm occurred from 14:00 UT on 23 April to 12:00 UT on 24 April, with the maximum MAE exceeding 0.8 m in the N direction. The maximum MAE was close to 0.6 m in the E direction, and the maximum MAE in the U direction exceeded 2.3 m. The GNSS positioning MAEs in all three latitude bands were affected by the geomagnetic storm, especially more significantly at high latitudes.

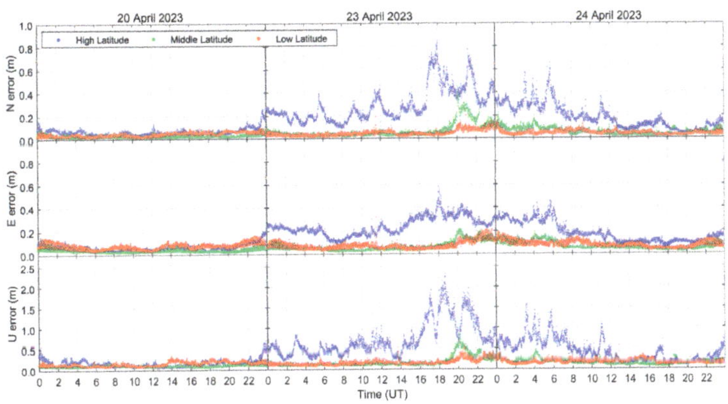

Figure 12. Mean absolute errors of the PPP in N, E, and U directions for GNSS stations at different latitudes on 20 April 2023 (**left**), 23 April 2023 (**middle**), and 24 April 2023 (**right**).

To further investigate the degradation of the PPP accuracy for GNSS stations in different temporal and spatial dimensions, the RMS of the PPP errors for 20 April, 23 April, and 24 April 2023 for the 3D direction are given in Figures 13–15. On geomagnetically quiet days, the larger RMS in the 3D direction is found mainly in the polar regions and in the low latitude EIA region. During the time period when no geomagnetic storms occurred on 23 April, the RMS exceeded 1.0 m only in the high latitude region of North America and the Antarctic region, while the RMS in all other regions was close to that of a geomagnetically quiet day. Starting at 08:00 UT, the RMS of GNSS stations in the high latitude region gradually increased as geomagnetic storms began to develop, and the number of GNSS stations with an RMS exceeding 1.0 m continued to increase. At 20:00 UT and 22:00 UT, the RMS of the GNSS stations in the high latitude region, the middle latitude region of North America, and the northern part of Africa were the largest, with the maximum exceeding 2.0 m. On 24 April, the RMS of the disturbed GNSS stations was smaller than the maximum on 23 April, which is in accordance with the above that the response of the ionosphere was smaller during the second step of the development of the geomagnetic storm. After 12:00 UT, the RMS in the disturbed region decreased further. The above variations in the RMS adequately illustrate the poor effects on GNSS positioning during the development of geomagnetic storms. The PPP error prior to 08:00 UT on 23 April may have been caused by the CME event that occurred on 21 April and will not be discussed in detail in this paper.

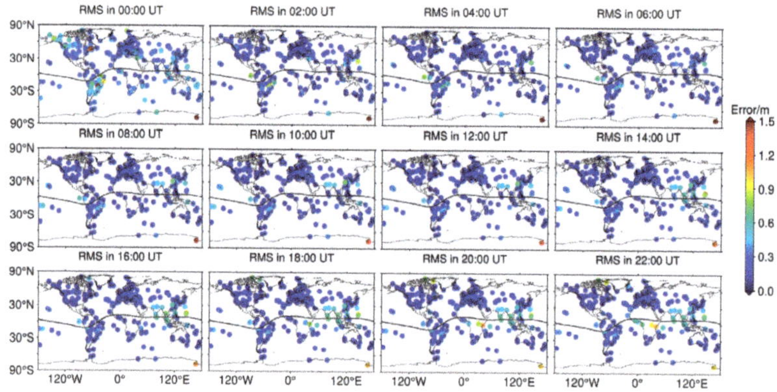

Figure 13. RMS of PPP errors for the 3D direction of global GNSS stations on 20 April 2023.

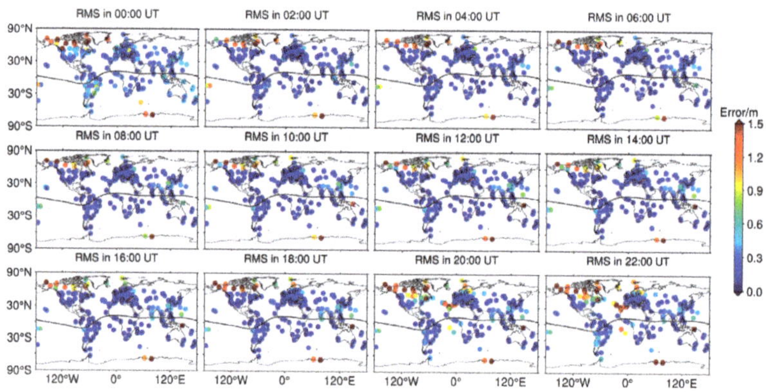

Figure 14. RMS of PPP errors for the 3D direction of global GNSS stations on 23 April 2023.

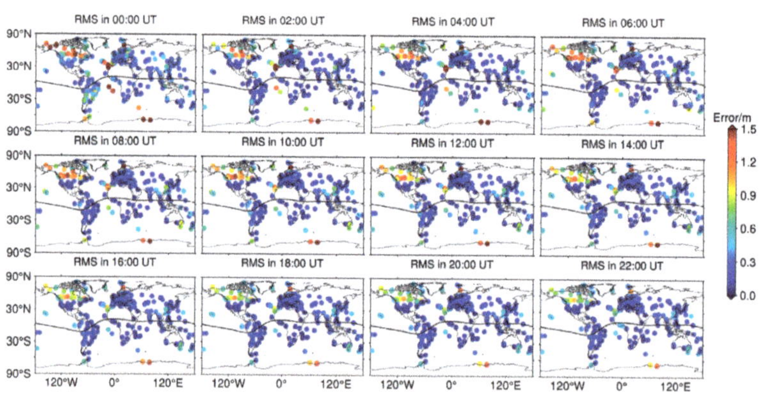

Figure 15. RMS of PPP errors for the 3D direction of global GNSS stations on 24 April 2023.

4. Discussion and Conclusions

In part of the analysis of ionospheric disturbances triggered by a geomagnetic storm, we found a significant positive ionospheric TEC anomaly over a wide range of low and middle latitudes during the main phase of the geomagnetic storm (i.e., from 20:00 UT on 23 April to 00:00 UT on 24 April), a phenomenon referred to as a positive ionospheric storm. Ionospheric storms caused by geomagnetic storms are categorized into positive and negative storms, and the physical factors affecting ionospheric parameter changes during geomagnetic storms include thermospheric circulation, disturbance dynamo electric fields, penetration electric fields, TIDs, and neutral component changes [40–42]. During geomagnetic storms, equatorward neutral winds may cause an eastward prompt penetration electric field (PPEP) to occur during the daytime, moving the EIA crests to higher than normal latitudes, and thus, positive ionospheric storms are likely to occur in the morning to noon local time [43]. Based on the global ionospheric TEC product provided by MIT and CODE, we found a significant enhancement of the EIA during the geomagnetic storms of 23 and 24 April 2023, and the EIA enhancement region was localized in the morning to noon time. This suggests that the cause of the EIA enhancement that occurred at low and middle latitudes in this geomagnetic storm was likely to be positive ionospheric storms caused by equatorward neutral winds. Typically, equatorward neutral winds lift the low and middle latitude F-layer to higher altitudes, where the recombination process is slower and therefore the TEC increases [44]. Therefore, we investigated the variations in the F layer of the ionosphere at low and middle latitudes using the FORMOSAT-7/COSMIC-2 ionospheric scintillation S4 index from 200 to 600 km. We found that ionospheric disturbances are more significant at 400~600 km altitude than at 200~400 km altitude during geomagnetic storms. This further demonstrated that the cause of this positive ionospheric storm is most likely the equatorward neutral wind.

Based on the observation data from more than 400 GNSS stations around the world, we calculated the ROTI during quiet days and geomagnetic storm days. We found that the ROTI during geomagnetic storms was significantly higher than on quiet days, and found that the ionospheric ROTI variations were consistent with the process of geomagnetic storms. The ROTI variation of GNSS stations in the high latitude region was the largest, and the maximum ROTI was more than 0.5 TECU/min. The maximum ROTI in the middle and low latitude was not more than 0.3 TECU/min. Phase scintillation data from four GNSS stations in CHAIN showed that ionospheric scintillation of varying levels were observed at all GNSS stations during geomagnetic storms in near high latitudes, and the largest mean phase scintillation indices exceeded 0.8.

The ionospheric disturbances induced by geomagnetic storms severely affected the performance of the GNSS PPP during this period. In particular, the error in the U direction was the largest, and the RMS exceeded 2.0 m, which led to an increase of the error in the 3D direction and the degradation of the PPP accuracy. Nie et al. [45] proposed that the reason for the increase in the PPP error during geomagnetic storms is related to the occurrence of cycle slips in the observing satellites, and that increasing the threshold value of the ionospheric scintillation detector can improve the PPP accuracy in the high latitude region. Luo, et al. [46] fused the ROTI into an elevation angle stochastic model (EAS), which also improved the performance of the PPP on a global scale during storms. This means that we can improve the GNSS data processing methods or incorporate perturbation models with a view to improving the accuracy of GNSS positioning during geomagnetic storms. This is the direction of our future investigations.

We have investigated in detail the ionospheric disturbances and the degradation of PPP accuracy during a severe geomagnetic storm that occurred during the rising year of the 25th solar active cycle. With the arrival of the peak solar activity year, we will consider the investigation of more cases of GNSS positioning problems caused by ionospheric disturbance phenomena, which will be of reference value for space weather and GNSS research during the upcoming peak solar activity year.

Author Contributions: Conceptualization, writing—original draft preparation, writing—review and editing, and methodology, Y.W., Y.Y., M.L., T.Z. and H.G.; validation, Y.W., T.Z. and G.W. (Guofang Wang); data curation, Y.Y., M.L. and G.W. (Gang Wen); formal analysis, Y.W. and T.Z.; funding acquisition, Y.W., Y.Y. and M.L. All authors have read and agreed to the published version of the manuscript.

Funding: This research was funded by the Science and technology project of China Southern Power Grid Yunnan Power Grid Co., Ltd. (grant number. YNKJXM20220033), Yunnan Fundamental Research Projects (grant number. 202301AU070101), the China Postdoctoral Science Foundation (grant number. 2021MD703896), and was partially supported by the National Natural Science Foundation of China (No. 42004027).

Data Availability Statement: The GNSS observations and CODE products are available from the IGS (https://cddis.nasa.gov/archive/gnss/data/, registration required, accessed on 25 October 2023). The FORMOSAT-7/COSMIC-2 occultation data are available from CDAAC (https://data.cosmic.ucar.edu/gnss-ro/cosmic2/, accessed on 25 October 2023). The MIT GNSS data products are stored in the Madrigal database (http://cedar.openmadrigal.org/openmadrigal, accessed on 25 October 2023). The GISTM Data are derived from CHAIN (http://chain.physics.unb.ca/data/, accessed on 25 October 2023), and the solar geomagnetic parameters are available from the GSFC/SPDF OMNIWeb (https://omniweb.gsfc.nasa.gov/form/dx4.html, accessed on 25 October 2023).

Acknowledgments: We acknowledge the use of MATrix LABoratory (MATLAB R2022b) software, the RTKLIB program package, and Generic Mapping Tools. We would also like to thank the anonymous reviewers for their helpful suggestions.

Conflicts of Interest: Authors Yifan Wang, Hao Geng, Guofang Wang and Gang Wen were employed by the company Yunnan Power Grid Company Ltd. The remaining authors declare that the research was conducted in the absence of any commercial or financial relationships that could be construed as a potential conflict of interest.

References

1. Yuan, Y. Study on Theories and Methods of Correcting Ionospheric Delay and Monitoring Ionosphere Based on GPS. Ph.D. Thesis, Institute of Geodesy and Geophysics, Chinese Academy of Sciences, Wuhan, China, 2002. (In Chinese)
2. Yuan, Y.; Tscherning, C.C.; Knudsen, P.; Xu, G.; Ou, J. The ionospheric eclipse factor method (IEFM) and its application to determining the ionospheric delay for GPS. *J. Geod.* **2008**, *82*, 1–8. [CrossRef]
3. Yuan, Y.; Wang, N.; Li, Z.; Huo, X. The BeiDou global broadcast ionospheric delay correction model (BDGIM) and its preliminary performance evaluation results. *Navigation* **2019**, *66*, 55–69. [CrossRef]
4. Feng, J.; Zhang, T.; Li, W.; Zhao, Z.; Han, B.; Wang, K. A new global TEC empirical model based on fusing multi-source data. *GPS Solut.* **2022**, *27*, 20. [CrossRef]
5. D'Amicis, R.; Bruno, R.; Bavassano, B. Is geomagnetic activity driven by solar wind turbulence? *Geophys. Res. Lett.* **2007**, *34*, 28896. [CrossRef]
6. Gonzalez, W.D.; Joselyn, J.A.; Kamide, Y.; Kroehl, H.W.; Rostoker, G.; Tsurutani, B.T.; Vasyliunas, V.M. What is a geomagnetic storm? *J. Geophys. Res. Space Phys.* **1994**, *99*, 5771–5792. [CrossRef]
7. Afraimovich, E.L.; Demyanov, V.V.; Kondakova, T.N. Degradation of GPS performance in geomagnetically disturbed conditions. *GPS Solut.* **2003**, *7*, 109–119. [CrossRef]
8. Astafyeva, E.; Yasyukevich, Y.; Maksikov, A.; Zhivetiev, I. Geomagnetic storms, super-storms, and their impacts on GPS-based navigation systems. *Space Weather* **2014**, *12*, 508–525. [CrossRef]
9. Peng, Y.; Scales, W.A.; Hartinger, M.D.; Xu, Z.; Coyle, S. Characterization of multi-scale ionospheric irregularities using ground-based and space-based GNSS observations. *Satell. Navig.* **2021**, *2*, 14. [CrossRef]
10. Lin, C.H.; Richmond, A.D.; Liu, J.Y.; Yeh, H.C.; Paxton, L.J.; Lu, G.; Tsai, H.F.; Su, S.Y. Large-scale variations of the low-latitude ionosphere during the October–November 2003 superstorm: Observational results. *J. Geophys. Res. Space Phys.* **2005**, *110*, 10900. [CrossRef]
11. Nishioka, M.; Saito, A.; Tsugawa, T. Super-medium-scale traveling ionospheric disturbance observed at midlatitude during the geomagnetic storm on 10 November 2004. *J. Geophys. Res. Space Phys.* **2009**, *114*, 13581. [CrossRef]
12. Park, J.; Lühr, H.; Jakowski, N.; Gerzen, T.; Kil, H.; Jee, G.; Xiong, C.; Wook Min, K.; Noja, M. A long-lived band of plasma density enhancement at mid-latitudes during the 2003 Halloween magnetic storm. *J. Atmos. Sol.-Terr. Phys.* **2012**, *80*, 100–110. [CrossRef]
13. Li, W.; Zhao, D.; He, C.; Hancock, C.M.; Shen, Y.; Zhang, K. Spatial-Temporal Behaviors of Large-Scale Ionospheric Perturbations During Severe Geomagnetic Storms on 7–8 September 2017 Using the GNSS, SWARM and TIE-GCM Techniques. *J. Geophys. Res. Space Phys.* **2022**, *127*, e2021JA029830. [CrossRef]

14. Dugassa, T.; Mezgebe, N.; Habarulema, J.B.; Habyarimana, V.; Oljira, A. Ionospheric response to the 23–31 August 2018 geomagnetic storm in the Europe-African longitude sector using multi-instrument observations. *Adv. Space Res.* **2023**, *71*, 2269–2287. [CrossRef]
15. Cherniak, I.; Zakharenkova, I. Dependence of the high-latitude plasma irregularities on the auroral activity indices: A case study of 17 March 2015 geomagnetic storm. *Earth Planets Space* **2015**, *67*, 151. [CrossRef]
16. Nava, B.; Rodríguez-Zuluaga, J.; Alazo-Cuartas, K.; Kashcheyev, A.; Migoya-Orué, Y.; Radicella, S.M.; Amory-Mazaudier, C.; Fleury, R. Middle- and low-latitude ionosphere response to 2015 St. Patrick's Day geomagnetic storm. *J. Geophys. Res. Space Phys.* **2016**, *121*, 3421–3438. [CrossRef]
17. Jacobsen, K.S.; Andalsvik, Y.L. Overview of the 2015 St. Patrick's day storm and its consequences for RTK and PPP positioning in Norway. *J. Space Weather Space Clim.* **2016**, *6*, 2016004. [CrossRef]
18. Zakharenkova, I.E.; Cherniak, I.V.; Shagimuratov, I.I.; Klimenko, M.V. Features of High-Latitude Ionospheric Irregularities Development as Revealed by Ground-Based GPS Observations, Satellite-Borne GPS Observations and Satellite In Situ Measurements over the Territory of Russia during the Geomagnetic Storm on March 17–18, 2015. *Geomagn. Aeron.* **2018**, *58*, 70–82. [CrossRef]
19. Aa, E.; Huang, W.; Liu, S.; Ridley, A.; Zou, S.; Shi, L.; Chen, Y.; Shen, H.; Yuan, T.; Li, J.; et al. Midlatitude Plasma Bubbles Over China and Adjacent Areas During a Magnetic Storm on 8 September 2017. *Space Weather* **2018**, *16*, 321–331. [CrossRef]
20. Habarulema, J.B.; Katamzi-Joseph, Z.T.; Burešová, D.; Nndanganeni, R.; Matamba, T.; Tshisaphungo, M.; Buchert, S.; Kosch, M.; Lotz, S.; Cilliers, P.; et al. Ionospheric Response at Conjugate Locations During the 7–8 September 2017 Geomagnetic Storm Over the Europe-African Longitude Sector. *J. Geophys. Res. Space Phys.* **2020**, *125*, e2020JA028307. [CrossRef]
21. Alfonsi, L.; Cesaroni, C.; Spogli, L.; Regi, M.; Paul, A.; Ray, S.; Lepidi, S.; Di Mauro, D.; Haralambous, H.; Oikonomou, C.; et al. Ionospheric Disturbances Over the Indian Sector During 8 September 2017 Geomagnetic Storm: Plasma Structuring and Propagation. *Space Weather* **2021**, *19*, e2020SW002607. [CrossRef]
22. Li, G.; Ning, B.; Wang, C.; Abdu, M.A.; Otsuka, Y.; Yamamoto, M.; Wu, J.; Chen, J. Storm-Enhanced Development of Postsunset Equatorial Plasma Bubbles Around the Meridian 120°E/60°W on 7–8 September 2017. *J. Geophys. Res. Space Phys.* **2018**, *123*, 7985–7998. [CrossRef]
23. Zhang, S.-R.; Erickson, P.J.; Coster, A.J.; Rideout, W.; Vierinen, J.; Jonah, O.; Goncharenko, L.P. Subauroral and Polar Traveling Ionospheric Disturbances During the 7–9 September 2017 Storms. *Space Weather* **2019**, *17*, 1748–1764. [CrossRef]
24. Liu, Y.; Li, Z.; Fu, L.; Wang, J.; Zhang, C. Studying the ionospheric responses induced by a geomagnetic storm in September 2017 with multiple observations in America. *GPS Solut.* **2019**, *24*, 3. [CrossRef]
25. Wan, X.; Xiong, C.; Gao, S.; Huang, F.; Liu, Y.; Aa, E.; Yin, F.; Cai, H. The nighttime ionospheric response and occurrence of equatorial plasma irregularities during geomagnetic storms: A case study. *Satell. Navig.* **2021**, *2*, 23. [CrossRef]
26. Zakharenkova, I.; Cherniak, I. Effects of storm-induced equatorial plasma bubbles on GPS-based kinematic positioning at equatorial and middle latitudes during the September 7–8, 2017, geomagnetic storm. *GPS Solut.* **2021**, *25*, 132. [CrossRef]
27. Duann, Y.; Chang, L.C.; Liu, J.-Y. Impact of the February 3–4, 2022 geomagnetic storm on ionospheric S4 amplitude scintillation index: Observations and implications. *Adv. Space Res.* **2023**, *72*, 4379–4391. [CrossRef]
28. Hapgood, M.; Liu, H.; Lugaz, N. SpaceX—Sailing Close to the Space Weather? *Space Weather* **2022**, *20*, e2022SW003074. [CrossRef]
29. Feng, J.; Zhang, Y.; Li, W.; Han, B.; Zhao, Z.; Zhang, T.; Huang, R. Analysis of ionospheric TEC response to solar and geomagnetic activities at different solar activity stages. *Adv. Space Res.* **2023**, *71*, 2225–2239. [CrossRef]
30. Cherniak, I.; Krankowski, A.; Zakharenkova, I. ROTI Maps: A new IGS ionospheric product characterizing the ionospheric irregularities occurrence. *GPS Solut.* **2018**, *22*, 69. [CrossRef]
31. Rideout, W.; Coster, A. Automated GPS processing for global total electron content data. *GPS Solut.* **2006**, *10*, 219–228. [CrossRef]
32. Jayachandran, P.T.; Langley, R.B.; MacDougall, J.W.; Mushini, S.C.; Pokhotelov, D.; Hamza, A.M.; Mann, I.R.; Milling, D.K., Kale, Z.C.; Chadwick, R.; et al. Canadian High Arctic Ionospheric Network (CHAIN). *Radio Sci.* **2009**, *44*, 4046. [CrossRef]
33. Heelis, R.A.; Stoneback, R.A.; Perdue, M.D.; Depew, M.D.; Morgan, W.A.; Mankey, M.W.; Lippincott, C.R.; Harmon, L.L.; Holt, B.J. Ion Velocity Measurements for the Ionospheric Connections Explorer. *Space Sci. Rev.* **2017**, *212*, 615–629. [CrossRef]
34. Schreiner, W.S.; Weiss, J.P.; Anthes, R.A.; Braun, J.; Chu, V.; Fong, J.; Hunt, D.; Kuo, Y.H.; Meehan, T.; Serafino, W.; et al. COSMIC-2 Radio Occultation Constellation: First Results. *Geophys. Res. Lett.* **2020**, *47*, e2019GL086841. [CrossRef]
35. Zumberge, J.F.; Heflin, M.B.; Jefferson, D.C.; Watkins, M.M.; Webb, F.H. Precise point positioning for the efficient and robust analysis of GPS data from large networks. *J. Geophys. Res. Solid Earth* **1997**, *102*, 5005–5017. [CrossRef]
36. Kouba, J.; Lahaye, F.; Tétreault, P. Precise Point Positioning. In *Springer Handbook of Global Navigation Satellite Systems*; Teunissen, P.J.G., Montenbruck, O., Eds.; Springer International Publishing: Cham, Switzerland, 2017; pp. 723–751.
37. Takasu, T.; Yasuda, A. Development of the low-cost RTK-GPS receiver with an open source program package RTKLIB. In Proceedings of the International Symposium on GPS/GNSS, Jeju, Republic of Korea, 4–6 November 2009.
38. Feng, J.; Yuan, Y.; Zhang, T.; Zhang, Z.; Meng, D. Analysis of Ionospheric Anomalies before the Tonga Volcanic Eruption on 15 January 2022. *Remote Sens.* **2023**, *15*, 4879. [CrossRef]
39. Prikryl, P.; Jayachandran, P.T.; Mushini, S.C.; Pokhotelov, D.; MacDougall, J.W.; Donovan, E.; Spanswick, E.; St.-Maurice, J.P. GPS TEC, scintillation and cycle slips observed at high latitudes during solar minimum. *Ann. Geophys.* **2010**, *28*, 1307–1316. [CrossRef]
40. Sastri, J.H.; Jyoti, N.; Somayajulu, V.V.; Chandra, H.; Devasia, C.V. Ionospheric storm of early November 1993 in the Indian equatorial region. *J. Geophys. Res. Space Phys.* **2000**, *105*, 18443–18455. [CrossRef]

41. Maruyama, T.; Nakamura, M. Conditions for intense ionospheric storms expanding to lower midlatitudes. *J. Geophys. Res. Space Phys.* **2007**, *112*, 12226. [CrossRef]
42. Zhao, B.; Wan, W.; Tschu, K.; Igarashi, K.; Kikuchi, T.; Nozaki, K.; Watari, S.; Li, G.; Paxton, L.J.; Liu, L.; et al. Ionosphere disturbances observed throughout Southeast Asia of the superstorm of 20–22 November 2003. *J. Geophys. Res. Space Phys.* **2008**, *113*, 13054. [CrossRef]
43. Balan, N.; Bailey, G.J. A Physical Mechanism of Positive Ionospheric Storms. In *Aeronomy of the Earth's Atmosphere and Ionosphere*; Abdu, M.A., Pancheva, D., Eds.; Springer: Dordrecht, The Netherlands, 2011; pp. 465–475.
44. Pedatella, N.M.; Liu, H.L. The Influence of Internal Atmospheric Variability on the Ionosphere Response to a Geomagnetic Storm. *Geophys. Res. Lett.* **2018**, *45*, 4578–4585. [CrossRef]
45. Nie, W.; Rovira-Garcia, A.; Li, M.; Fang, Z.; Wang, Y.; Zheng, D.; Xu, T. The Mechanism for GNSS-Based Kinematic Positioning Degradation at High-Latitudes Under the March 2015 Great Storm. *Space Weather* **2022**, *20*, e2022SW003132. [CrossRef]
46. Luo, X.; Du, J.; Galera Monico, J.F.; Xiong, C.; Liu, J.; Liang, X. ROTI-Based Stochastic Model to Improve GNSS Precise Point Positioning Under Severe Geomagnetic Storm Activity. *Space Weather* **2022**, *20*, e2022SW003114. [CrossRef]

remote sensing

MDPI

Article

Analysis of Ionospheric Disturbances during X-Class Solar Flares (2021–2022) Using GNSS Data and Wavelet Analysis

Charbeth López-Urias [1,*], G. Esteban Vazquez-Becerra [1], Karan Nayak [1] and Rebeca López-Montes [2]

1 Department of Earth and Space Sciences, Autonomous University of Sinaloa, Culiacán 80040, Mexico; gvazquez@uas.edu.mx (G.E.V.-B.); nayakkaran.facite@uas.edu.mx (K.N.)
2 Department of Applied Geophysics, Center for Scientific Research and Higher Education of Ensenada, La Paz 23050, Mexico; rebecamont@gmail.com
* Correspondence: charbethlopez@uas.edu.mx; Tel.: +52-6672-433-822

Abstract: The influence of solar activity on the ionosphere, a critical area of investigation due to its relevance to the Sun–Earth relationship, has been extensively examined through various methodologies. The ability of solar events to induce disturbances in both the ionosphere and the geomagnetic field is widely acknowledged. This specific study focused on sporadic incidents resulting from X-class solar flares that occurred between 2021 and 2022. Utilizing a methodology that involved analyzing data at 5Hz intervals using wavelet algorithms, the data from the GNSS stations of the National Autonomous University of Mexico (UNAM) were investigated. The primary emphasis was on deducing the Total Electron Content (TEC) within the ionosphere. Subsequently, this parameter for each satellite during instances of solar flares was analyzed. The approach uncovered disruptions in the ionosphere triggered by solar flares, even in cases where events transpired at the periphery of the solar disk and were of magnitudes smaller than X2.

Keywords: ionosphere; GNSS; wavelets; solar flare; total electron content

1. Introduction

The ionosphere, situated in the upper region of Earth's atmosphere, maintains a continuous state of ionization primarily caused by solar radiation. This layer spans from approximately 60 km to 1000 km above the Earth's surface [1]. The Earth's ionosphere plays a crucial role in facilitating high-frequency communication and navigation systems, serving as a critical medium for the propagation of radio waves. It exhibits complicated spatial and temporal changes that can disrupt the transmission of radio signals, adversely impacting satellites, communication, navigation, and geographic location [2–6]. The ionosphere's Total Electron Content (TEC) is a key parameter that characterizes the integrated electron density along the signal's path. The TEC is the total number of electrons in the path from the satellite to the receiver, in a column with a 1 m^2 cross-section. Its unit is the TECU, where $1TECU = 10^{16}\ e/m^{-2}$ is a parameter that describes the behavior of the ionosphere [7]. During radio wave propagation, the electron density in the Earth's atmosphere is expected to vary over a nominal range of $10^{16}\ e/m^{-2}$ to $10^{19}\ e/m^{-2}$, being the minimum and maximum values reached around midnight and mid-afternoon, respectively [8], and is obtained from the signals L1 (1.57542 GHz) and L2 (1.2276 GHz) of the Global Positioning Systems [9]. However, this region is not impervious to the influence of solar activity, particularly solar flares—intense bursts of radiation and energy released from the Sun's surface. Solar flares can lead to significant disturbances in the ionosphere, impacting TEC and affecting the reliability and performance of high-frequency communication and navigation systems.

A solar flare, also referred to as a chromospheric burst, is a significant release of high-energy radiation frequently observed from the Sun. It holds immense importance in the realm of space weather due to its powerful and energetic nature [10,11]. Solar flares stem from the active region of the Sun, typically near a cluster of sunspots, and can endure anywhere from minutes to hours. These solar flares represent the most powerful and abrupt explosive events during which a substantial number of high-energy protons, electrons, and intense radiation spanning the entire electromagnetic spectrum are emitted. Consequently, there is a notable surge in extreme ultraviolet (EUV) and X-ray emissions, leading to additional ionization in the dayside ionosphere [12].

Numerous studies have indicated a complex relationship between solar flares and ionospheric perturbations, with the potential to cause disruptions in TEC. Variations in electron density during solar flares can affect the phase and amplitude of radio signals propagating through the ionosphere, leading to signal fading, scintillation, and navigation inaccuracies [13–17]. Such effects can have substantial consequences for global navigation satellite systems (GNSS), satellite-based communication links, and other applications that rely on ionospheric propagation. Yasyukevich et al. [13] investigated the influence of X-class solar flares on GNSS signals during the solar maximum of Cycle 24. Their study utilized GNSS data to analyze the TEC variations caused by solar flares and assess the impact on positioning accuracy. Sreeja [14] reviewed the space weather effects on high-accuracy GNSS-based navigation, discussing the implications of solar flares on GNSS signal propagation, TEC estimation, and navigation performance. Yadav et al. [15] analyzed the impacts of the St. Patrick's Day storm on the high-latitude ionosphere using GPS and CHAMP data. Their study investigated the TEC variations induced by solar flares during this geomagnetic storm. Reddybattula et al. [16] examined the ionospheric response to the solar flare of 10 September 2017 and its impact on GNSS positioning in the Asia–Oceania region. Their research highlighted the regional variability of TEC disturbances caused by solar flares. Currently, much research is being carried out using GPS technology, because the impact of the ionosphere on GPS signals provides the information necessary to understand how the ionosphere varies temporally and spatially, and how this affects navigation and navigation. satellite communication systems [18].

This study investigates the noteworthy influence of solar flares on TEC at high frequencies, with a focus on their subsequent implications for space weather. Solar flares, powerful bursts of high-energy radiation emitted by the Sun, have been recognized as influential space weather events that disrupt the ionospheres electron density and cause variations in TEC. This research methodically analyzes data from multiple Global Navigation Satellite System (GNSS) stations during various solar flare events. By applying wavelet analysis and careful consideration of sampling intervals, this study establishes the optimal conditions for detecting and characterizing ionospheric disturbances induced by solar flares. This research not only highlights the complex interplay between solar flares and the ionosphere but also establishes a robust methodology for identifying and analyzing these disturbances, contributing to our understanding of space weather impact on Earth's technological infrastructure.

2. Data and Methods

2.1. TEC Data

GNSS systems utilize two or three frequencies to assess the characteristics of the ionosphere. These frequencies aid in the assessment of the ionosphere itself rather than its influence on the signals. Having access to dual frequencies allows for a more accurate examination of relative phase delays between signals, such as L1 and L2. A greater electron density in the ionosphere results in a more pronounced difference between the signals, as the impact of density on signals varies depending on their frequency [19]. Signals are affected by the ionosphere in terms of both amplitude and phase [20]. The extent of the dispersive effect is directly proportional to the total electron content [21].

In the conducted study, data from six GNSS stations affiliated with the Institute of Geophysics at the National Autonomous University of Mexico were utilized (see Table 1). The RINEX files were processed to exclusively retain the necessary parameters for the calculation of Total Electron Content (TEC). For slant TEC, particularly in the case of TEC calculated from pseudorange, Equation (1) [22] was employed, characterized by noise yet maintaining ambiguity-free characteristics [23]. To derive phase-based TEC, Equation (2) was applied, yielding results with less noise albeit being ambiguous.

$$TEC_R = 9.52(R_1 - R_2) \tag{1}$$

$$TEC_\Phi = 9.52(\lambda_1 \cdot \Phi_1 - \lambda_2 \cdot \Phi_2) \tag{2}$$

The values of slant TEC depend on the satellite's elevation angle, making it necessary to compute the vertical TEC. The vertical TEC is obtained using the mapping function (Equation (3)), which assumes an ionospheric point of interest at a height of 350 km [24].

$$VTEC = STEC \cdot cos\left[arcsin\frac{R_e cos(\theta)}{R_e + h}\right] \tag{3}$$

Here, R_e represents the radius of the Earth (6371.2 km), h represents the height of the ionosphere relative to the Earth's surface (350 km), and θ denotes the elevation angle between the satellite and the receiver, measured in degrees.

Table 1. GNSS station coordinates.

Station	Latitude (N)	Longitude (W)	Height (m)
SPIG	31°02′45.87″	115°27′57.68″	2751.110
UCOE	19°48′47.47″	101°41′39.86″	1977.427
UNPM	20°52′06.66″	86°52′05.43″	1.800
OXUM	15°39′44.09″	96°29′56.50″	82.267
SSNX	19°19′37.78″	99°10′36.37″	2275.850
UTEO	19°44′28.46″	99°11′19.00″	2267.452

2.2. *Wavelets*

The Discrete Wavelet Transform (DWT) is a method that iteratively decomposes an input signal, $S_0(n)$, into two lower resolution subsequences, known as approximation and detail. The subsequences $S_i(n)$ and $W_i(n)$ represent the approximation and detail, respectively, of the signal at level i [25]. The calculation of the approximation of the signal at level $i+1$ is performed according to Equations (4) and (5).

$$S_{i+1}(n) = \sum_k g(k) S_i(2n-k) \tag{4}$$

The detail of the signal at level $i+1$ can be calculated as follows:

$$W_{i+1}(n) = \sum_k h(k) S_i(2n-k) \tag{5}$$

This calculation process, known as the Mallat pyramid algorithm [26], is employed for the DWT.

According to Osorio et al. [27], in certain cases, it is advantageous to limit the amount of data for signal processing involving wavelet coefficients. Consequently, the number of analyzed samples is reduced by half through compression. In our scenario, even though we had time windows ranging from 2 to 4 h, it was feasible to detect ionospheric disturbances using wavelets at levels 7 and 8, where levels 7 and 8 are the seventh and eighth application of the wavelet algorithms, respectively. Figure 1 displays the results of the wavelet analysis for one of the satellites, where it can be observed that levels 7 and 8 exhibit a noticeable disturbance at the time of the event. In addition to this, we calculated the correlation between each of the levels and TEC, aiming to have a more solid foundation for selecting the wavelet levels to use. This led to the conclusion that the best levels are 7 and 8. It is worth mentioning that the application of each wavelet level is similar to applying filtering for high and low frequencies, with one being implemented for detail and the other for approximation.

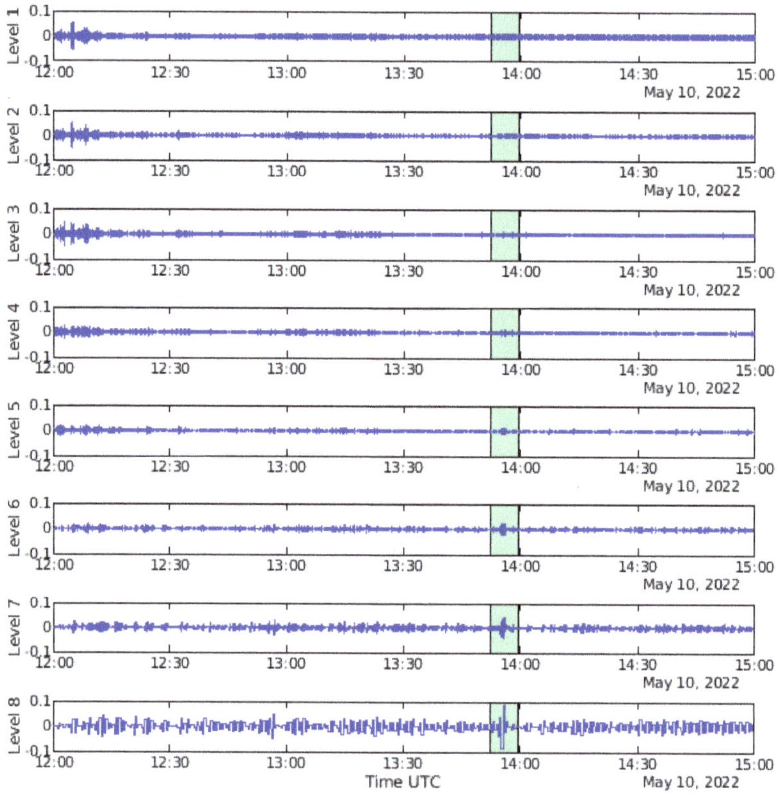

Figure 1. The figure illustrates the differentiation of each level's application from 1 to 8 in wavelet algorithms used for solar flare detection. The green rectangle is the solar flare being more distinct in level 7 and 8.

2.3. *Analysis Methods*

For over thirty years, the Geostationary Operational Environmental Satellite (GOES) series from the National Oceanic and Atmospheric Administration (NOAA) has been responsible for measuring meteorological and space weather phenomena. One of the essential components of the GOES-R instrument suite is the EUV and X-Ray Irradiance Sensor System (EXIS), which consists of two instruments designed to monitor solar spectral irradiance: the Extreme Ultraviolet Sensor (EUVS) and the X-Ray Sensor (XRS) [28]. For our study, we utilize data from the GOES 17 satellite to select and analyze six solar flare events that occurred during the years 2021 and 2022 (details provided in Table 2).

Figure 2 displays the positions of these solar flares on the solar disk, revealing occurrences both in the central region and at the periphery. Recognizing the spatial distribution of these solar eruptions is crucial as it offers pivotal insights into solar activity dynamics. Additionally, it is important to note that, in accordance with the Dst index (Figure 3), the Earth's magnetic field remained calm during this period. This calm magnetic field condition stands as an essential factor for our study, as it ensures that disturbances arising from magnetic field fluctuations are minimal. The data used from the dst index were obtained from the Center for Geomagnetism and Space Magnetism Data Analysis at Kyoto University.

Table 2. Event Features.

Date	Day GPS	Start Time (UT)	Peak Time (UT)	End Time (UT)	Flare Class
03 July 2021	184	14:18	14:29	14:34	X1.5
28 October 2021	301	15:17	15:35	15:48	X1
30 March 2022	89	17:21	17:37	17:46	X1.3
30 April 2022	120	13:37	13:47	13:52	X1.1
10 May 2022	130	13:50	13:55	13:59	X1.5
2 October 2022	275	19:53	20:25	20:34	X1.0

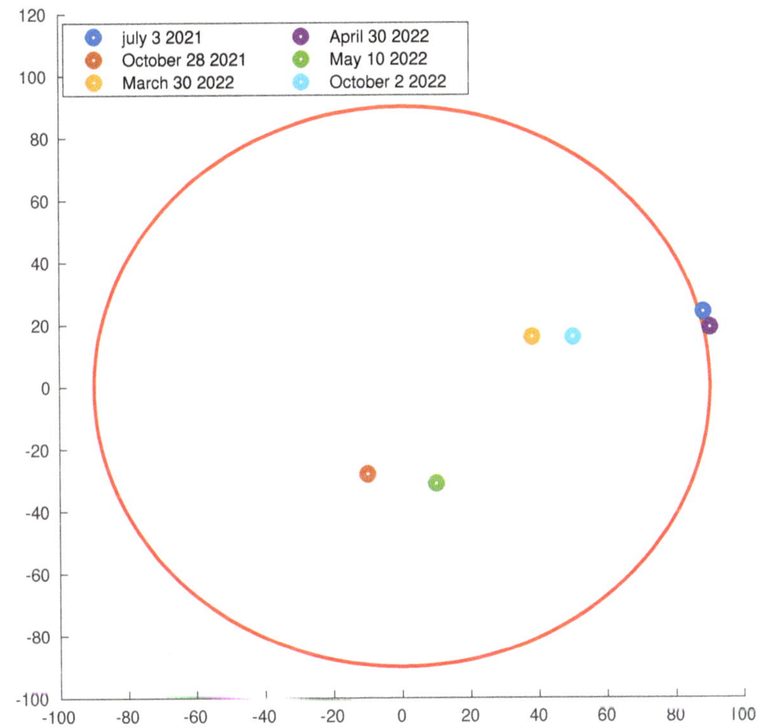

Figure 2. Location of the solar flare on the solar disk.

When investigating the Total Electron Content (TEC) on a per-station basis, we risk overlooking significant insights offered by analyzing TEC data for individual satellites. The computation of TEC for each station mandates the application of diverse filters to counter temporal mean effects that arise as the number of visible satellites observed by the station fluctuates. By individually scrutinizing TEC for each satellite, a more intricate evaluation of TEC conditions becomes achievable. To further enhance our understanding of ionospheric perturbations, we calculated TEC at 5 Hz intervals, allowing us to capture ionospheric fluctuations every 1/5 of a second.

The deployment of multiple monitoring stations played a pivotal role in mitigating distortions arising solely from the peculiarities of a single station's receiver. Employing wavelet analysis for TEC examination enabled us to distinctly and concisely pinpoint the influence of solar flares on TEC while simultaneously highlighting distinct frequency components evident over time.

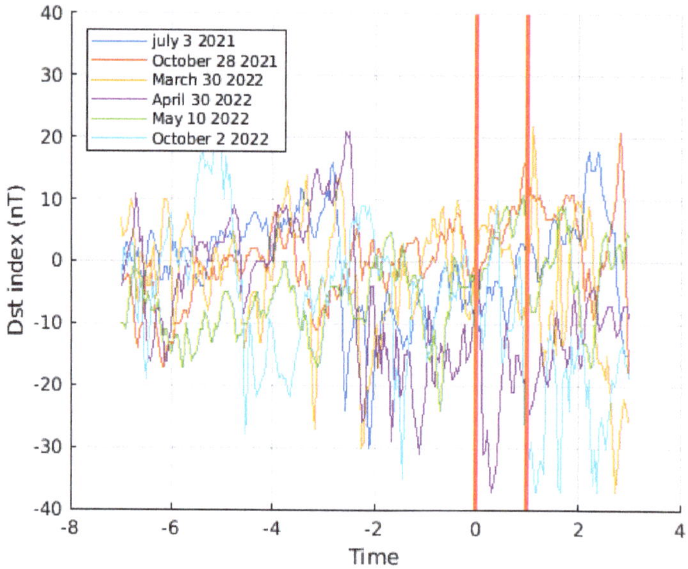

Figure 3. Variations of the Dst index during the events.

3. Results and Discussion

3.1. Event 1 (3 July 2021)

The event on 3 July 2021 took place in region AR12838, located in zone N24 W88, reaching a higher intensity in the X1.59 type flare, starting at 14:18 UTC and reaching its maximum intensity at 14:29 UTC, ending at 14:34 UTC. Although the event occurred in a location that was not favorable for its study, alterations in the ionosphere can still be found due to the implemented methodology. During the event time period, we have information from six GNSS stations (UTEO, UNPM, OXUM, SPIG, SSNX and UCOE), with an average of five satellites' information for each station during the event hour, with a window from 13:00 UTC to 15:00 UTC. Figure 4 shows the results of the event for only OXPE station (PRN 14), including X-ray irradiance measurements, vTEC, and wavelet details at different levels over a 2-hour period in UTC time. On the other hand, Figure 5 shows the wavelet results for the same event utilizing all six stations. In these, it is possible to identify the event in question; however, at the SSNX station, despite identifying the event, there are signal alterations that are not shown in other receivers, even though they interact with the same satellites. Therefore, for this station, it could be said that this is due to disturbances independent of the solar flare, or even noise from the receiver. Satellite 30 presented alterations at the beginning of the time window for all receivers; nevertheless, it is possible to identify the alterations that occurred due to the solar flare. Satellites 15 and 19 of the SPIG station had a poorer response, making it difficult to clearly identify the event. A possible cause of this could be the location of the station, which is further north than the other stations studied.

The solar flare occurred on 3 July 2021, but according to the magnetic field and GOES 17 satellite data, the previous day (2 July 2021) was calm (without external disturbances in the ionosphere). Therefore, for comparative purposes, we applied the same algorithms to this quiet day, yielding random and nonsensical alterations as results. This is in contrast to when a solar flare occurs, resulting in disturbances across all stations at a specific time (during the solar flare event). For each of the events, we followed the same procedure, generating a graph for a calm day and another for the event day. In the case of Event 1, Figure 6 presents the obtained results.

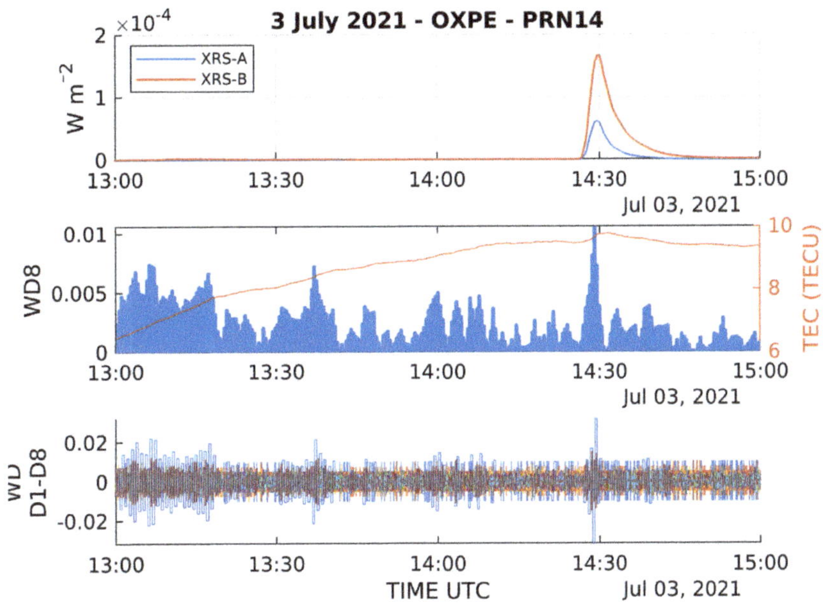

Figure 4. Results from the X1.59 solar flare event that occurred on 3 July 2021, in Region AR12838 (Zone N24 W88) for the OXPE station (PRN 14). The upper graph shows two irradiance measurements within the X-ray part of the electromagnetic spectrum, which were obtained through solar observations. These measurements originate from the XRS-A and XRS-B channels. The XRS-A channels detect wavelengths ranging from 0.05 to 0.4 nm, represented by the solid blue line, while the XRS-B channels cover wavelengths from 0.1 to 0.8 nm, represented by the solid red line. In the middle graph, the vTEC (vertical total electron content) is depicted as a solid red line measured in TECU, and it is accompanied by wavelet details at level 8 (WD8), represented by blue column bars. The last graph displays the wavelet details at all levels (WD D1-D8) together. The X-axis represents the time in UTC for a 2-hour period, with each grid corresponding to 30 min.

3.2. Event 2 (28 October 2021)

On 28 October 2021, there were two intense solar flares observed. One of these flares was categorized as type M with an intensity level of M8.77, while the other was of type X with an intensity of X1.06. The M-type flare was observed during the nighttime hours. As a result, the focus of the study is directed towards the X-type flare, which attained its peak intensity at 15:35 UTC. The initiation of this flare took place at 15:17 UTC, and its conclusion was marked at 15:48 UTC. The origin of this X-type flare was traced back to the solar region known as AR12887, situated at coordinates S28 E10.

The impact of this event was visible at the three designated stations, namely, UNPM, SPIG, and UCOE. These stations encountered notable disturbances during the time of the flare. It is worth noting that Satellite 24 displayed an unusual noise pattern at 17:48 UTC. This occurrence is potentially linked to noise originating from the mentioned satellite, especially considering the considerable geographical separation between the stations where the noise was detected.

Event 2 exhibited comparable features to the previous one, even though, in this instance, we had fewer satellites accessible across all stations. Nonetheless, a pronounced disturbance was evident, coinciding with the solar flare's occurrence. However, during the calm day, the alterations appeared random, lacking a consistent pattern attributable to flare-induced disruptions (Figure 7).

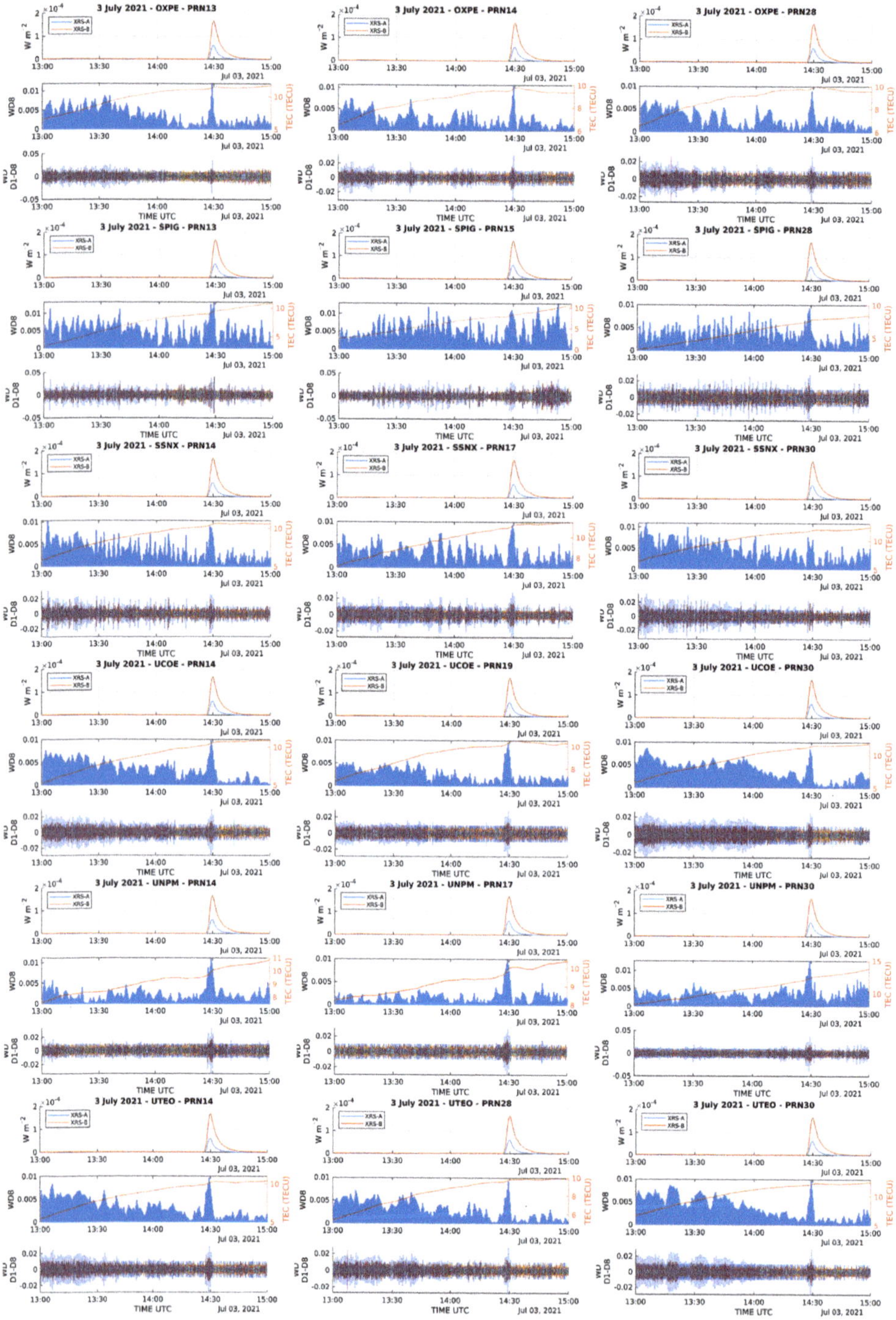

Figure 5. The wavelet results for the same event with accordance to all the six stations (OXPE, SPIG, SSNX, UCOE, UNPM, UTEO), each with three PRNs. The specific details for each station align with those depicted in Figure 4.

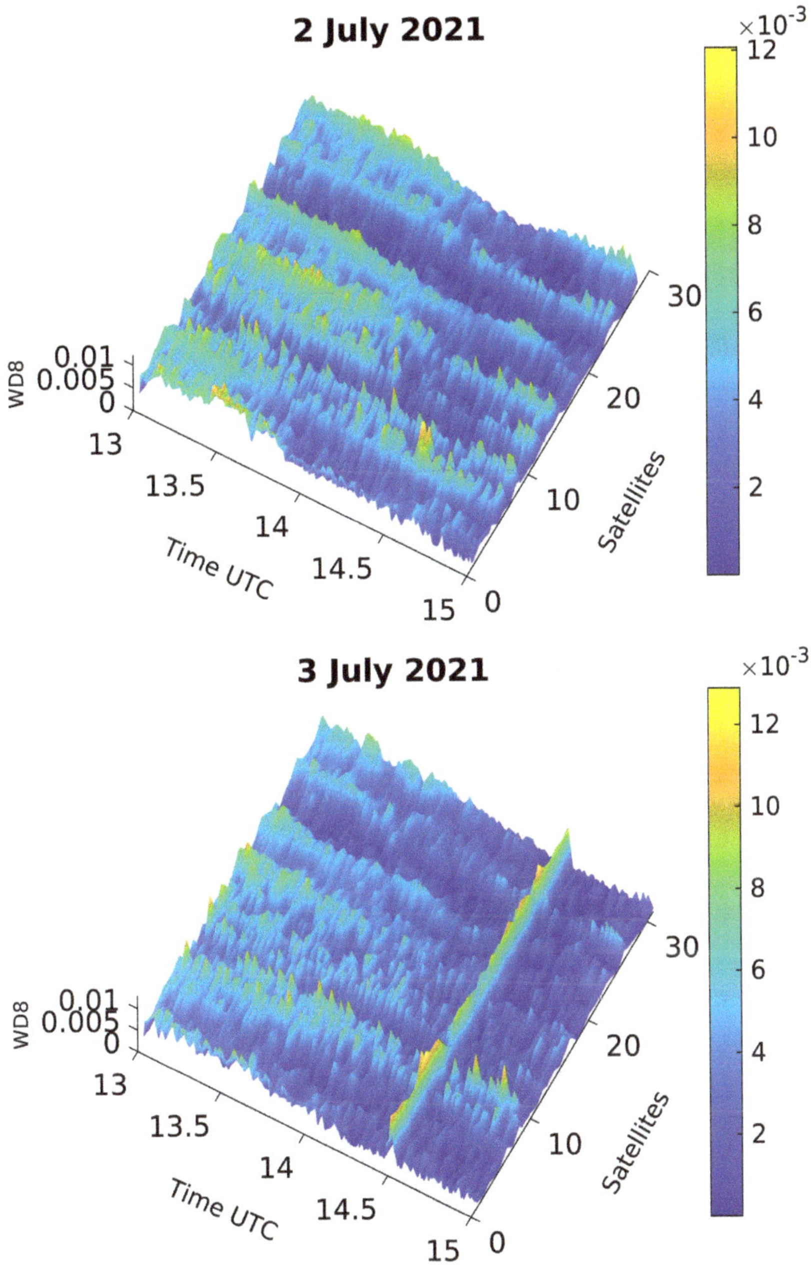

Figure 6. Visualization of Total Electron Content Wavelet Analysis from GNSS Signals for the Event on 3 July 2021

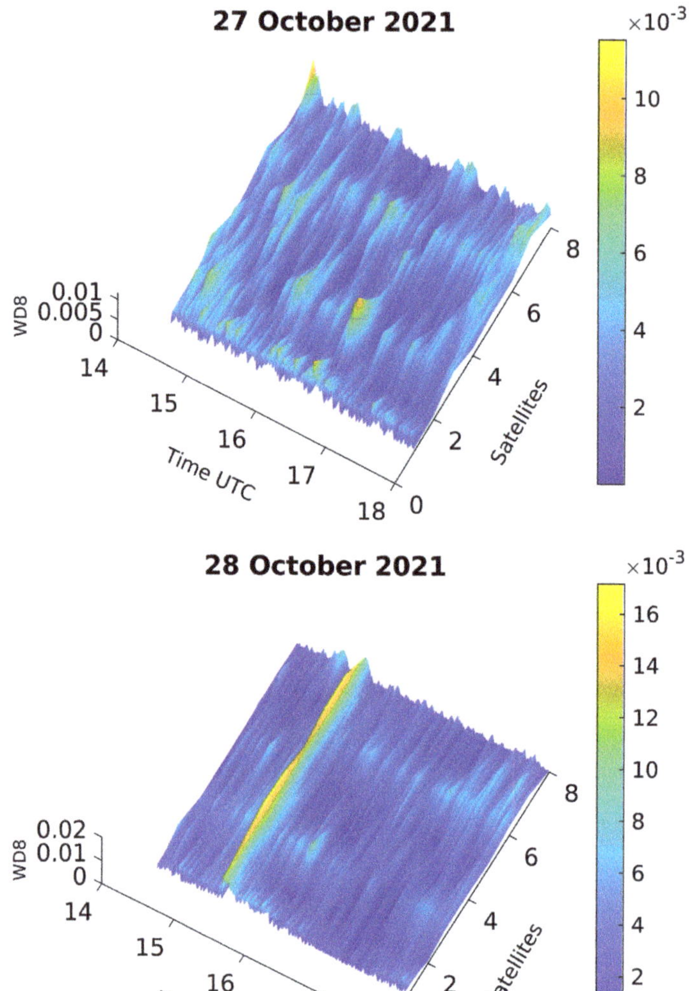

Figure 7. Visualization of Total Electron Content Wavelet Analysis from GNSS Signals for the Event on 28 October 2021.

3.3. Event 3 (30 March 2022)

On 30 March 2022, the Sun emitted a solar flare in the AR12975 region, which reached an X1.38 type intensity. This flare originated at location N16 W38, starting at 17:21 UTC, reaching its maximum intensity at 17:37 UTC, and ending at 17:46 UTC. This event caused disturbances in the ionosphere that were visible at all analyzed stations (UNPM, OXUM, SPIG, SSNX, and UCOE). In addition, for satellites 7 and 30, signal noise increased after the event.

In this instance, the alterations were also noticeable, although not as pronounced as the previous ones. Nonetheless, it remains possible to identify the moment of the solar flare. The results from the calm day display alterations that would not be associated with the flare but rather with the daily ionospheric cycle (Figure 8).

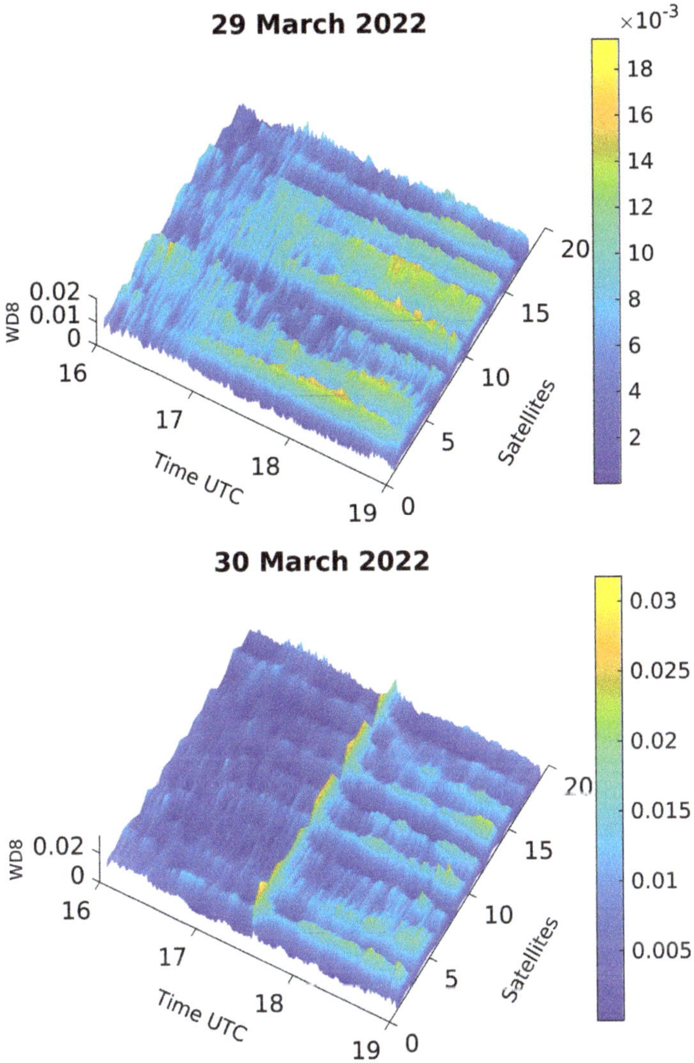

Figure 8. Visualization of Total Electron Content Wavelet Analysis from GNSS Signals for the Event on 30 March 2022.

3.4. Event 4 (30 April 2022)

The event of 30 April 2022 occurred in the solar spot region AR12994, located on the solar disk at N19 W90. Despite being in this location, it is possible to analyze the impact of this flare on the Earth's ionosphere, which was generated at 13:37 UTC, reached its maximum at 13:47 UTC, and ended at 13:52 UTC. The results of this event show a considerable increase in the wavelets after the event (Figure 9). This is a visible pattern where the flare did not have much impact. In this case, it is mainly noticeable in satellite 4. Additionally, it is possible to appreciate that there is not as much impact for the SPIG station as in the others, because this station is further north. The same happened in the other case where the location of the flare on the solar disk was very close to the edge.

During this event, we observed more distinct patterns of the diurnal cycle during the calm day. As the ionosphere began to charge, frequencies that impact the wavelet results were detected. Once the increase subsided, the wavelets also decreased. However, concerning the event detection, the day of the event was clearly and concisely distinguishable (Figure 9).

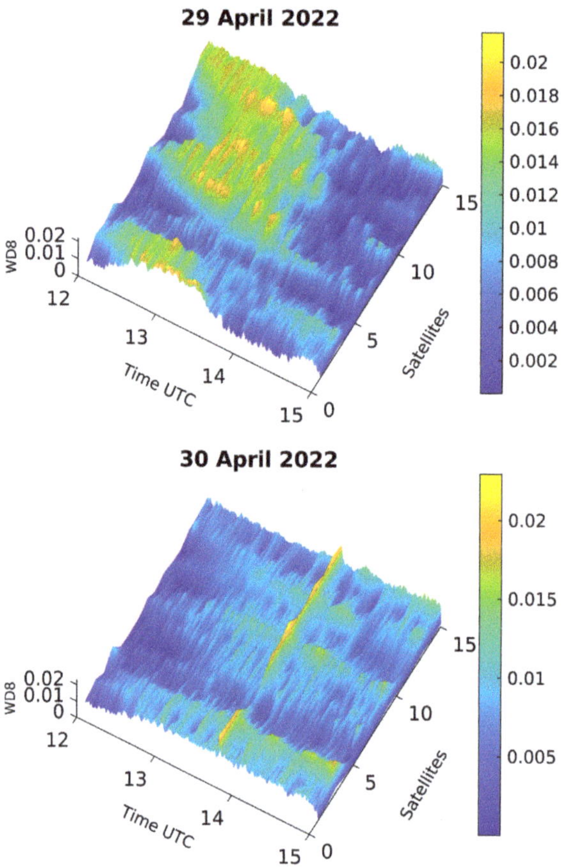

Figure 9. Visualization of Total Electron Content Wavelet Analysis from GNSS Signals for the Event on 30 April 2022.

3.5. Event 5 (10 May 2022)

On 10 May 2022, a solar flare of X1.5 type occurred in the AR13006 region located at S31 W10. The peak intensity was reached at 13:55 UTC, while the onset and end of the flare were recorded at 13:50 UTC and 13:59 UTC, respectively. During this event, all stations (UNPM, OXUM, SPIG, SSNX, and UCOE) experienced disturbances in the ionosphere, and the disturbances could also be observed in all satellites.

This event exhibited the most significant alterations in the wavelet results (Figure 10), displaying a pronounced disturbance with values nearing 0.05 at the level of detail 8. Considering that the average for the other events is around 0.02, this event experienced a more substantial alteration according to the wavelets. One possible reason for such pronounced disturbances in this event could be its solar flare's location on the solar disk, originating from the central part of the Sun. It is worth mentioning that, in the case of the calm day, we obtained maximum wavelet values less than 0.01, demonstrating the correlation between flare-induced disturbances and wavelet outcomes.

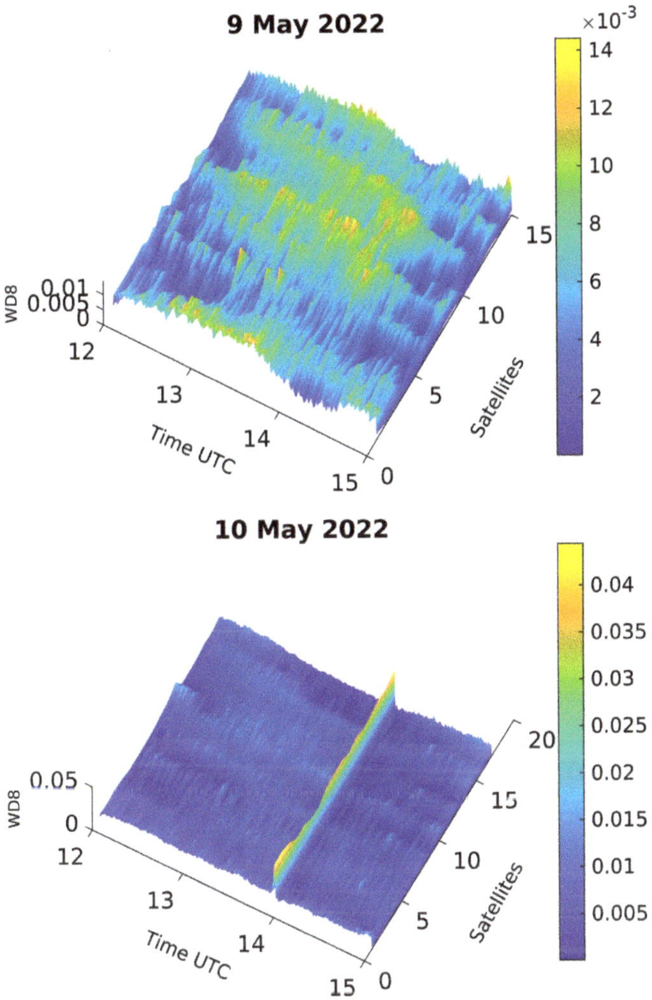

Figure 10. Visualization of Total Electron Content Wavelet Analysis from GNSS Signals for the Event on 10 May 2022.

3.6. Event 6 (2 October 2022)

The event of 2 October 2022 began at 19:53 UTC, reached its peak intensity at 20:25 UTC, and ended at 20:34 UTC. The genesis of this event can be traced back to the coordinates N16 W50, pinpointing the origin within the solar region AR13110. The impact of the solar flare on each of the satellites for all stations can be observed in the wavelet results. While some satellites show disturbances in the wavelets independent of the solar flare's impact, the maximum values of the wavelets occur precisely when the solar flare happens (see Figure 11).

The wavelet outcomes for this event revealed subtle alterations during the solar flare. Despite their subtlety, it was feasible to discern the moment when the flare impacted the ionosphere and distinguish between the results of a calm day and a day affected by the flare-induced disturbances (Figure 11).

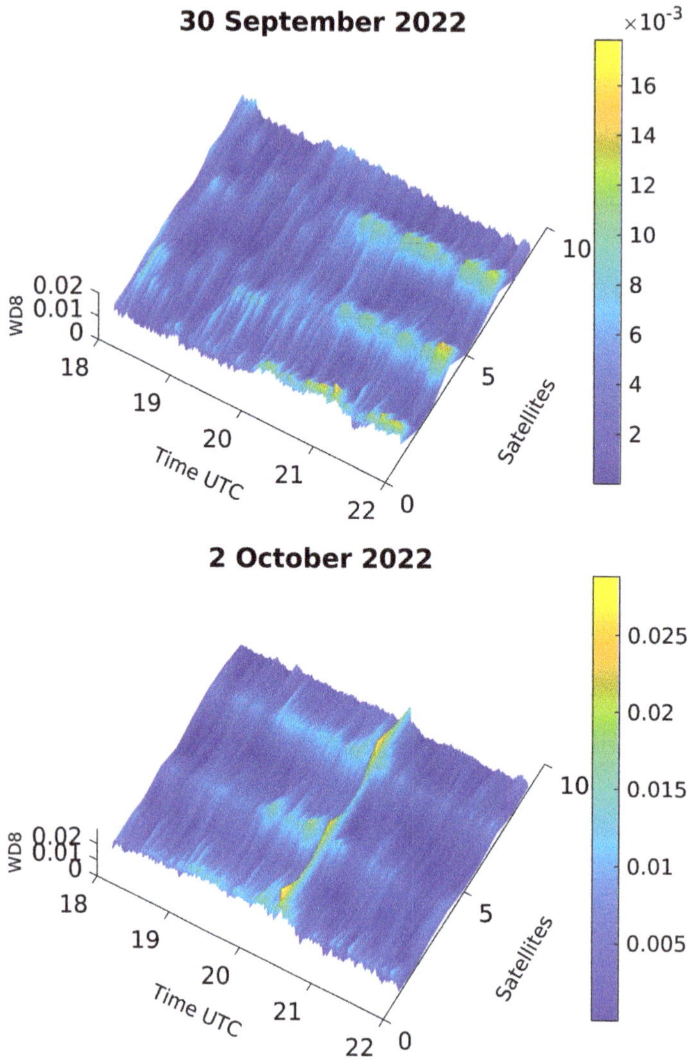

Figure 11. Visualization of Total Electron Content Wavelet Analysis from GNSS Signals for the Event on 2 October 2022.

Conducting the methodology's implementation on days characterized by tranquil Total Electron Content (TEC) conditions, unaffected by storms or phenomena that atypically disrupt the TEC, is pivotal. This approach allows for the comparison of results between normal days and days influenced by solar flares, as seen in prior events. However, by conducting an aggregation of wavelet results for each station, a coherent and distinctive pattern emerges, revealing the solar flare's consistent impact on the total electron content within the ionosphere. This visual representation is showcased in Figure 12, where the average wavelet trends per station are depicted. This illustration unmistakably underscores that all events exhibit a discernible surge in wavelet values across all the stations employed in the analysis.

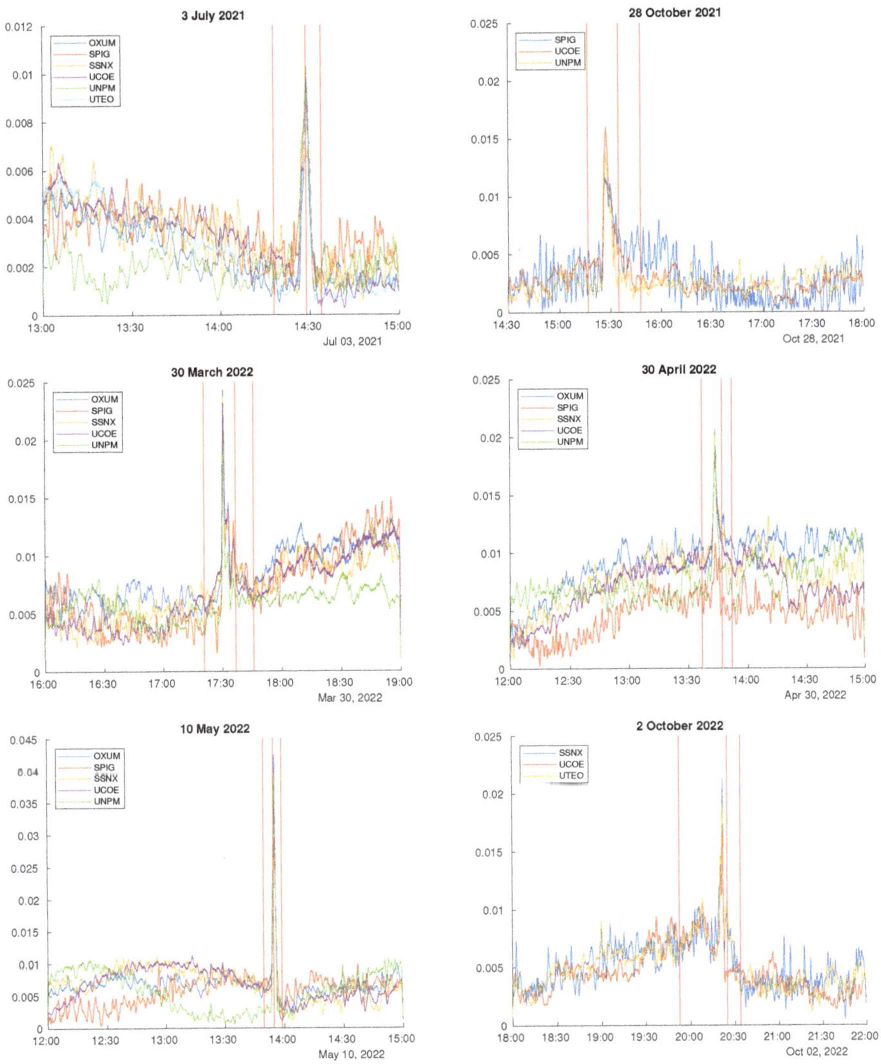

Figure 12. Averaging the wavelets per station of the analyzed events.

4. Conclusions

The comprehensive investigation into the interplay between solar flares and the ionosphere has yielded profound insights that hold vast implications for both scientific knowledge and practical applications. Through meticulous data analysis, strategic methodology selection, and rigorous validation, a nuanced understanding of the intricate relationship between these space weather events has emerged. A pivotal discovery arising from this study is the significant influence of sampling intervals on the detection and characterization of ionospheric disturbances induced by solar flares. The identification of previously unnoticed variations at high frequencies within the Total Electron Content (TEC) during solar flare events underscores the critical importance of selecting appropriate sampling intervals. The establishment of a 5 Hz sampling interval as optimal not only enhances the capture of sporadic events but also ensures efficient data processing. This choice was determined through comprehensive testing, which demonstrated that higher sampling rates led to increased data volume without significantly improving event detection accuracy, while lower rates risked missing important fluctuations in TEC during solar flare events.

Furthermore, the implementation of wavelet analysis, particularly at levels 7 and 8, has provided unprecedented accuracy in portraying ionospheric perturbations caused by solar flares. This refined methodology, capable of detecting minute anomalies at 0.1 TECU and 2 Hz intervals, reaffirms its effectiveness in unraveling even subtle ionospheric fluctuations. The investigation reveals the influence of solar flares, specifically those of X-class intensity up to X2, on the ionosphere. Disturbances originating from solar flares even at the solar disk periphery emphasize the far-reaching consequences of these events. The convergence of anomalies across multiple monitoring stations solidifies their authenticity and establishes them as significant events with substantial implications for ionospheric dynamics. The study findings have been robustly validated through application to event-free days, confirming their accuracy and reliability. The alignment of detected perturbations with the onset and termination phases of solar flares further supports the causal link between solar flares and ionospheric disturbances, reinforcing the importance of these findings in the context of space weather phenomena.

In summary, this intricate study, involving thorough data analysis and accurate techniques, provides a novel approach for detecting solar influences on the ionosphere. While it represents an important step, further research is needed to deepen our comprehension of the complex interaction between solar flares and the ionosphere. Examining the ionosphere at these specific time points reveals certain frequencies that go unnoticed when investigating Total Electron Content (TEC) through individual stations and at lower frequencies.

Author Contributions: Conceptualization, C.L.-U. and G.E.V.-B.; methodology, C.L.-U. and K.N.; software, C.L.-U.; validation, C.L.-U., G.E.V.-B. and R.L.-M.; formal analysis, C.L.-U., R.L.-M. and K.N.; investigation, C.L.-U.; resources, C.L.-U. and K.N.; data curation, C.L.-U.; writing—original draft preparation, C.L.-U. and K.N.; writing—review and editing, C.L.-U. and K.N.; visualization, C.L.-U. and K.N.; supervision, G.E.V.-B. and R.L.-M. All authors have read and agreed to the published version of the manuscript.

Funding: This work was carried out with the support (CVU: 859540) of the National Council of Science and Technology (CONACyT) in Mexico.

Data Availability Statement: The data can be made available on request.

Acknowledgments: The authors would like to express sincere gratitude to the UNAM Geophysical Institute for their invaluable contribution to this research through the provision of essential data. The availability of this data was crucial in conducting a comprehensive analysis and achieving the outcomes presented in this manuscript.The authors also impart their deepest thanks to NASA for providing GOES satellite data. The authors further extend their heartfelt thanks to CONACYT (National Council for Science and Technology) for their financial support that made this research possible.

Conflicts of Interest: The authors declare no conflict of interest.

References

1. Budden, K.G. *The Propagation of Radio Waves: The Theory of Radio Waves of Low Power in the Ionosphere and Magnetosphere*; Cambridge University Press: Cambridge, UK , 1988.
2. Davies, K.; Smith, E.K. Ionospheric effects on satellite land mobile systems. *IEEE Antennas Propag. Mag.* **2002**, *44*, 24–31. [CrossRef]
3. Yuan, Y.; Tscherning, C.; Knudsen, P.; Xu, G.; Ou, J. The ionospheric eclipse factor method (IEFM) and its application to determining the ionospheric delay for GPS. *J. Geod.* **2008**, *82*, 1–8. [CrossRef]
4. Klobuchar, J.A. Ionospheric time-delay algorithm for single-frequency GPS users. *IEEE Trans. Aerosp. Electron. Syst.* **1987**, *AES-23*, 325–331. [CrossRef]
5. Prieto-Cerdeira, R.; Orús-Pérez, R.; Breeuwer, E.; Lucas-Rodriguez, R.; Falcone, M. Performance of the Galileo single-frequency ionospheric correction during in-orbit validation. *GPS World* **2014**, *25*, 53–58.
6. Yuan, Y.; Wang, N.; Li, Z.; Huo, X. The BeiDou global broadcast ionospheric delay correction model (BDGIM) and its preliminary performance evaluation results. *Navigation* **2019**, *66*, 55–69. [CrossRef]

7. Feng, J.; Zhang, Y.; Li, W.; Han, B.; Zhao, Z.; Zhang, T.; Huang, R. Analysis of ionospheric TEC response to solar and geomagnetic activities at different solar activity stages. *Adv. Space Res.* **2023**, *71*, 2225–2239. [CrossRef]
8. Bilitza, D.; McKinnell, L.A.; Reinisch, B.; Fuller-Rowell, T. The international reference ionosphere today and in the future. *J. Geod.* **2011**, *85*, 909–920. [CrossRef]
9. Bhuyan, P.; Borah, R.R. TEC derived from GPS network in India and comparison with the IRI. *Adv. Space Res.* **2007**, *39*, 830–840. [CrossRef]
10. Eastwood, J.; Biffis, E.; Hapgood, M.; Green, L.; Bisi, M.; Bentley, R.; Wicks, R.; McKinnell, L.A.; Gibbs, M.; Burnett, C. The economic impact of space weather: Where do we stand? *Risk Anal.* **2017**, *37*, 206–218. [CrossRef]
11. Baker, D.; Daly, E.; Daglis, I.; Kappenman, J.G.; Panasyuk, M. Effects of space weather on technology infrastructure. *Space Weather* **2004**, *2*. [CrossRef]
12. Nishimoto, S.; Watanabe, K.; Kawai, T.; Imada, S.; Kawate, T. Validation of computed extreme ultraviolet emission spectra during solar flares. *Earth Planets Space* **2021**, *73*, 79. [CrossRef]
13. Yasyukevich, Y.; Astafyeva, E.; Padokhin, A.; Ivanova, V.; Syrovatskii, S.; Podlesnyi, A. The 6 September 2017 X-class solar flares and their impacts on the ionosphere, GNSS, and HF radio wave propagation. *Space Weather* **2018**, *16*, 1013–1027. [CrossRef]
14. Sreeja, V. Impact and mitigation of space weather effects on GNSS receiver performance. *Geosci. Lett.* **2016**, *3*, 24. [CrossRef]
15. Yadav, S.; Sunda, S.; Sridharan, R. The impact of the 17 March 2015 St. Patrick's Day storm on the evolutionary pattern of equatorial ionization anomaly over the Indian longitudes using high-resolution spatiotemporal TEC maps: New insights. *Space Weather* **2016**, *14*, 786–801. [CrossRef]
16. Reddybattula, K.D.; Panda, S.K.; Sharma, S.K.; Singh, A.K.; Kurnala, K.; Haritha, C.S.; Wuyyuru, S. Anomaly effects of 6–10 September 2017 solar flares on ionospheric total electron content over Saudi Arabian low latitudes. *Acta Astronaut.* **2020**, *177*, 332–340. [CrossRef]
17. Yuan, Y.; Ou, J. Auto-covariance estimation of variable samples (ACEVS) and its application for monitoring random ionospheric disturbances using GPS. *J. Geod.* **2001**, *75*, 438–447. [CrossRef]
18. Olwendo, O.; Baki, P.; Mito, C.; Doherty, P. Characterization of ionospheric GPS Total Electron Content (GPS–TEC) in low latitude zone over the Kenyan region during a very low solar activity phase. *J. Atmos. Sol.-Terr. Phys.* **2012**, *84–85*, 52–61. [CrossRef]
19. Eftaxiadis, K.; Cervera, M.A.; Thomas, R.M. *A Global Positioning System Receiver for Monitoring Ionospheric Total Electron Content*; Technical Report; Defence Science and Technology Organisation: Canberra, Australia, 1999.
20. Turel, N.; Arikan, F. Probability density function estimation for characterizing hourly variability of ionospheric total electron content. *Radio Sci.* **2010**, *45*, 1–10. [CrossRef]
21. Sardon, E.; Rius, A.; Zarraoa, N. Estimación del contenido total de electrones en la ionosfera usando datos del Sistema de Posicionamiento Global. *Física Tierra* **1993**, *5*, 167–182.
22. Araujo-Pradere, E. GPS-derived total electron content response for the Bastille Day magnetic storm of 2000 at a low mid-latitude station. *Geofísica Int.* **2005**, *44*, 211–218. [CrossRef]
23. Rodríguez, M. Estudio de Perturbaciones Ionosféricas a Través del Contenido Total de Electrones en Europa Meridional. Ph.D. Thesis, Universidad Complutense de Madrid, Madrid, Spain, 2017.
24. Brunini, C.; Camilion, E.; Azpilicueta, F. Simulation study of the influence of the ionospheric layer height in the thin layer ionospheric model. *J. Geod.* **2011**, *85*, 637–645. [CrossRef]
25. Colom, R.J.; Gadea, R.; Sebastia, A.; Martinez, M.; Ballester, F.; Herrero, V. Implementación de la Transformada Wavelet Discreta 2D con filtros no separables. In Proceedings of the I Jornadas Sobre Computación Reconfigurable y Aplicaciones, Alicante, España, 19 September 2001 .
26. Mallat, S. Multifrequency channel decompositions of images and wavelet models. *IEEE Trans. Acoust. Speech Signal Process.* **1989**, *37*, 2091–2110. [CrossRef]
27. Osorio Sánchez, A. Algoritmo Para Detección de Vibraciones Anormales en Maquinarias Utilizando la Transformada Wavelet. *Repos. Nac. Conacyt*. 2006. Available online: http://catarina.udlap.mx/u_dl_a/tales/documentos/meie/osorio_s_a (accessed on 21 August 2023).
28. Eparvier, F.G.; Crotser, D.; Jones, A.R.; McClintock, W.E.; Snow, M.; Woods, T.N. The extreme ultraviolet sensor (EUVS) for GOES-R. In Proceedings of the Solar Physics and Space Weather Instrumentation III, San Diego, CA, USA, 4–6 August 2009; Volume 7438, pp. 31–38.

remote sensing

Article

Comparative Analysis of the H2PT Ionosphere Model

Paulina Gajdowska *, Anna Świątek, Łukasz Tomasik and Mariusz Pożoga

Centrum Badań Kosmicznych Polskiej Akademii Nauk, Bartycka 18A, 00-716 Warsaw, Poland; ana@cbk.waw.pl (A.Ś.); tomasik@cbk.waw.pl (Ł.T.); pozoga@cbk.waw.pl (M.P.)
* Correspondence: pwozniak@cbk.waw.pl

Abstract: The ionosphere stands in the path of signals emitted by Global Navigation Satellite System (GNSS) satellites to receivers located on the Earth's surface. Many factors affect the accuracy of satellite positioning, but error due to ionospheric refraction is the largest among them. For this reason, it is important to minimize the impact of ionospheric refraction, and ionospheric models are one of the methods used. As the intensity of the processes taking place in the ionosphere is variable because of solar activity, the influence on satellite observations is also not constant; it varies by location and time of day and year. Therefore, models focusing on the region of interest to users are especially useful in precise GNSS applications. In this research, the H2PT model covering the region of Europe was examined at a temporal resolution of 15 min and two spatial resolutions (latitude × longitude) of 1° × 1° and 5° × 5°. This study aimed to compare the H2PT model with the solution obtained from the International GNSS Service (IGS) in the context of vertical total electron content (VTEC). The H2PT values in high-latitude regions turned out to be overestimated compared to IGS VTEC maps, while, in low-latitude regions, the situation was the opposite. Although the differences between the analyzed maps were usually a few TECUs, it was observed that, during the course of a day, they could increase to several dozen TECUs. Furthermore, the data from selected days characterized by high or low activity of the ionosphere were subjected to a detailed analysis (in relation to quiet days, as well as to the median). The data available with a 15-min interval allowed the identification of short-term disturbances appearing in the ionosphere. The analyzed model, which is of a regional nature and has a relatively high resolution, allows improvement to be made to the quality of the determined ionospheric correction in GNSS positioning.

Keywords: total electron content; ionospheric models; ionospheric regional maps; empirical ionosphere modeling; ionospheric disturbances

1. Introduction

The ionosphere is a medium that affects radio signal propagation in two ways. While the ionosphere makes global radio communication possible, it is a dispersion source for electromagnetic waves in the microwave band, including ultra-high-frequency (UHF) waves transmitted by satellites of Global Navigation Satellite Systems (GNSSs). On the way to a receiver located on the surface of the Earth, the electromagnetic signal emitted by a satellite encounters a heterogeneous layer of ionized gas and free electrons of variable density, in which its radio wave is refracted [1,2]. The passage of a radio wave through the ionosphere slows down its modulation (decrease in group velocity) and increases its carrier phase velocity [3–5]. Ionospheric refraction is an important factor that needs to be taken into account when developing satellite observations, especially in precision applications [6,7].

A measure of the ionosphere state is the total electron content (TEC) coefficient, expressed in the total electron content units (TECUs), where 1 TECU refers to the number of 10^{16} electrons contained in a cylinder which has a base with an area of 1 m^2 and whose height is the distance between a receiver and a satellite [8–11]. As a result of the complexity of issues determining the state of the ionosphere, approximation of its parameters in the form of a mathematical model requires various simplifications, which can affect the reliability of a given pattern [12]. The condition of the ionosphere is constantly being monitored by numerous research institutions that prepare mathematical models at various scales, both globally, e.g., the International Reference Ionosphere (IRI) of the Committee on Space Research (COSPAR), the International Union of Radio Science (URSI) [13], and NeQuick-G [14] (which in turn is based on NeQuick [15]) developed by the European Space Agency (ESA), as well as regionally for various parts of the world, e.g., for Europe, the University of Warmia and Mazury in Olsztyn (UWM) [6], and, for Africa, the National Space Research and Development Agency (NASRDA) in Nigeria [16].

Regarding the subject model of this manuscript, the International Space Environment Service Regional Warning Center (ISES/RWC) in Warsaw incorporates the HELioGEOphysical (HELGEO) prediction service in Poland and the software [17] used in space weather monitoring and forecasting activities. HELGEO was designed to perform data analysis tasks and create alerts and predictions for end users. The system's next generation is named "HELGEO2 Pożoga Tomasik" (H2PT), and it aims to enhance the system operator's awareness of ongoing space weather events and provide processed data with minimal modifications.

The subject of this analysis is the regional H2PT ionosphere model generated by the employees of the Laboratory of Heliogeophysical Forecasts in Space Research Centre of the Polish Academy of Sciences. This study aimed (a) to compare the H2PT model with the maps provided by the International GNSS Service (IGS), which simulate VTEC measurement maps for the purpose of this study, and (b) to compare disturbed days with quiet or median data as a reference. We created tools for downloading and developing data, as well as for generating visualizations and lists of statistical parameters. Time series were also prepared to analyze the variability of the TEC parameter during a specific day for nodes of the same longitude (for both H2PT model and IGS VTEC maps). We also analyzed the time series for the surroundings of Warsaw for days characterized by positive or negative disturbance of the ionosphere, with reference to a quiet day or the median for a given day in a month (for the H2PT model). The data from January to March 2019 were processed and analyzed at an interval of 15 min for the H2PT model, as well as at an interval of 2 h for the H2PT model in reference to the IGS VTEC maps.

2. The H2PT Model

The discussed model is of a regional type and covers the European region (latitude range: 30°N to 70°N; longitude range: 10°W to 40°E). Data for 15-min intervals are saved in files that resemble IONEX (IONosphere Map Exchange) in two spatial resolutions of latitude × longitude: $1^\circ \times 1^\circ$ and $5^\circ \times 5^\circ$.

The H2PT model is an empirical thin-layer ionospheric model (TLIM), for which the main data source is 15-min Receiver INdependent EXchange Format (RINEX) files for the Regional Reference Frame Sub-Commission for Europe (EUREF) Permanent Network (EPN). Most TLIMs, such as the H2PT model, use an infinitely thin fixed-height layer [18,19]. The intersection of the pseudo line from the satellite (line of sight, LOS) to the receiver occurs at a given height, determining the ionospheric pierce point (IPP). This scheme is shown in Figure 1 below.

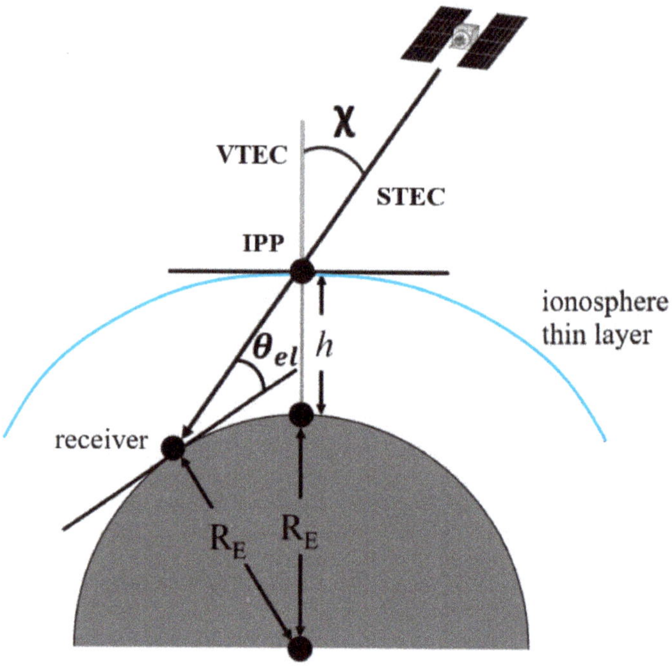

Figure 1. Conversion model from slant total electron content (STEC) to vertical total electron content (VTEC) in a thin-layer ionospheric model (TLIM) [20].

Using this approximation, a two-dimensional ionospheric map can be created by converting the slant total electron content (STEC) to vertical total electron content (VTEC) using a formula from [21]:

$$\mathrm{VTEC} = \mathrm{STEC} \cdot \cos\chi, \tag{1}$$

where χ is the angle between the zenith direction and the satellite direction from the IPP.

The disadvantages of this method are the means of errors generated by model simplifications and the fixed IPP height. The global ionosphere TEC maps provided by the IGS Ionospheric Analysis Center use an IPP height of 450 km [18,22,23]. Some researchers proposed the use of different heights in different ionospheric conditions or applications [24,25]. The new feature of the H2PT model is the possibility to generate new maps based on user-defined IPP heights (such a request would have to be submitted to the authors of the model).

The current version of the software allows us to use support procedures to perform calculation and generate maps of IPP heights with the following options:

Fixed height of the ionosphere—typically 350 or 450 km;

- Hourly height estimated from the NeQuick-G electron density model as the balanced height for which the top and bottom densities are equal;
- Hourly height of combined electron densities from the bottom electron density based on ionosonde measurement plus the scaled top profile from the NeQuick-G model; in case of less informative ionograms, the full NeQuick-G electron density profile can be used through the injections of ionospheric parameters;
- Hourly *hm*F2 height plus correction factor, where the correction factor is defined as a constant by the researchers.

For this work, the calculation was performed with a fixed IPP height of 350 km for the Europe region.

The computational procedure of the H2PT model consists of the following steps:

1. Data acquisition:
    a. RINEX observation files;
    b. RINEX navigation files (Almanac data, Galileo ionospheric message);
    c. Ionospheric parameters from recognized ionograms.
2. Computation of GNSS satellite position and data reviewing:
    a. Generation of IPP height based on user options;
    b. Calculating STEC [21,26,27] (which does not take phase correction into account) for continuous observational arcs longer than 15 min—elimination of discontinuities;
    c. Elimination of carrier phase shifts (whereby the ambiguity estimation is based on the Least-squares AMBiguity Decorrelation Adjustment (LAMBDA) method [28]; data are also checked for outlier observations, and jumps are defined if the TEC value changes in subsequent observation epochs by more than 0.1 TECU; in this case, the arc is split at the jump point. If the length of the resulting arc is at least 15 min, it is still included in the calculation).
3. Calculation of observation geometry assuming the estimated IPP height (according to the diagram shown in Figure 1).
4. Computation of VTEC separately for each available station (every 30 s) and evaluation of phase correction (as a sum of satellite and receiver phase biases) for each arc using the least-squares method (LSM). LSM fitting is performed for all arches within the 12-h period. The time window is shifted by 6 h and the operation is repeated.
5. Computation of VTEC values in all IPPs using Formula (1) according to Figure 1.
6. Binning the obtained VTEC values according to the given grid ($1^\circ \times 1^\circ$ or $5^\circ \times 5^\circ$) and time resolution (for a given grid node, all IPPs with a margin of $\pm$ requested spatial resolution and $\pm$ time resolution are collected).
7. Computation of VTEC as a median value from the binned data.

The process runs automatically every 15 min with a 1-day delay, which is necessary to obtain all available RINEX files from the European region. It is also possible to build a near-real-time local model based on the stream data from permanent GNSS stations.

Discontinuities sometimes occur in the model, with no data for some nodes in some observation epochs. These discontinuities result from too few calculated VTEC values from a station in a given epoch to determine the median for a given node. The reason may be too few observation arcs for a given station, or delays in providing data by the station. It can be considered as a H2PT disadvantage, but the purpose of this procedure is to ensure a high quality of the resulting maps.

The model is still in development, and future features will allow it to perform additional analysis related to the area variation in TEC values, which can be compared with the local rate of TEC index (ROTI). It is planned to make the final version of the H2PT software available to the ionospheric community for public use.

3. Comparison Data

The ionospheric maps generated by the IGS cover the entire globe, and solutions in the final version are available at a delay of about 11 days. The files in the IONEX format contain data with a time resolution of 2 h, while their spatial resolution (latitude $\times$ longitude) is $2.5^\circ \times 5^\circ$. The data were downloaded from the National Aeronautics and Space Administration (NASA) Crustal Dynamics Data Information System (CDDIS) archive (https://urs.earthdata.nasa.gov/, accessed on 20 October 2022), which contains datasets based on space and satellite geodesy techniques. The solution created by the Geodynamics Research Laboratory, University of Warmia and Mazury (GRL/UWM) [29,30] was chosen. In this study, IGS VTEC maps were used as reference VTEC measurements through which the performance of the H2PT model was evaluated.

The information on ionosphere disturbances was obtained from the ISES/RWC in the Warsaw website, where catalogues of disturbances for Warsaw (http://rwc.cbk.waw.pl/

rwc/disturb.war, accessed on 11 May 2023), as well as disturbed and quiet days for Europe (http://rwc.cbk.waw.pl/rwc/q_d_days.ctl, accessed on 22 August 2023), are available.

All computations and visualizations were made in the MATLAB environment and involved data collection, processing, cataloguing, completeness control, performing calculations, determining statistics, and generating maps and plots.

4. Comparative Analysis between Models

The TEC values of the H2PT model and IGS maps in Europe for the period from January to March 2019 with an interval of 2 h were developed using the 5° × 5° grid version, then increased to 1° × 1°, while the IGS VTEC map data were interpolated as its nominal resolution is 2.5° × 5°. Differences between the models were determined and visualized, and their statistical parameters were calculated for the specific epochs under consideration. In the next stage, time series covering the period of a specific day were prepared, showing that the TEC parameter is variable according to latitude.

4.1. TEC Maps

The data were visualized on the map of Europe, and a common, uniform scale was used for the H2PT model and IGS VTEC maps (from 0 to 20 TECU). A TEC map of the H2PT model, the IGS, and the differences between them was generated for each epoch. Figures 2 and 3 show an example set for a relatively quiet day, 10 March 2019, at 14:00 UTC, at a resolution of 5° × 5° and 1° × 1°, respectively.

4.2. Statistical Parameters

As an addition to the plots, the statistical parameters of the models and the difference between them were computed for each of the considered epochs (Tables 1 and 2, for a 5° and 1° grid, respectively). The differences Δ between the H2PT model values (TEC_{H2PT}) and the IGS VTEC values (TEC_{IGS}) for subsequent epochs i were determined as follows:

$$TEC_{\Delta i} = TEC_{H2PTi} - TEC_{IGSi}. \quad (2)$$

Table 1. Statistical parameters of the H2PT model, the IGS VTEC map, and their differences at a spatial resolution of 5° × 5° (date: 10 March 2019, at 14:00 UTC).

Name	Min (TECU)	Max (TECU)	Range (TECU)	Mean (TECU)	Standard Deviation (TECU)	RMSE (TECU)	NRMSE (-)	Correlation Coefficient (-)
H2PT	6.3	13.5	7.2	8.28	2.09	2.93	0.35	0.96
IGS	3.9	19.8	15.9	9.71	4.50			
Δ	−6.4	2.4	8.8	−1.43	2.57			

Table 2. Statistical parameters of the H2PT model, the IGS VTEC map, and their differences at a spatial resolution of 1° × 1° (date: 10 March 2019, at 14:00 UTC).

Name	Min (TECU)	Max (TECU)	Range (TECU)	Mean (TECU)	Standard Deviation (TECU)	RMSE (TECU)	NRMSE (-)	Correlation Coefficient (-)
H2PT	4.0	15.8	11.8	8.23	2.22	2.65	0.32	0.90
IGS	3.9	19.8	15.9	9.55	4.07			
Δ	−8.5	3.3	11.8	−1.35	2.28			

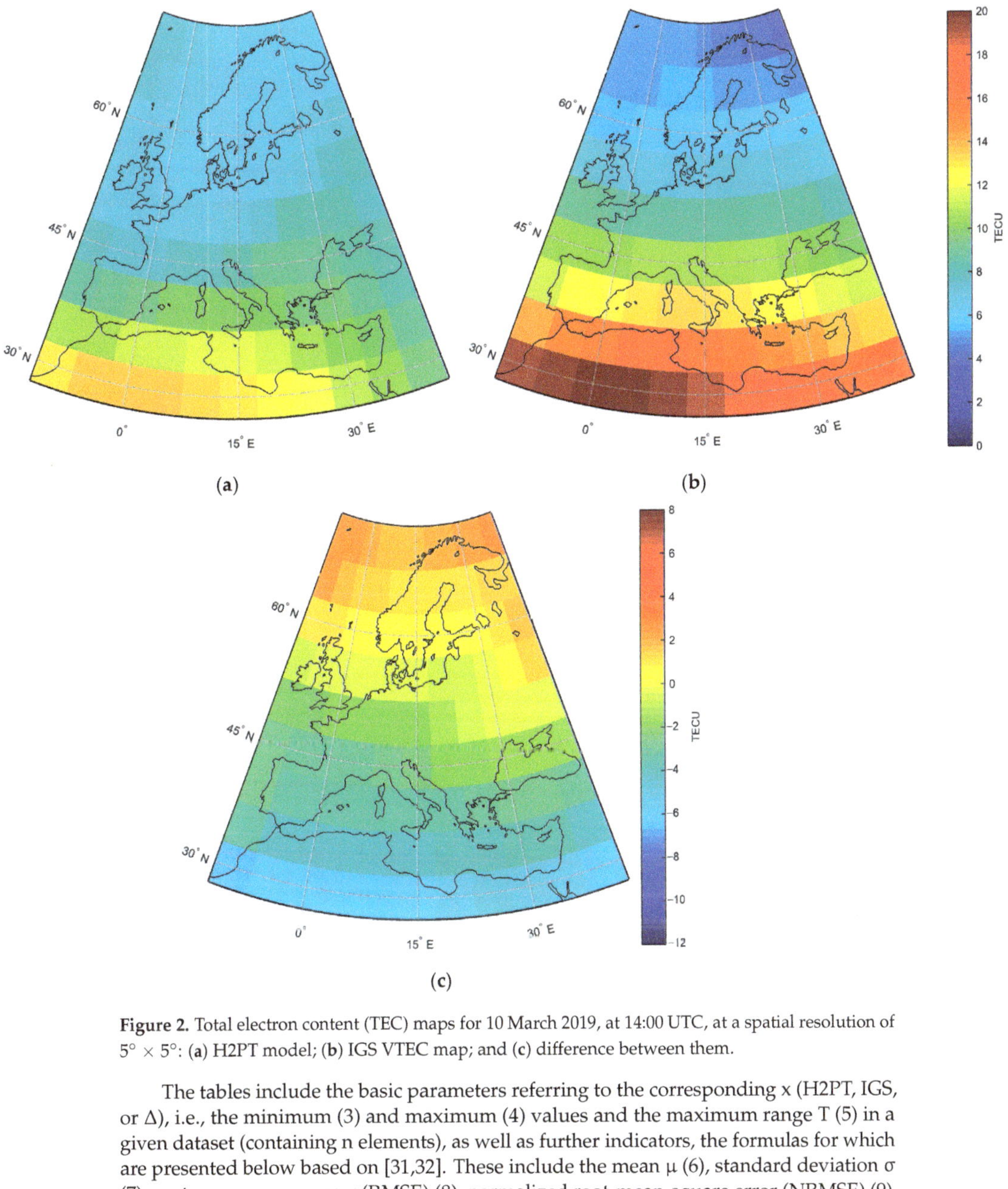

Figure 2. Total electron content (TEC) maps for 10 March 2019, at 14:00 UTC, at a spatial resolution of 5° × 5°: (**a**) H2PT model; (**b**) IGS VTEC map; and (**c**) difference between them.

The tables include the basic parameters referring to the corresponding x (H2PT, IGS, or Δ), i.e., the minimum (3) and maximum (4) values and the maximum range T (5) in a given dataset (containing n elements), as well as further indicators, the formulas for which are presented below based on [31,32]. These include the mean μ (6), standard deviation σ (7), root-mean-square error (RMSE) (8), normalized root-mean-square error (NRMSE) (9), and Pearson correlation coefficient ρ (10):

$$\min_x = \min(\mathrm{TEC}_{x_1}, \mathrm{TEC}_{x_2}, \ldots, \mathrm{TEC}_{x_n}) \quad (3)$$

$$\max_x = \max(\mathrm{TEC}_{x_1}, \mathrm{TEC}_{x_2}, \ldots, \mathrm{TEC}_{x_n}) \quad (4)$$

$$T_x = max_x - min_x \tag{5}$$

$$\mu_x = \frac{1}{n}\sum_{i=1}^{n} TEC_{xi} \tag{6}$$

$$\sigma_x = \sqrt{\sum_{i=1}^{n}(TEC_{xi} - \mu_x)^2} \tag{7}$$

$$RMSE = \sqrt{\frac{1}{n}\sum_{i=1}^{n}|TEC_{H2PT} - TEC_{IGS}|^2} \tag{8}$$

$$NRMSE = \frac{RMSE}{\mu_{H2PT}} \tag{9}$$

$$\rho = \frac{1}{n-1}\sum_{i=1}^{n}\left(\frac{TEC_{H2PTi} - \mu_{H2PT}}{\sigma_{H2PT}}\right)\left(\frac{TEC_{IGSi} - \mu_{IGS}}{\sigma_{IGS}}\right) \tag{10}$$

Furthermore, analogous indices were determined for individual latitudes (Tables 3 and 4, for the 5° and 1° grid, respectively). Consistent with the results presented in Figures 2 and 3, the information contained in the tables are related to 10 March 2019, at 14:00 UTC.

Table 3. Statistical parameters of the differences between the H2PT model and the IGS VTEC map for individual latitudes at a spatial resolution of 5° × 5° (date: 10 March 2019, at 14:00 UTC).

φ (°)	Min (TECU)	Max (TECU)	Range (TECU)	Mean (TECU)	Standard Deviation (TECU)	RMSE (TECU)	NRMSE (-)	Correlation Coefficient (-)
70	1.5	2.4	0.9	2.01	0.29	2.03	0.25	0.79
65	1.0	2.2	1.2	1.29	0.38	1.34	0.16	0.12
60	0.3	1.5	1.2	0.75	0.34	0.82	0.10	−0.21
55	−0.5	1.2	1.7	−0.13	0.53	0.53	0.06	−0.27
50	−2.2	−0.2	2.0	−1.27	0.80	1.48	0.18	−0.29
45	−3.2	−1.5	1.7	−2.39	0.70	2.48	0.30	−0.02
40	−3.4	−2.8	0.6	−3.15	0.24	3.15	0.38	0.94
35	−4.5	−3.3	1.2	−4.09	0.36	4.11	0.50	0.94
30	−6.4	−5.1	1.3	−5.92	0.44	5.93	0.72	0.97

Table 4. Statistical parameters of the differences between the H2PT model and the IGS VTEC map for individual latitudes at a spatial resolution of 1° × 1° (date: 10 March 2019, at 14:00 UTC).

φ (°)	Min (TECU)	Max (TECU)	Range (TECU)	Mean (TECU)	Standard Deviation (TECU)	RMSE (TECU)	NRMSE (-)	Correlation Coefficient (-)
...								
43	−4.1	−1.6	2.5	−2.62	0.71	2.71	0.33	0.53
42	−4.0	−1.8	2.3	−2.89	0.63	2.96	0.36	0.64
41	−5.8	−1.9	3.9	−3.30	0.90	3.42	0.42	0.57
40	−5.1	−2.1	3.0	−3.28	0.60	3.34	0.41	0.75
39	−5.1	−1.4	3.7	−3.48	0.66	3.54	0.43	0.81
38	−6.0	−2.0	4.0	−3.56	0.71	3.62	0.44	0.81
37	−7.0	−2.3	4.7	−3.73	0.88	3.83	0.47	0.73
36	−7.6	−2.4	5.1	−3.96	0.96	4.07	0.49	0.77
35	−6.3	−1.4	4.9	−3.76	0.81	3.84	0.47	0.77
...								

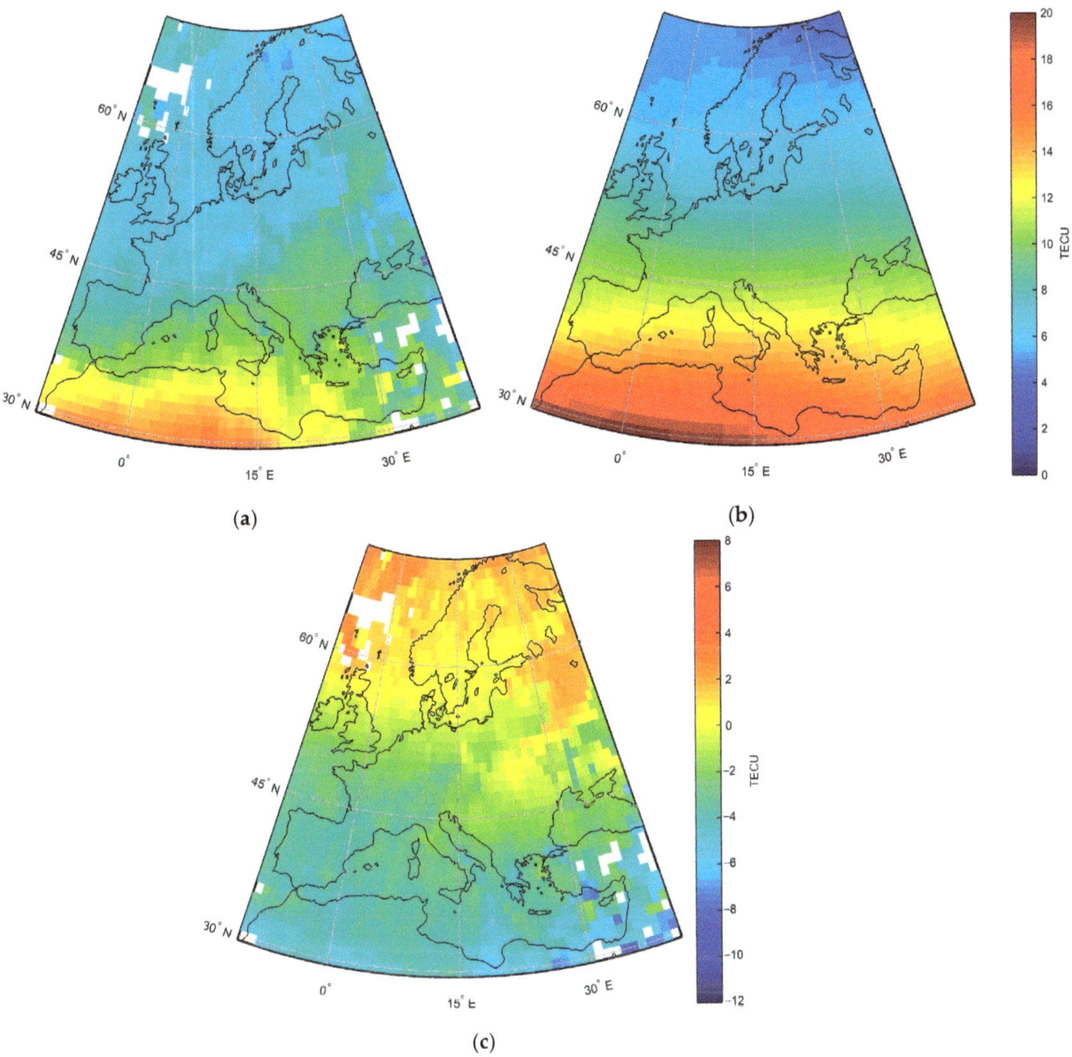

Figure 3. TEC maps for 10 March 2019, at 14:00 UTC, at a spatial resolution of 1° × 1°: (**a**) H2PT model; (**b**) IGS VTEC map; and (**c**) difference between them.

4.3. Time Series

In the next stage, time series were generated, showing variation in the TEC parameter along individual meridians depending on the latitude and time of day. Figure 4 shows the data for the 20°E meridian for 10 March 2019.

4.4. Discussion

In this section, available data for the period from January to March 2019 were examined and visualized, with a time interval of two hours (imposed by the resolution of the IGS VTEC map) at resolutions of 5° × 5° and 1° × 1°. The graphs and tables presented for 10 March 2019 are in line with the general trend that we noticed when compiling the H2PT model and IGS maps for other dates. The comparison indicates the occurrence of differences in the values of the TEC parameter, which are particularly depending on the latitude. At high latitudes, the values of the H2PT model are higher than those of the IGS VTEC maps (overestimated), whereas they are lower in tropical regions (underestimated). As the differences between the models generally amount to several TECUs, this may be related to the accuracy of the model itself. When considering the correlation of the models along a given parallel, no constant relationship is observed, whereas when considering the time of day, a larger correlation coefficient is observed in daytime. The RMSE values at successive parallels indicate that much smaller errors occur at high latitudes. However, when moving closer to the equator, where the activity of the ionosphere is increasing and more variable over the course of a day, the RMSE increases more.

The availability of the H2PT model data depends on the time of day (there are more missing nodes at nighttime), and there is also less availability in its edge fragments.

A comparison of the usefulness of the two different spatial resolutions of 5° × 5° and 1° × 1° reveals that, while the 5° × 5° grid allows general analyses showing trends of ionosphere behavior in Europe, the 1° × 1° grid gives the opportunity to carry out detailed interpretations, especially in the context of ionospheric disturbances.

(a)

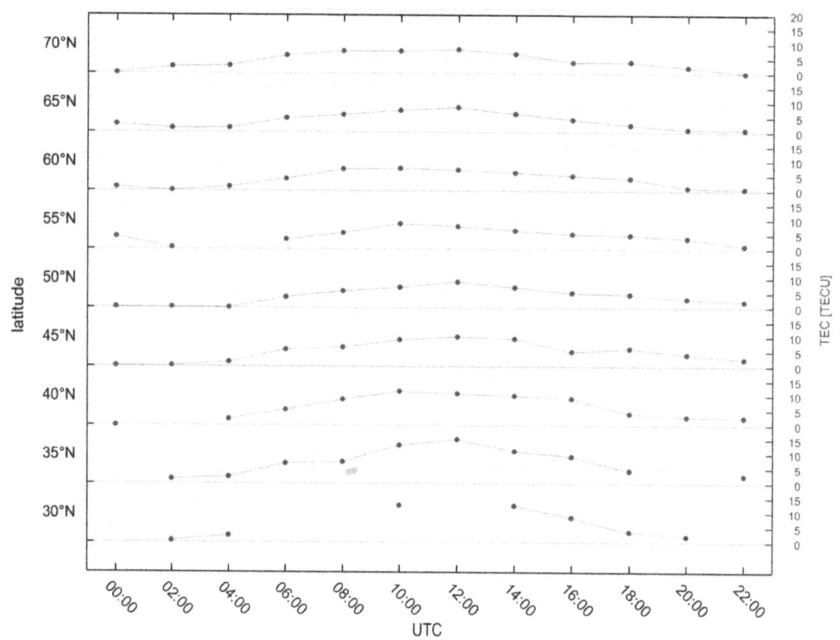

Figure 4. *Cont.*

(b)

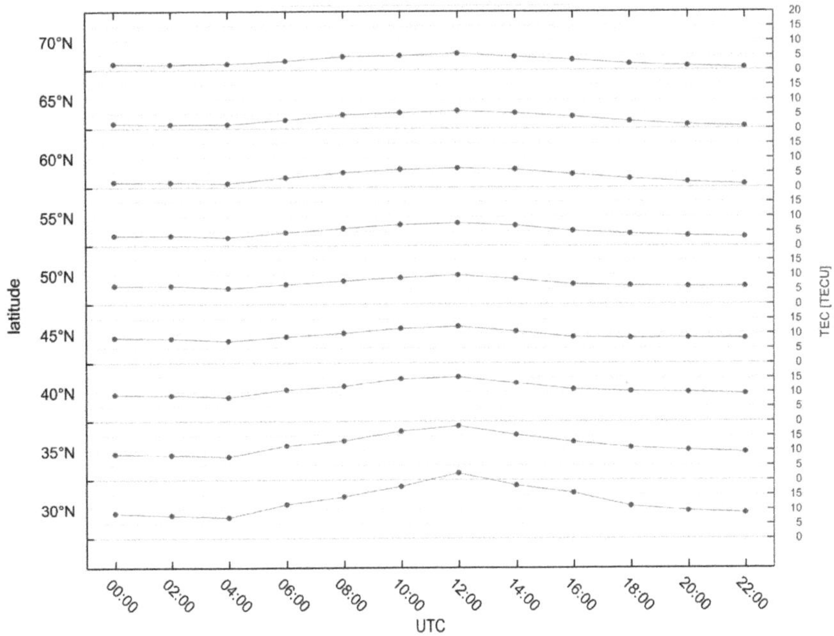

(c)

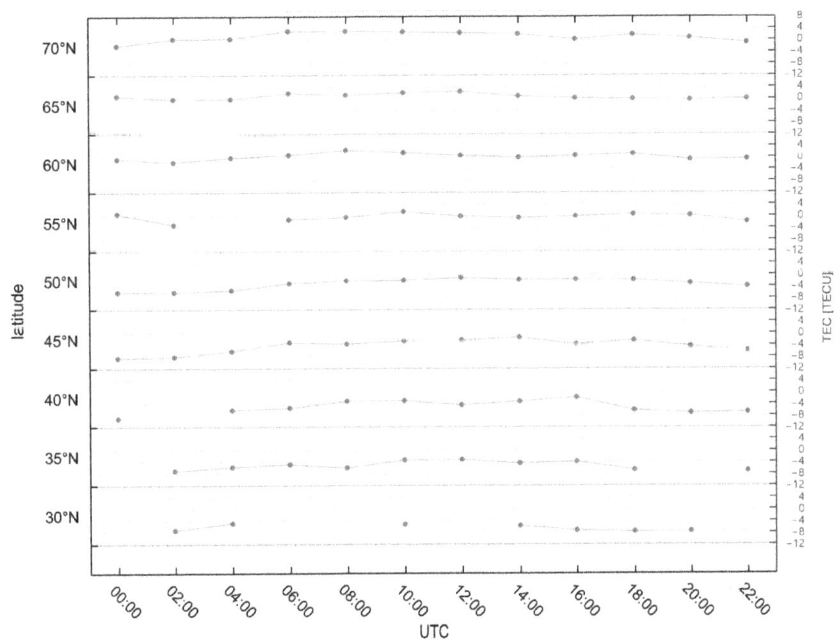

Figure 4. Time series for the 20°E meridian for 10 March 2019: (**a**) H2PT model; (**b**) IGS VTEC map; and (**c**) differences between them.

5. Analysis of H2PT Model Solutions

In the next stage, the H2PT model data were developed using the 1° × 1° grid with an interval of 15 min, which is undoubtedly an advantage of this version compared to the commonly used IGS solution. Similar to the previous analysis phase, the data were visualized on the map of Europe, and basic statistical parameters were determined for each of the datasets. Daily time series were then generated for the 21°E meridian (approximate longitude of Warsaw) and the parallel Warsaw station nearby (52°N), taking into account a 3° buffer to the south and the north. The datasets present data for a day during which both positive and negative ionospheric disturbances were recorded, as well as for a quiet day or the daily median determined for the month in question.

5.1. TEC Maps

For each epoch, a map of the TEC parameter was generated, which was recorded in the .png format as well as the .gif format, producing 96 maps per day and creating a visualization in the form of a movie. The sample data are presented in Figure 5, again for 10 March 2019, at 00:00, 4:00, 8:00, 12:00, 16:00, and 20:00 UTC.

5.2. Statistical Parameters

Similar to the comparison of the H2PT model and IGS VTEC maps discussed above, a statistical summary was generated showing the higher time resolution for the H2PT model at the selected epochs on 10 March 2019 (Table 5).

Table 5. Statistical parameters of the H2PT model for selected epochs (date: 10 March 2019).

UTC	Min (TECU)	Max (TECU)	Range (TECU)	Mean (TECU)	σ (TECU)
00:00	0.0	6.2	6.2	1.33	1.03
4:00	0.0	5.1	5.1	1.77	0.88
8:00	1.7	12.6	10.9	6.58	1.73
12:00	5.8	18.2	12.4	9.54	1.66
16:00	2.5	12.7	10.2	6.34	1.93
20:00	0.0	7.9	7.9	2.53	1.30

5.3. Time Series

Using the collections posted on the RWC Warsaw website, disturbance catalogues for the Warsaw location (52°N, 21°E) were downloaded, as well as a set of data on positive/negative days and quiet days in Europe. The catalog was used to select the potential days when there would be a chance to observe disturbances on the maps. However, these dates are indicated in relation to the median for the whole month; therefore, a similar tactic was used in our analyses, and potentially disturbed days (positively and negatively) were also compared with the median calculated in two ways. Using a number of possible combinations, time series were generated to produce visualizations for disturbed and calm days (examples are shown in Figures 6–8) or the daily median (examples are shown in Figures 9–11). The median was determined in two ways (for each node independently): as a common value for the entire month, and separately for each of the 15-min epochs.

5.3.1. Positively Disturbed, Negatively Disturbed, and Quiet Days

The first discussed set contains the time series for positively and negatively disturbed days, in reference to a quiet day.

Although the catalogue of disturbances dedicated to Warsaw lacks information on the course of disturbances, 5 January 2019 was among the days characterized by an increase in ionospheric activity in Europe. Figure 6 shows a clear increase in the TEC parameter from 8:00 to 13:00 UTC. In contrast, on 17 January 2019, a negative disturbance, according to the disturbance catalogue, occurred at 20:00 UTC; therefore, a decrease in TECU units can be seen. The disturbance did not appear at the same time on all the parallels shown, but moved from south to north over time, and simultaneously decreased and returned to normal values.

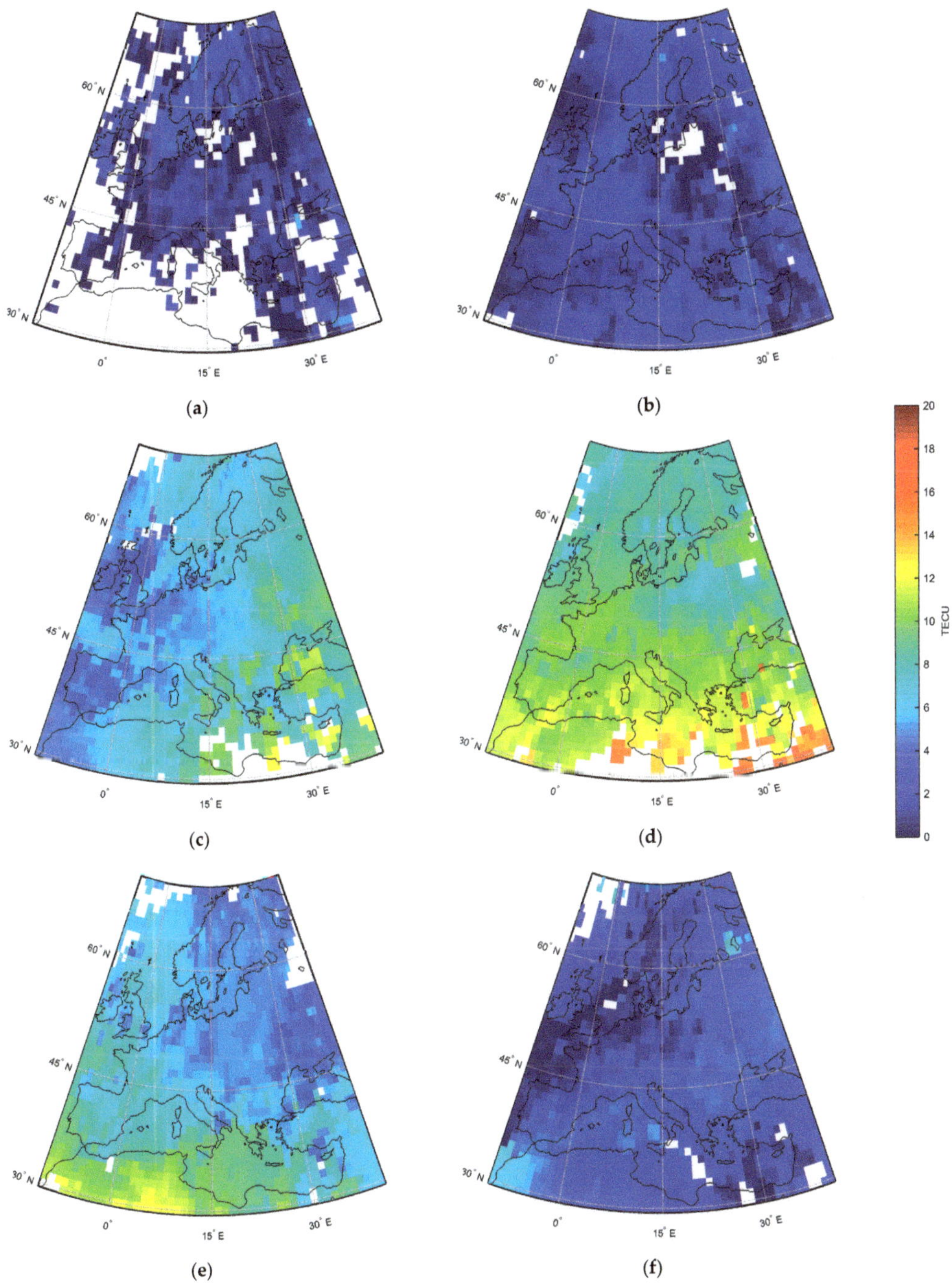

Figure 5. TEC maps for the H2PT model for 10 March 2019, at (**a**) 00:00; (**b**) 4:00; (**c**) 8:00; (**d**) 12:00; (**e**) 16:00; and (**f**) 20:00 UTC.

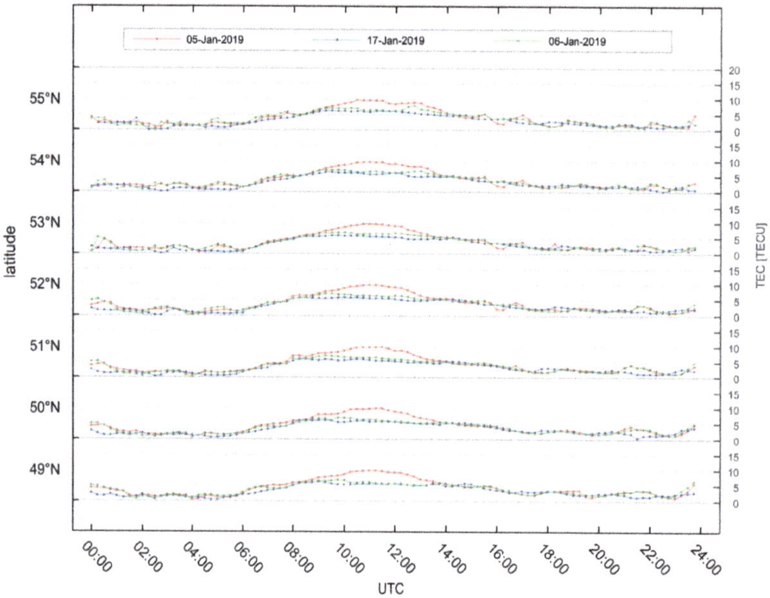

Figure 6. TEC time series at a longitude of 21°E for a positively disturbed day (5 January 2019), a negatively disturbed day (17 January 2019), and a quiet day (6 January 2019).

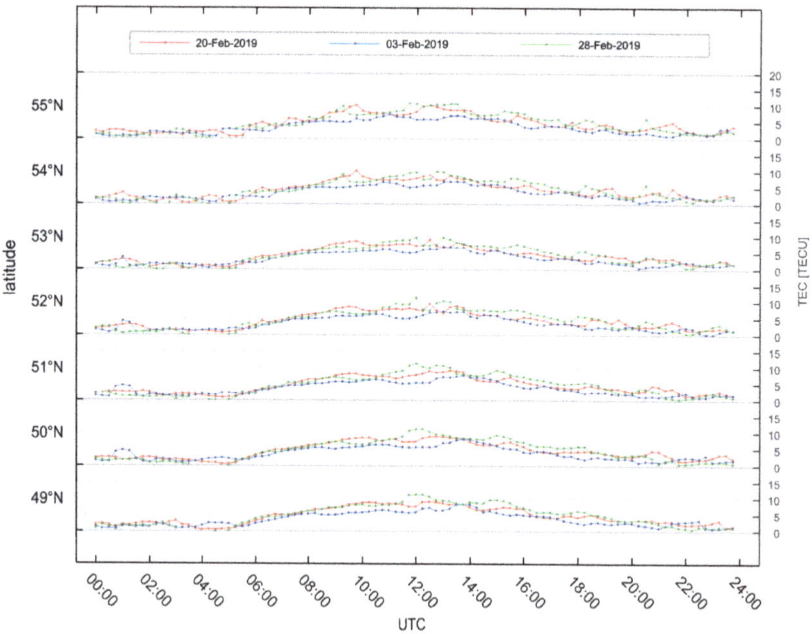

Figure 7. TEC time series at a longitude of 21°E on a positively disturbed day (22 February 2019), a negatively disturbed day (3 February 2019), and a quiet day (28 February 2019).

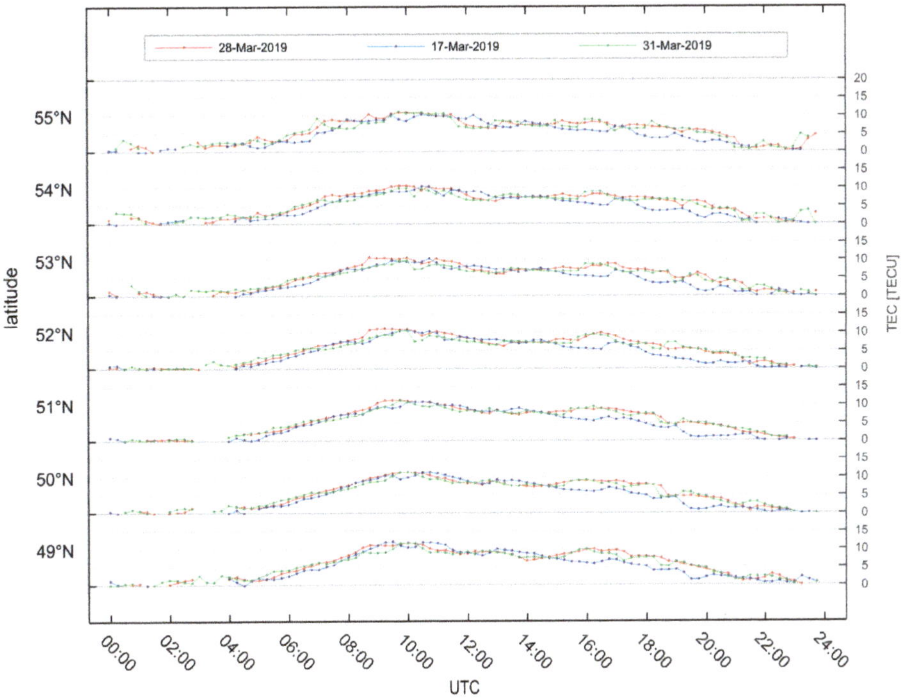

Figure 8. Time series at a longitude of 21°E for a positively disturbed day (28 March 2019), a negatively disturbed day (17 March 2019), and a quiet day (31 March 2019).

The last day of February was placed in the ranking as the calmest of the month, but the largest disturbances that occurred (119% and 138% from 16:00 to 19:00 and at 21:00 UTC, respectively) are not reflected in the TEC. Moreover, the variability in the parameter on the quiet day indicates the occurrence of disturbances, in particular at around 12:00 and 15:00 UTC. For 3 February 2019, Figure 7 shows an underestimation in TEC values from 7:00 to 13:00 UTC, and further underestimation from 17:00 to 22:00 UTC, although the catalogue data indicate that disturbances occurred from 3:00 to 10:00 and from 18:00 to 00:00 UTC.

On 17 March 2019 (nominally negatively disturbed), it is difficult to observe any clear deviations from the series (Figure 8), illustrating the course of the TEC on a quiet day during a disturbance (from 22:00 UTC); however, a clear decrease in the parameter can be seen a little earlier, from 15:00 to 21:00 UTC. During the positively disturbed day (28 March 2019), the TEC is similar to that expected for a quiet day, so no apparent increase can be seen from 18:00 to 20:00 UTC, as indicated by the information in the catalogue.

5.3.2. Positively Disturbed, Negatively Disturbed, and Median

In this section, sets containing two figures are presented. In each case, the first set shows positively and negatively disturbed days, with the median calculated for subsequent 15-min epochs, and the second set shows the positively and negatively disturbed days with the median for the whole day.

(a)

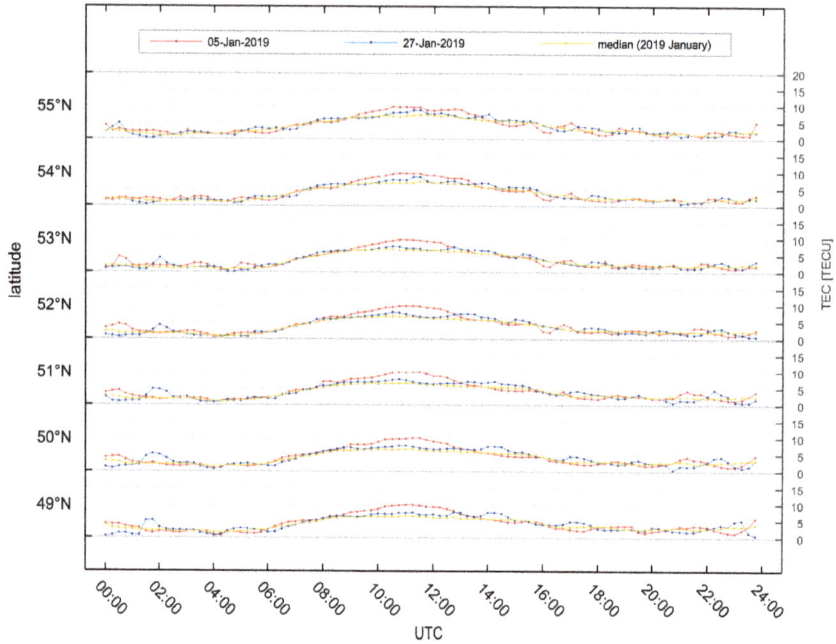

(b)

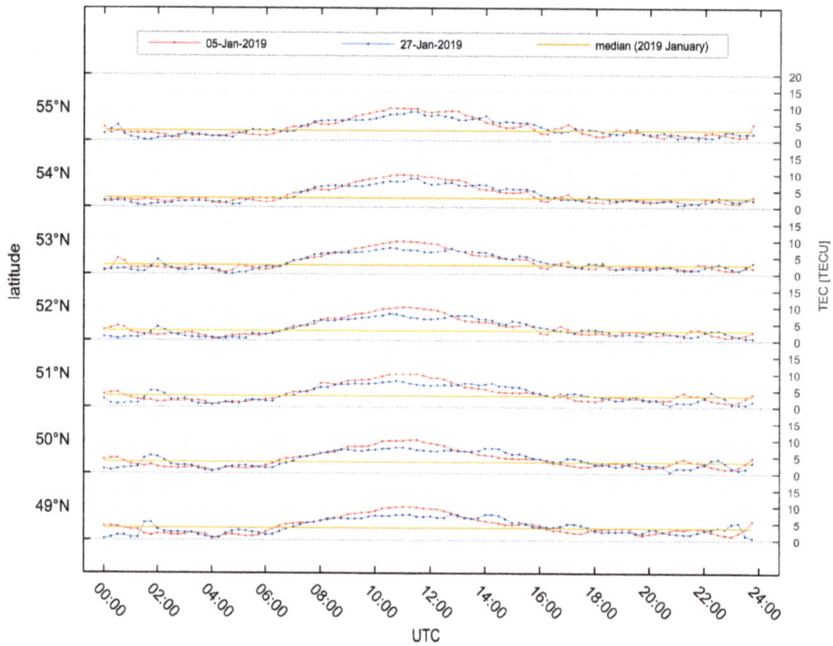

Figure 9. Time series at a longitude of 21°E for a positively disturbed day (5 January 2019) and a negatively disturbed day (17 January 2019), along with the median of (**a**) particular epochs and of (**b**) the whole day.

(a)

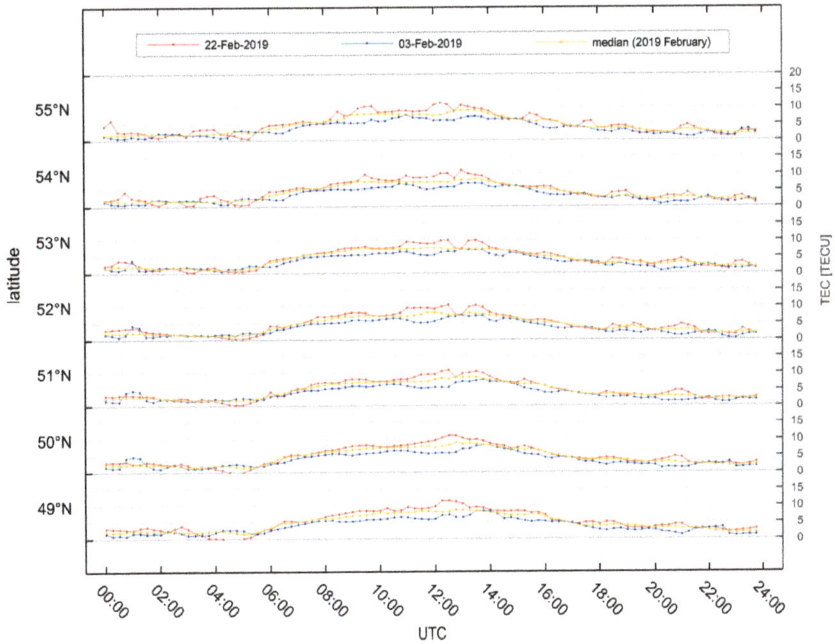

(b)

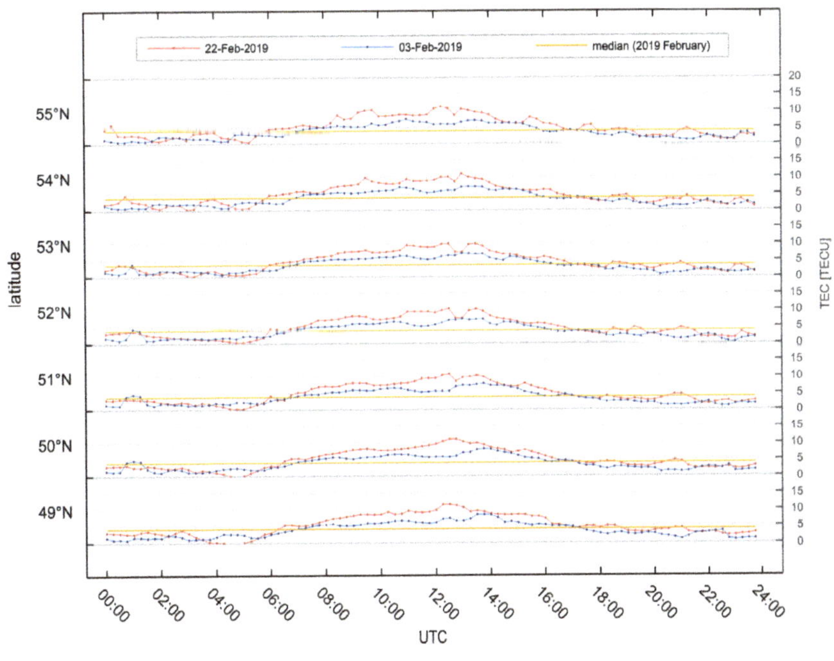

Figure 10. Time series at a longitude of 21°E for a positively disturbed day (22 February 2019) and a negatively disturbed day (3 February 2019), along with the median of (**a**) particular epochs and of (**b**) the whole day.

(a)

(b)

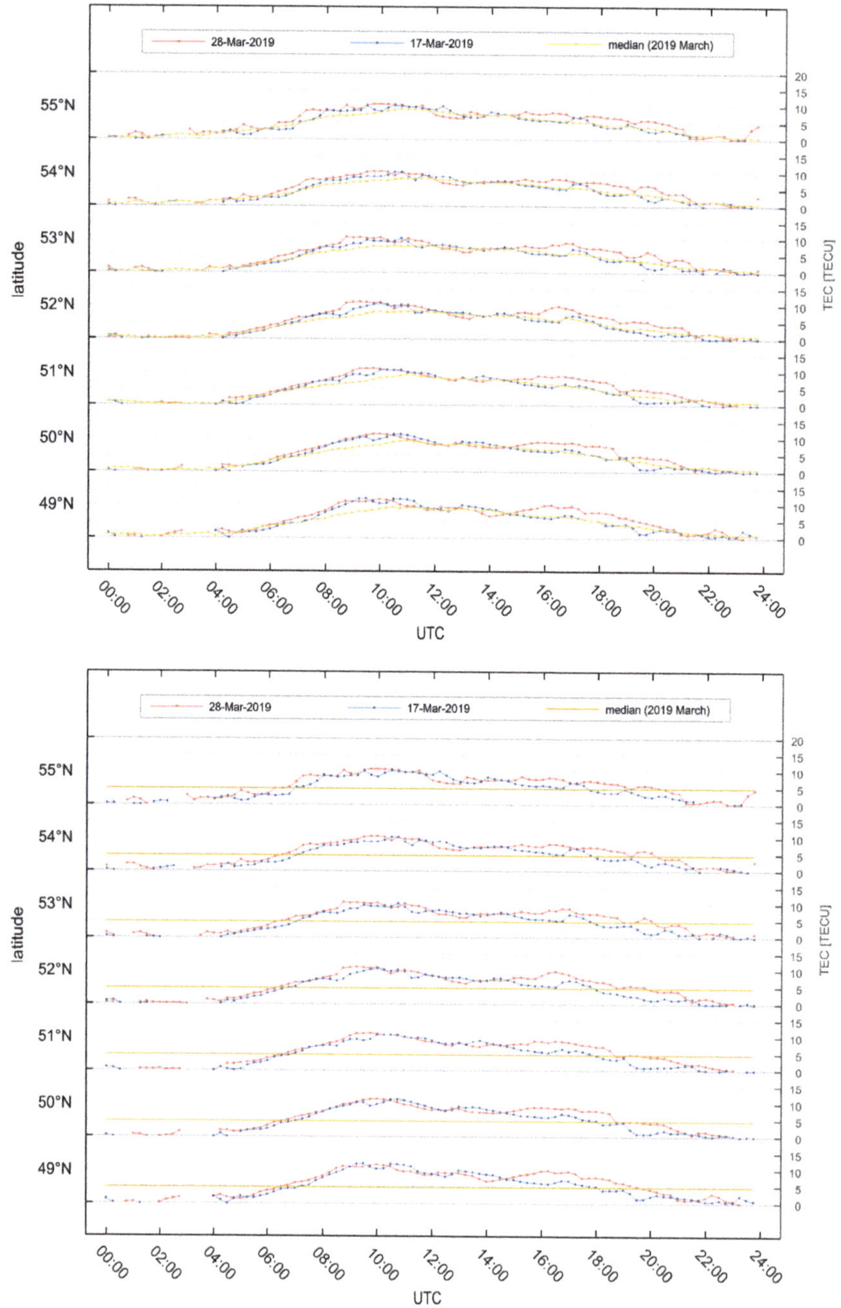

Figure 11. Time series at a longitude of 21°E for a positively disturbed day (28 March 2019) and negatively disturbed day (17 March 2019), along with the median of (**a**) particular epochs and of (**b**) the whole day.

For January (Figure 9), the TEC values for a calm day and the median are very similar. On 5 January 2019, there is a clear increase in TEC around noon, while on 17 January there is a slight decrease in the parameter relative to the median. Considering the next month (Figure 10), it should be noted that the TEC size of a quiet day (28 February 2019, Figure 7) differs considerably from the median. Correspondingly, the positive disturbance on 22 February 2019 is clear, as is the negative disturbance on 3 February 2019. For March, there is a significant deviation between the TEC of a non-disturbed day (31 March 2019, Figure 7) and the median (Figure 11), especially in the afternoon. Comparing this with 28 March 2019, when a positive disturbance was recorded, its occurrence can be confirmed from 15:00 to 21:00 UTC (although the catalogue includes the period from 18:00 to 20:00 UTC).

5.4. Discussion

The H2PT model data were processed at 15-min intervals and a resolution of $1^\circ \times 1^\circ$ for the entire period in question. The generated maps provide the basis for analyzing individual disturbances in the state of the ionosphere, while the set of statistical parameters allows the recording of general trends to characterize the variability in the activity of the ionized atmosphere layer during a daily course.

Similar to the examples of the time series presented, the TEC values during many disturbances (both positive and negative), as well as numerous combinations between them (within a given month), were analyzed. The information contained in the catalogue of disturbed days is not always reflected in the joint combination with the median or the time series of a quiet day.

Considering the two approaches presented, as a result of the discrepancy between the illustrative catalogues for Europe and for the Warsaw station, the median is a more objective and neutral reference than a time series for a quiet day. Considering the method of quiet-day determination, unifying it for the whole month, regardless of the time of day, introduces far too much simplification; reducing data with a repetitive, variable course to a single value results in the loss of some information. Therefore, subsequent analyses should choose the option of determining the median for each epoch during the day separately.

6. Conclusions

The H2PT model is of a regional nature and is dedicated to Europe, which allows it to focus on a relatively small area with high resolution. Its undoubted advantage is resolution; maps are generated for epochs of every 15 min, while its grid (in a more accurate version) has dimensions of $1^\circ \times 1^\circ$ (latitude $\times$ longitude). As the IGS VTEC maps have an interval of two hours and a spatial resolution of $2.5^\circ \times 5^\circ$, it is necessary to interpolate the data. Artificial smoothing can lead to many irregularities and eliminate ongoing ionospheric disturbances. The duration of such disturbances is usually just a few hours, and they cannot be expected to be clearly reflected on maps or time series presenting data with a 2-h interval.

The exemplary results presented in this study, as illustrated by the H2PT and IGS VTEC map sets, show the diversity of model characteristics. The $5^\circ \times 5^\circ$ H2PT model significantly blurs locally occurring disturbances; thus, the $1^\circ \times 1^\circ$ version is definitely more useful for detailed analysis of changes in the state of the ionosphere. In the case of the IGS VTEC maps with a lower spatial resolution, adaptation to a $1^\circ \times 1^\circ$ grid via the interpolation of data obviously causes a forced result. However, the most visible differences between the two solutions are the differences related to latitude: there are higher TEC values from the H2PT model than the IGS VTEC maps in the north of the study area and lower in the southern fragment, with these differences being greater at night than during the day.

The analysis of the H2PT model data with a higher time resolution (15 min) and a $1^\circ \times 1^\circ$ grid allows the detection of disturbances that last only a few hours and, therefore, cannot be reliably displayed using 2-h data. The course of the TEC parameter for the days during which a positive or negative disturbance occurred gives a reliable view compared with the median calculated independently for each epoch during the day. The median for the entire month introduces too much data degradation because it eliminates the typical variability in ionospheric activity during a daily course. A detailed examination of the disturbances listed in the catalogues indicates the occurrence of deviations between the presumably occurring disturbances and the median. Correspondingly, some of the positive or negative disturbances turn out to be not different from the median, and in other cases, the TEC values of positive disturbances are smaller than the mentioned parameter and, analogously, they are larger during negative disturbances. In addition, in many cases, the duration of the disturbance differs from the result of the comparison with the reference (median), and uncatalogued positive and negative deviations are observed.

In summary, these analyses have made it possible to become familiar with the characteristics of the H2PT ionosphere model, as well as to study its advantages and disadvantages. The relatively high temporal and spatial resolution is undoubtedly an advantage of the H2PT model because, unlike global models, this regional model allows conscientious analysis of ionospheric characteristics for the region of Europe. These analyses also indicated in which circumstances the H2PT model best reflects the behavior of the ionosphere and allowed us to find the model's shortcomings in some conditions. Unfortunately, the incompleteness of the maps, especially at night and in higher spatial resolution, is a significant limitation of the model, indicating a direction for its improvement. On the other hand, a key advantage of the H2PT model lies in its numerical transparency and operability. This entails that within a single cell of either $5^\circ \times 5^\circ$ or $1^\circ \times 1^\circ$, raw VTEC data from the receivers are present. The final value in a specific node is computed as the median of the existing measurement data only. This approach does mean that occasionally, due to insufficient data, the value for a certain node may not be determined. Nevertheless, it is important to consider that other models attempt to forcefully supplement data, and such an approach might result in erroneous conclusions when assessing the ionosphere's state based on GNSS measurements. The developed model serves as an analytical tool for consolidating data sourced from GNSS measurements, which will also allow analysis if the correctness of other ionospheric models, such as NeQuick-G, IRI, Klobuchar, or other IGS solutions, based on real measurement datasets. The other considered issue is improving the efficiency of calculation; additionally, it is planned to generate maps of ROTI scintillations.

However, relatively dense surface coverage enables detailed analyses of passing disturbances in the ionosphere, while a small interval allows the detection of short-term changes. Certainly, it is necessary to expand the analysis of the H2PT model, which is why there are plans to extend the relevant period to include all seasons and the changing ionospheric activity during the course of the 25th solar cycle. The H2PT model can also potentially be used as a new solution for determining the ionospheric correction in the course of GNSS observation data processing for the European region. In this context, an important point for further research on the H2PT model will be its analysis compared to the navigation models: NeQuick-G and Klobuchar.

Author Contributions: Conceptualization, P.G., A.Ś., Ł.T. and M.P.; methodology, P.G. and A.Ś.; software, P.G., Ł.T. and M.P.; validation, P.G., Ł.T. and M.P.; formal analysis, P.G., A.Ś., Ł.T. and M.P.; investigation, P.G. and A.Ś.; resources, Ł.T., M.P. and P.G.; data curation, Ł.T., M.P. and P.G.; writing—original draft preparation, P.G. and Ł.T.; writing—review and editing, P.G., A.Ś., Ł.T. and M.P.; visualization, P.G.; supervision, A.Ś.; project administration, A.Ś.; funding acquisition, A.Ś. All authors have read and agreed to the published version of the manuscript.

Funding: This research was funded by the Polish Ministry of Education and Science from funds for science in 2020 allocated to the implementation of a co-financed international project.

Data Availability Statement: Publicly available IONEX datasets were analyzed in this study. These data can be found at https:/urs.earthdata.nasa.gov (accessed on 20 October 2022). Information on ionosphere disturbances from the ISES RWC Warsaw can be found at http://rwc.cbk.waw.pl/rwc/disturb.war (accessed on 11 May 2023). Catalogue of ionospheric disturbed and quiet days in Europe can be found at http://rwc.cbk.waw.pl/rwc/q_d_days.ctl (accessed on 22 August 2023). The H2PT data are available from the authors, Ł.T. and M.P., upon request.

Conflicts of Interest: The authors declare no conflict of interest.

References

1. Teunissen, P.J.G.; Montenbruck, O. *Springer Handbook of Global Navigation Satellite Systems*; Springer International Publishing AG: Cham, Switzerland, 2017; ISBN 9783319429267.
2. Savchuk, S.; Hiryak, I. Evaluation of Static GNSS Positioning Accuracy during Selected Normal and High Ionospheric Activity Periods. *Tech. Sci./Univ. Warm. Maz. Olsztyn* **2015**, *18*, 299–311.
3. Seeber, G. *Satellite Geodesy*; Walter de Gruyter: Berlin, Germany, 2003; Volume 2, ISBN 3110175495.
4. Mainul, M.; Jakowski, N. Ionospheric Propagation Effects on GNSS Signals and New Correction Approaches. In *Global Navigation Satellite Systems: Signal, Theory and Applications*; IntechOpen: Rijeka, Croatia, 2012. [CrossRef]
5. Narkiewicz, J. *GPS i Inne Satelitarne Systemy Nawigacyjne*; Wydawnictwa Komnikacji i Łączności: Warszawa, Poland, 2007; ISBN 978-83-206-1642-2.
6. Krypiak-Gregorczyk, A.; Wielgosz, P.; Borkowski, A. Ionosphere Model for European Region Based on Multi-GNSS Data and TPS Interpolation. *Remote Sens.* **2017**, *9*, 1221. [CrossRef]
7. Wielgosz, P.; Krypiak-Gregorczyk, A.; Borkowski, A. Regional Ionosphere Modeling Based on Multi-GNSS Data and TPS Interpolation. In Proceedings of the 2017 Baltic Geodetic Congress (BGC Geomatics), Gdansk, Poland, 22–25 June 2017; pp. 287–291. [CrossRef]
8. Czarnecki, K. *Geodezja Współczesna*; Wydawnictwo Naukowe PWN SA: Warszawa, Poland, 2014; ISBN 978-83-01-18011-9.
9. Specht, C. *System GPS*; Wydawnictwo Bernardinum Sp. z o. o.: Pelplin, Poland, 2007; ISBN 978-83-7380-469-2.
10. Davies, K.; Hartmann, G.K. Studying the Ionosphere with the Global Positioning System. *Radio Sci.* **1997**, *32*, 1695–1703. [CrossRef]
11. Hofmann-Wellenhof, B.; Lichtenegger, H.; Wasle, E. *GNSS—Global Navigation Satellite Systems—GPS, GLONASS, Galileo, and More*; SpringerWienNewYork: Wien, Austria, 2007; ISBN 9783211730126.
12. Hernández-Pajares, M.; Wielgosz, P.; Paziewski, J.; Krypiak-Gregorczyk, A.; Krukowska, M.; Stepniak, K.; Kaplon, J.; Hadas, T.; Sosnica, K.; Bosy, J.; et al. Direct MSTID Mitigation in Precise GPS Processing. *Radio Sci.* **2017**, *52*, 321–337. [CrossRef]
13. Bilitza, D.; Altadill, D.; Truhlik, V.; Shubin, V.; Galkin, I.; Reinisch, B.; Huang, X. International Reference Ionosphere 2016: From Ionospheric Climate to Real-Time Weather Predictions. *Space Weather* **2017**, *15*, 418–429. [CrossRef]
14. European Union. European GNSS (Galileo) Open Service—Ionospheric Correction Algorithm for Galileo Single Frequency Users. 2016. Available online: https://www.gsc-europa.eu/sites/default/files/sites/all/files/Galileo_Ionospheric_Model.pdf (accessed on 23 May 2023).
15. Nava, B.; Coïsson, P.; Radicella, S.M. A New Version of the NeQuick Ionosphere Electron Density Model. *J. Atmos. Sol.-Terr. Phys.* **2008**, *70*, 1856–1862. [CrossRef]
16. Okoh, D.; Seemala, G.; Rabiu, B.; Habarulema, J.B.; Jin, S.; Shiokawa, K.; Otsuka, Y.; Aggarwal, M.; Uwamahoro, J.; Mungufeni, P.; et al. A Neural Network-Based Ionospheric Model Over Africa From Constellation Observing System for Meteorology, Ionosphere, and Climate and Ground Global Positioning System Observations. *J. Geophys. Res. Space Phys.* **2019**, *124*, 10512–10532. [CrossRef]
17. Klos, Z.; Stanislawska, I.; Dziak-Jankowska, B. Heliogeophysical Prediction Service in Poland:Past, Present and Future. *Hist. Geo Space Sci.* **2019**, *10*, 193–199. [CrossRef]
18. Mannucci, A.J.; Wilson, B.D.; Yuan, D.N.; Ho, C.H.; Lindqwister, U.J.; Runge, T.F. A Global Mapping Technique for GPS-Derived Ionospheric Total Electron Content Measurements. *Radio Sci.* **1998**, *33*, 565–582. [CrossRef]
19. Brunini, C.; Camilion, E.; Azpilicueta, F. Simulation Study of the Influence of the Ionospheric Layer Height in the Thin Layer Ionospheric Model. *J. Geod.* **2011**, *85*, 637–645. [CrossRef]
20. Hein, W.Z.; Kashiva, Y.; Goto, Y.; Kasahara, Y. Estimation Method of Ionospheric TEC Distribution from Single Frequency GPS Measurements Using a Slant Effect Model. In Proceedings of the 2016 URSI Asia-Pacific Radio Science Conference (URSI AP-RASC), Seoul, Republic of Korea, 21–25 August 2016; IEEE: Seoul, Republic of Korea, 2016; pp. 104–106.
21. Hernández-Pajares, M.; Juan, J.M.; Sanz, J.; Aragón-Àngel, À.; García-Rigo, A.; Salazar, D.; Escudero, M. The Ionosphere: Effects, GPS Modeling and the Benefits for Space Geodetic Techniques. *J. Geod.* **2011**, *85*, 887–907. [CrossRef]
22. Hernández-Pajares, M.; Juan, J.M.; Sanz, J.; Orús, R. Second-Order Ionospheric Term in GPS: Implementation and Impact on Geodetic Estimates. *J. Geophys. Res. Solid Earth* **2007**, *112*, 1–16. [CrossRef]
23. Jiang, H.; Jin, S.; Hernández-Pajares, M.; Xi, H.; An, J.; Wang, Z.; Xu, X.; Yan, H. A New Method to Determine the Optimal Thin Layer Ionospheric Height and Its Application in the Polar Regions. *Remote Sens.* **2021**, *13*, 2458. [CrossRef]
24. Klobuchar, J. Ionospheric Time-Delay Algorithm for Single-Frequency GPS Users. *IEEE Trans. Aerosp. Electron. Syst.* **1987**, *AES-23*, 325–331. [CrossRef]
25. Li, Z.; Yuan, Y.; Wang, N.; Hernandez-Pajares, M.; Huo, X. SHPTS: Towards a New Method for Generating Precise Global Ionospheric TEC Map Based on Spherical Harmonic and Generalized Trigonometric Series Functions. *J. Geod.* **2015**, *89*, 331–345. [CrossRef]
26. Wielgosz, P.; Milanowska, B.; Krypiak-Gregorczyk, A.; Jarmołowski, W. Validation of GNSS-Derived Global Ionosphere Maps for Different Solar Activity Levels: Case Studies for Years 2014 and 2018. *GPS Solut.* **2021**, *25*, 103. [CrossRef]
27. Zhang, Y.; Wu, F.; Kubo, N.; Yasuda, A. TEC Measurement By Single Dual-Frequency GPS Receiver. In Proceedings of the 2003 International Symposium on GPS/GNSS, Tokyo, Japan, 15–18 November 2003.

28. Teunissen, P.J.G. The Least-Squares Ambiguity Decorrelation Adjustment: A Method for Fast GPS Integer Ambiguity Estimation. *J. Geod.* **1995**, *70*, 65–82. [CrossRef]
29. Hernández-Pajares, M.; Juan, J.M.; Sanz, J.; Orus, R.; Garcia-Rigo, A.; Feltens, J.; Komjathy, A.; Schaer, S.C.; Krankowski, A. The IGS VTEC Maps: A Reliable Source of Ionospheric Information since 1998. *J. Geod.* **2009**, *83*, 263–275. [CrossRef]
30. Hernández-Pajares, M.; Roma-Dollase, D.; Krankowski, A.; García-Rigo, A.; Orús-Pérez, R. Methodology and Consistency of Slant and Vertical Assessments for Ionospheric Electron Content Models. *J. Geod.* **2017**, *91*, 1405–1414. [CrossRef]
31. Janssen, P.H.M.; Heuberger, P.S.C. Calibration of Process-Oriented Models. *Ecol. Modell.* **1995**, *83*, 55–66. [CrossRef]
32. Cohen, J. *Statistical Power Analysis for the Behavioral Sciences*, 2nd ed.; New York University: New York, NY, USA, 1988; ISBN 0-8058-0283-5.

remote sensing

MDPI

Article

Assimilating GNSS TEC with an LETKF over Yunnan, China

Jun Tang [1,2,3], Shimeng Zhang [2], Dengpan Yang [3] and Xuequn Wu [1,*]

1 Faculty of Land Resources Engineering, Kunming University of Science and Technology, Kunming 650093, China; jtang@kust.edu.cn
2 School of Information Engineering, Nanchang University, Nanchang 330031, China; shimeng@email.ncu.edu.cn
3 School of Transportation Engineering, East China Jiaotong University, Nanchang 330013, China; yangdengpan@ecjtu.edu.cn
* Correspondence: xuequnwu@kust.edu.cn

Abstract: A robust ionospheric model is indispensable for providing the atmospheric delay corrections for global navigation satellite system (GNSS) navigation and positioning and forecasting the space environment. The accuracy of ionospheric models is limited due to the simplified model structures. Complicated spatiotemporal variations in total electron content (TEC) biases between GNSS and international reference ionosphere (IRI) suggest a robust strategy to optimally combine GNSS and IRI TEC for high-precision modeling. In this paper, we propose a novel ionospheric data assimilation method, which is a local ensemble transform Kalman filter (LETKF), to construct an ionospheric model over Yunnan in southwestern China. We used the LETKF method to assimilate the ionospheric TEC extracted from GNSS observations in Yunnan into the IRI-2016 model. The experimental results indicate that the ionospheric data assimilation has a more pronounced improvement effect on the IRI empirical model during periods of geomagnetic quiet than during periods of geomagnetic disturbance. On quiet magnetic days, the skill score (SKS) of the assimilation is 0.60 and the root mean square error (RMSE) values before and after assimilation are 5.08 TECU and 2.02 TECU, respectively. The correlation coefficient after assimilation increases from 0.94 to 0.99. On magnetic storm days, the SKS of the assimilation is 0.42 and the RMSE values before and after assimilation are 5.99 TECU and 3.46 TECU, respectively. The correlation coefficient after assimilation increases from 0.98 to 0.99. The results suggest that the LETKF algorithm can be considered an effective method for ionospheric data assimilation.

Keywords: ionospheric TEC; LETKF; GNSS; data assimilation

1. Introduction

In the near-Earth space environment, the ionosphere has an important impact on human activities as a medium for radio wave propagation. Dramatic spatial and temporal variations in the ionosphere can seriously reduce the quality of global short-wave communications and the accuracy of satellite navigation and positioning [1,2]. The ionospheric total electron content (TEC) is one of the important parameters used to characterize the spatial and temporal variability of the ionosphere [3]. The question of how to accurately obtain the characteristics and variation patterns of ionospheric TEC distribution and establish a system that both contains intrinsic physics and reflects real observations while satisfying the real-time monitoring of the space environment and providing high-precision quasi-real-time ionospheric error correction information has become one of the key focuses of research in the study of the ionosphere.

In recent years, ionospheric data assimilation has become an important method for obtaining high-precision ionospheric TEC with the improvement of the global navigation satellite system (GNSS) and the increase in the number of ground-based observation networks [4–13]. The commonly used assimilation methods are mainly divided into two categories: the variational method and the Kalman filter method. For the variational method, Bust et al. [14] used a three-dimensional variational assimilation technique to assimilate GNSS TEC into an empirical ionospheric model and constructed an ionospheric data assimilation three-dimensional (IDA3D) model, which can handle a wide range of observations such as of the altimeter, ground-based GNSS, and occultation data. The model can process a variety of observations such as of the altimeter, ground-based GNSS, and occultation data to obtain information on ionospheric variability in a uniform global grid. Mengist et al. [15,16] investigated the performance of the four-dimensional technique for ionospheric data assimilation using the empirical model of the international reference ionosphere (IRI) as a background model in order to develop a regional ionospheric forecast model. Performance was improved in regions with no observational data and also improved by adding multi-source data. The system could operate in both calm and storm conditions. Jeong et al. [17] evaluated the performance of the data assimilation model (IDA4D) by introducing multiple types of data into the empirical model around the Korean Peninsula and found that the IDA4D model can be used as a reliable nowcast model. For the Kalman filter method, Schunk et al. [18] developed a data assimilation model, the Global Assimilation of Ionospheric Measurements (GAIM), based on a physical model of the ionospheric plasma, that uses Kalman filtering as the data assimilation algorithm. Lin et al. [19] proposed a Gauss–Markov Kalman filter method to assimilate GPS receiver and space radio occultation instrument observations into the IRI model. It was shown that this method could better improve the accuracy of the data assimilation analysis, and it was confirmed that this method could be used to reconstruct the three-dimensional ionospheric electron density. Yue et al. [20] used the ensemble Kalman filter (EnKF) algorithm to assimilate the electron density data observed by the Millstone Hill incoherent scattering radar. The results showed that the EnKF technique outperformed the 3DVAR technique, especially in the region of data gaps. He et al. [21] used the EnKF algorithm to evaluate the effectivenesses of various types of information for the nowcasting and forecasting of ionospheric parameters in China and adjacent areas and found that the algorithm can be considered a useful tool for the accurate specification of ionospheric nowcasts and predictions. Kosary et al. [22] presented a sequential calibration approach, based on the EnKF needed to calibrate model parameters, that uses a short period of GNSS network measurements to improve TEC estimations. Tang et al. [23] presented a data assimilation model for a regional ionosphere based on the local EnKF and found that the method could allow effective regional ionospheric data assimilation. Forootan et al. [24] presented a data assimilation method based on principal component analysis and the EnKF. The method can be used to improve VTEC estimates for globally available ionospheric models using IGS GIM products and can be easily applied to regional data assimilation by varying the background model and the domain of the observed field. However, the deviation between IRI and GNSS is not very clear at present, and the ionosphere varies greatly across different times and spaces. It is difficult to accurately represent prior noise with simple functional models.

The LETKF is an integrated square root filter that uses a low-rank estimate of the forecast covariance matrix to compute the relevant analysis [25]. The analysis is computed independently, grid point by grid point, by simultaneously absorbing nearby observations, which allows the LETKF to be implemented efficiently on a cluster of parallel computers [26]. The local ensemble Kalman filter, an early version of the LETKF, has been applied to global forecast models and has proven to be an effective data assimilation scheme in the troposphere, especially in regions with sparse observation densities [27]. Yunnan is located in a low latitude region, where the ionosphere overhead is very active. Establishing a high-precision ionospheric assimilation model is beneficial to improve the performance of satellite navigation and positioning. In this paper, we present the LETKF assimilation method to assimilate GNSS ionospheric TEC observations into the IRI-2016 model over Yunnan in China. Observations from 27 ground-based GNSS stations are applied to construct the Yunnan regional ionospheric TEC assimilation model. Using the self-modelled data, a preliminary assessment of the assimilation results of the ionospheric TEC data from day 121 to day 151 in 2017 is carried out.

2. Local Ensemble Transform Kalman Filter Assimilation Algorithm

Since the prediction error covariance matrix is not explicitly computed in the ETKF method, but is implemented by passing the prediction ensemble perturbation matrix, the matrix dimension will be inconsistent when using the Covariance Localization (CL) method in the ETKF and performing the Schur product operation [28]. To solve this problem, the CL method can approximate the Schur product operation between the prediction error covariance matrix and the localization function by requiring a Schur product operation in the square root matrix of the prediction ensemble perturbation matrix and the localization function in the ETKF [29]. In this way, the CL method can also be used in the ETKF method for localization.

The LETKF, like the LEnKF method, also roughly consists of three processes: prediction, updating, and localization. Since the background model used in this experiment was the empirical model IRI model, as in Experiments 1 and 2, the Gauss–Markov method was used for temporal updating, and the predicted value at the current moment t was calculated from the background and analytical values at moment $t-1$.

The LETKF prediction process is as follows:

$$X_t^f = X_t^b + \left(X_{t-1}^a - X_{t-1}^b\right)e^{-\Delta t/\tau} = \left(x_1^f, x_2^f, \cdots, x_N^f\right) \tag{1}$$

$$\overline{x}^f = \frac{1}{N}\sum_{i=1}^{N} x_i^f \tag{2}$$

$$X'^f = \frac{1}{\sqrt{N-1}}\left(x_1^f - \overline{x}^f, x_2^f - \overline{x}^f, \cdots, x_N^f - \overline{x}^f\right) = (u_1, u_2, \cdots, u_N) \tag{3}$$

$$P^f = X'^f\left(X'^f\right)^T \tag{4}$$

where X_t^f is the set of predictions at the current moment. X^b is given by the background model IRI. X_{t-1}^a is the set of state analysis at the previous moment. N denotes the number of ensemble members. $\overline{x}^f$ denotes the average state of x_i^f $(i = 1 \sim N)$. X'^f is the prediction ensemble perturbation matrix. P^f is the prediction error covariance matrix.

The LETKF localization process is as follows:

$$\rho_{ij} = e^{-\alpha(\varphi_{ij}^2/L_x^2)}e^{-\alpha(\theta_{ij}^2/L_y^2)} \tag{5}$$

$$\rho = WW^T = (w_1, w_2, \cdots, w_L)(w_1, w_2, \cdots, w_L)^T \tag{6}$$

$$W = \rho_{eigenvectors}\left(\rho_{eigenvalue}\right)^{1/2} \tag{7}$$

$$P_{loc}^f = P^f \circ \rho = Z^f\left(Z^f\right)^T \tag{8}$$

$$Z^f = WX'^f = [(w_1 \circ u_1, w_1 \circ u_2, \cdots, w_1 \circ u_N), \cdots, (w_L \circ u_1, w_L \circ u_2, \cdots, w_L \circ u_N)] \tag{9}$$

$$v_k^f = \overline{x}^f + \sqrt{M}z_k,\ k = 1, 2, \cdots, M,\ M = N \times L \tag{10}$$

$$V^f = \left(v_1^f, v_2^f, \cdots, v_M^f\right) \tag{11}$$

$$P_{loc}^f = \frac{1}{M}\sum_{i=1}^{M}\left(v_i^f - \overline{v}^f\right)\left(v_i^f - \overline{v}^f\right)^{\mathrm{T}} = Z^f\left(Z^f\right)^{\mathrm{T}} \tag{12}$$

Here, ρ_{ij} is the correlation coefficient between point i and point j and α is the distance correlation coefficient. φ_{ij} and θ_{ij} are the distances between two points in the longitude and dimensional directions, respectively. L_x and L_y denote the correlation distances in the longitude and dimensional directions, respectively. In this paper, the correlation coefficient $e^{-\alpha}$ is taken to be 0.75, the correlation distance in the dimensional direction is 10°, and the correlation distance in the longitude direction varies from about 40° in mid-latitudes to about 20° in the equatorial region. W denotes the square root matrix of ρ, which is obtained by decomposing the eigenvalues of ρ and arranging the obtained eigenvalues in descending order, taking the first 10 eigenvalues and the eigenvector pair composition; L is taken as 10. P_{loc}^f is the prediction error covariance matrix after localization by the CL method. Z^f denotes the square root matrix of P_{loc}^f, obtained by using the ensemble expansion technique, with a matrix size of $n \times (N \times L)$ where n denotes the number of state values in a single ensemble. z_k denotes the k column of matrix Z^f. M is the size of the set after localization expansion. V^f is the reconstructed prediction ensemble after localization. Equation (11) gives the reconstructed prediction error covariance matrix P_{loc}^f and its square root matrix Z^f.

The LETKF update process is as follows:

$$\hat{Y}^f = R^{-\frac{1}{2}}HZ^f \tag{13}$$

$$\hat{Y}^f = U\Sigma V^T \tag{14}$$

$$Z^a = Z^f U\left(\Sigma\Sigma^T + I\right)^{-\frac{1}{2}}U^T \tag{15}$$

$$\overline{v}^a = \overline{v}^f + Z^f U\Sigma\left(\Sigma^T\Sigma + I\right)^{-1}V^T R^{-\frac{1}{2}}\left[y - H\left(\overline{v}^f\right)\right] \tag{16}$$

$$X^a = \overline{v}^a 1_N^T + \sqrt{N-1}Z^a randn(M,N) \tag{17}$$

Here, $\hat{Y}^f$ is the annotated prediction–observation ensemble perturbation matrix. R is the observation error covariance matrix. U and V are orthogonal matrices and Σ is the singular value diagonal matrix of $\hat{Y}^f$; all three are derived from a singular value decomposition of $\hat{Y}^f$. Z^a is the square root matrix of the analysis error covariance matrix. $\overline{v}^a$ is the mean of the assimilation analysis set. X^a is the assimilation analysis ensemble. 1_N^T denotes an $1 \times N$ dimensional row vector in which all elements are 1s. $randn(M,N)$ denotes an $M \times N$ dimensional random matrix with elements obeying the standard normal distribution, which completes the selection of the N column analysis set; the matrix Z^a is randomly transformed into an $n \times N$ dimensional matrix. The implementation steps are based on those used by Hunt et al. [25].

3. Experimental Data Accuracy Assessment

3.1. Experimental Data

In this experiment, data from 27 GNSS stations in the Crustal Movement Observation Network of China (CMONOC) were used for a 31-day ionospheric assimilation experiment that included the annual cumulative days from day 121 to day 151 in 2017 (i.e., 1 to 31 May 2017). The longitude of the region is from 97°E to 107°E and the latitude is from 21°N to

29°N, with a latitude and longitude resolution of 0.5° × 0.5°. As shown in Figure 1, the black triangles indicate the GNSS station distribution.

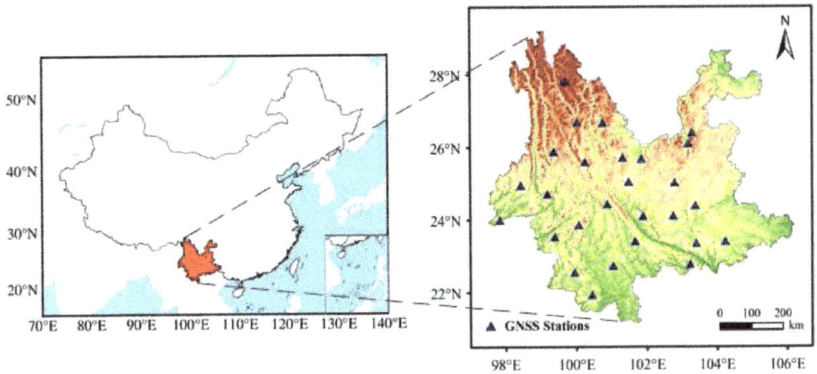

Figure 1. Geographical distribution of GNSS stations in Yunnan region.

Figure 2 shows the trends of the Dst index and Kp index during the period comprising days 121 to 151 in 2017. It can be seen that a magnetic storm occurred from day 147 to day 148 in 2017. The Dst index reached a minimum value of −125 nT at 08:00 UT on day 148 in 2017, and the Kp index reached a maximum value of 7 between 04:00 UT and 06:00 UT on day 148 in 2017.

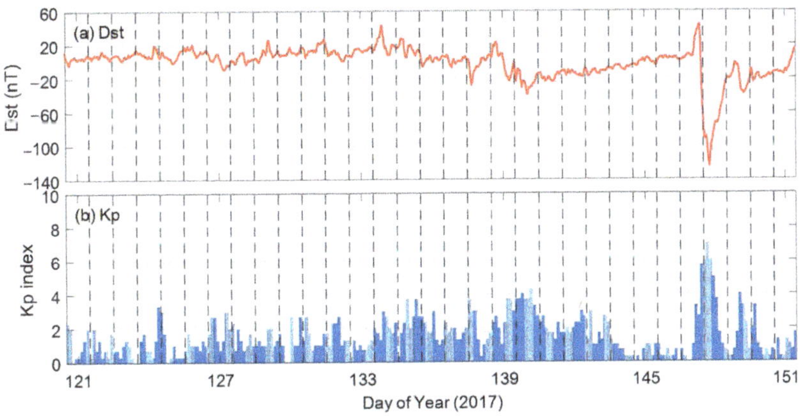

Figure 2. Geomagnetic indices observed during days 121 to 151, 31 days in 2017. (**a**) Dst; (**b**) Kp.

In this experiment, it was difficult to include comparative data in the experimental analysis because of the large spatial resolution of Center for Orbit Determination in Europe (CODE) TEC data and the small distribution of Massachusetts Institute of Technology (MIT) TEC data in the region of Yunnan. In this study, TEC maps established by using the low-degree spherical harmonic function were used as true values for comparative analysis [30]. In the following sections, we refer to the TEC maps as self-modeling TECs.

3.2. Accuracy Assessment

To assess the accuracy of assimilation results, the RMSE can be used as a criterion for accuracy assessment; it is defined as follows:

$$\mathrm{RMSE} = \sqrt{\frac{1}{N}\sum_{n=1}^{N}(\mathrm{TEC}_n^{\mathrm{ass}} - \mathrm{TEC}_n^{\mathrm{rea}})^2} \quad (18)$$

where N is the total number, $\mathrm{TEC}_n^{\mathrm{ass}}$ is the TEC value at point n obtained by data assimilation, and $\mathrm{TEC}_n^{\mathrm{rea}}$ is the truth data.

SKS is a common method for the quantitative assessment of assimilation effects, allowing a comprehensive assessment of the performance of ionospheric data assimilation. Three types of data are required for the calculation of SKS, namely the present or forecast value to be compared, the reference value, and the true value, which are calculated as follows [31]:

$$\mathrm{RMSE}_{ass} = \mathrm{RMSE}(\mathrm{TEC}_{ass} - \mathrm{TEC}_{truth}) \quad (19)$$

$$\mathrm{RMSE}_{ref} = \mathrm{RMSE}\left(\mathrm{TEC}_{ref} - \mathrm{TEC}_{truth}\right) \quad (20)$$

$$\mathrm{SKS} = 1 - \mathrm{RMSE}_{ass}/\mathrm{RMSE}_{ref} \quad (21)$$

where RMSE_{ass} and RMSE_{ref} denote the root mean square errors of the present or forecast and reference values to be compared, respectively, and TEC_{ass}, TEC_{ref}, and TEC_{truth} denote the present or ionospheric TEC data to be compared with the forecast, reference, and true values, respectively. From the equation, it can be derived that $\mathrm{SKS} \leq 1$, and the larger SKS is, the better the value to be compared is. When $\mathrm{SKS} = 1$, it means that the value to be compared is exactly the same as the true value. When $\mathrm{SKS} = 0 - 1$, it indicates that the value to be compared is better than the reference value. When $\mathrm{SKS} < 0$, it indicates that the value to be compared is worse than the reference value.

4. Results and Analysis

4.1. Ionospheric Geomagnetic Calm Conditions

To verify the reliability of the assimilation method, a comparison of the true TEC and the TECs before and after assimilation for two GNSS stations, KMIN and XIAG, for the 10-day period from day 125 to day 134 in 2017 is given in Figure 3. From Figure 3, it can be seen that the assimilated TEC data is closer to the true value with good agreement compared to the background model IRI data.

The daily average RMSE of the ionospheric TECs before and after the assimilation of the data and the daily average SKS of the assimilated data from day 125 to day 134 in 2017—i.e., for a 10-day period—are given in Figure 4, with the self-modeled data included as true values for comparison. As can be seen in Figure 4, the background data IRI TEC RMSE before assimilation deviates more from the true value. After assimilation, it is closer to the true value and the daily average RMSE decreases significantly. In Figure 4, all SKS values are greater than 0 during the aforementioned 10 days. This result indicates a certain improvement of IRI TEC relative to the background data.

The SKS values of the data-assimilated ionospheric TEC and the RMSEs before and after assimilation over time are given in Figure 5 for 24 h on day 126, 2017. It can be seen from Figure 5 that after data assimilation, the SKS values of the assimilated TEC are greater than 0, which indicates that after data assimilation, the accuracy of the assimilated TEC has a significant improvement compared to the IRI TEC. The greatest improvement relative to the IRI model for this day occurred at 06:00 UT with an SKS of 0.92 and RMSEs of 8.19 TECU and 0.67 TECU for IRI TEC and the assimilated TEC, respectively. The average SKS for this day was 0.72 and the average RMSE for the TECs before and after assimilation was 4.19 TECU and 1.18 TECU, respectively.

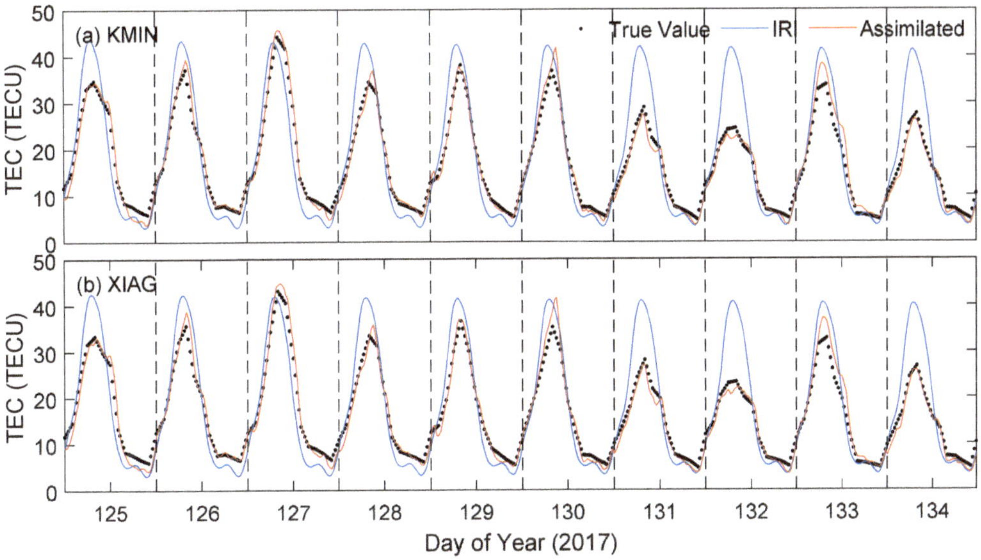

Figure 3. Comparison of true TEC and TECs before and after assimilation for KMIN and XIAG observatories during 10 days from day 125 to 134, 2017.

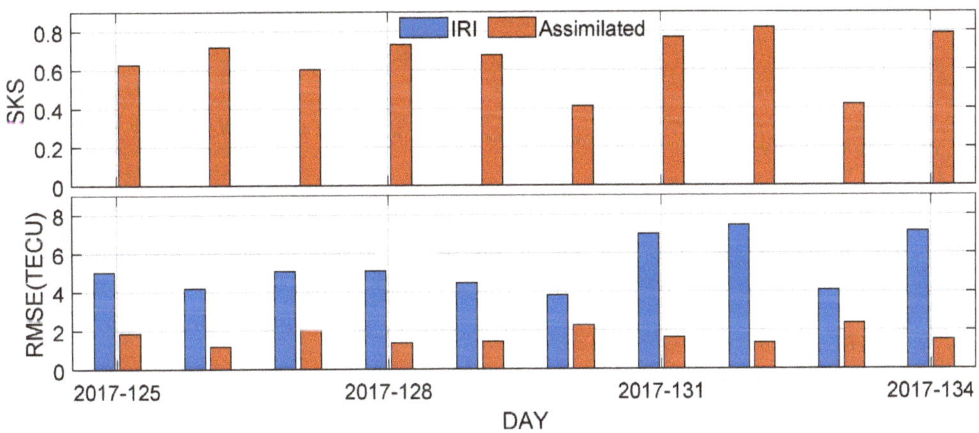

Figure 4. Average daily SKS of assimilated TEC and average daily RMSE of TECs before and after assimilation during the 10 days from day 125 to day 114, 2017.

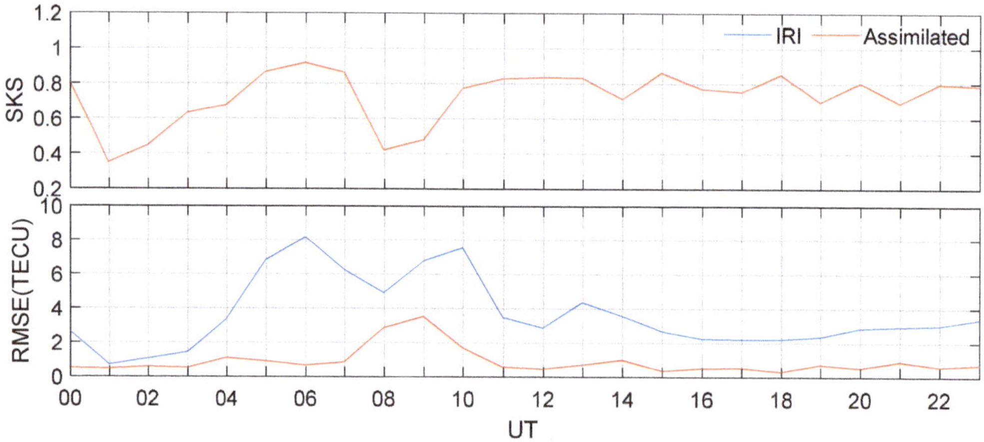

Figure 5. SKS values of assimilated TEC and RMSE of TEC before and after assimilation during 24 h on day 126, 2017.

Figure 6 shows the distribution of ionospheric TEC data in Yunnan and the surrounding regions for ionospheric pierce point (IPP) TEC, background model IRI TEC, assimilated TEC, and true TEC for the observed data at 09:00 UT on day 132, 2017. As can be seen in Figure 6, overall, the data in the background model IRI TEC is significantly overestimated compared to the true value while the assimilated TEC is closer to the IRI TEC than the IRI TEC is to both the true and IPP TECs. This suggests that the assimilated TEC has better agreement with the true value compared to the IRI TEC.

Figure 7 gives the TECs before and after data assimilation compared to the true data on day 127 in 2017. The histograms of the residuals of the background model IRI TEC and assimilated TEC compared to the true values are given in Figure 7a,b, respectively, which show that the residual distribution of the assimilated TEC data is closer to the unbiased Gaussian distribution compared to the IRI TEC. The mean and RMSE values of the residuals between the IRI TEC and the true value are −2.16 TECU and 5.08 TECU, respectively, while the mean and RMSE values of the residuals between the assimilated TEC and the true value drop to 0.25 TECU and 2.02 TECU, respectively. Figure 7c,d give scatter plots of the TECs before and after assimilation compared with the true value, respectively, and it can be seen that the correlation coefficient with the true value increases from 0.94 to 0.99 before assimilation.

Figure 8 shows the background model IRI TEC, true TEC, and assimilated TEC distributions for ionospheric TEC in Yunnan and the surrounding regions for 132 full days in 2017. As can be seen in Figure 8, the background model gives an obvious overestimation of the ionospheric TEC compared to the true TEC distribution, especially at the 06:00 UT and 10:00 UT moments, where the difference is obvious, with a maximum difference of 12 TECU, while the TEC distribution after assimilation is more consistent with the true TEC distribution.

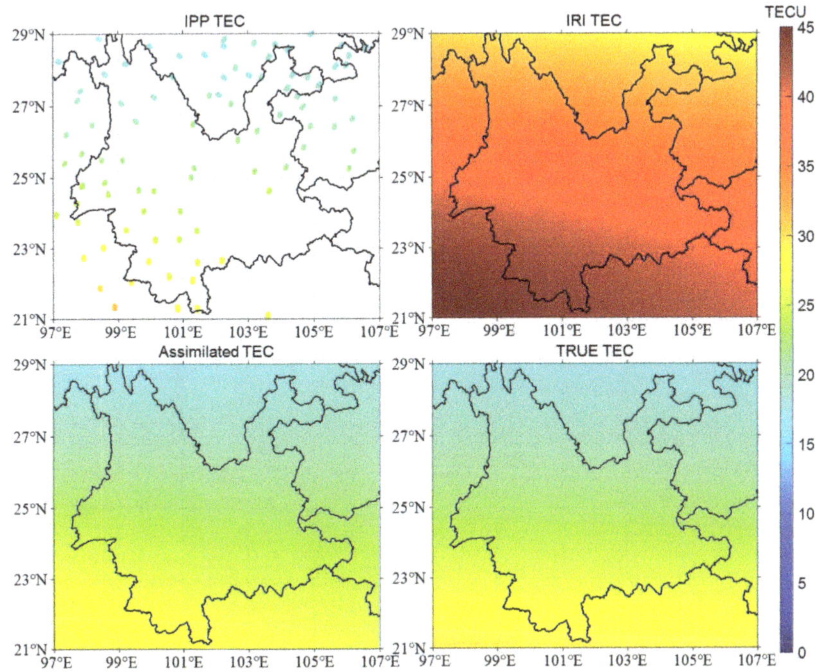

Figure 6. Distribution of ionospheric TEC data for IPP, IRI, assimilation, and true values at 09:00 UT on day 132, 2017.

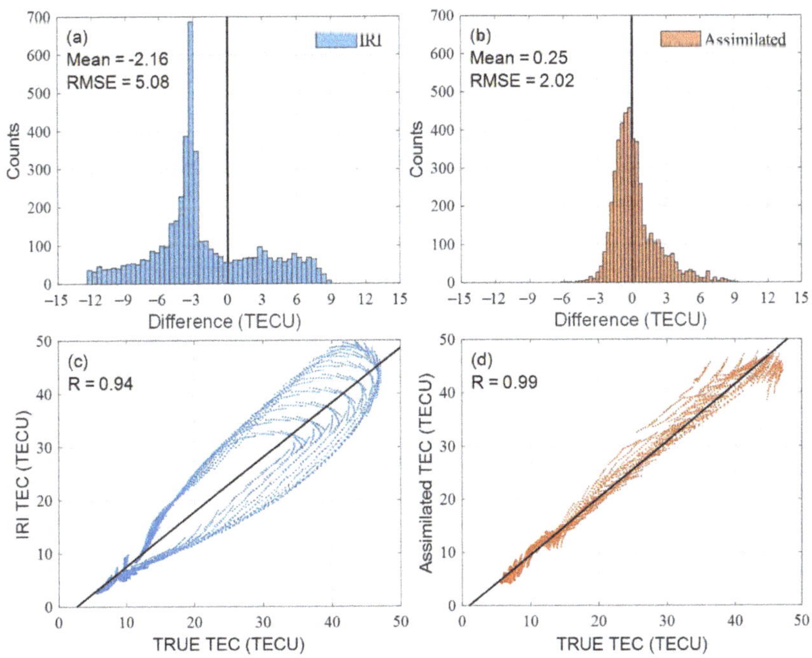

Figure 7. Residual histograms and scatter plots of IRI and assimilated TEC compared to true values, respectively, for day 127 in 2017. (**a**,**b**) are difference distribution; (**c**,**d**) are correlation.

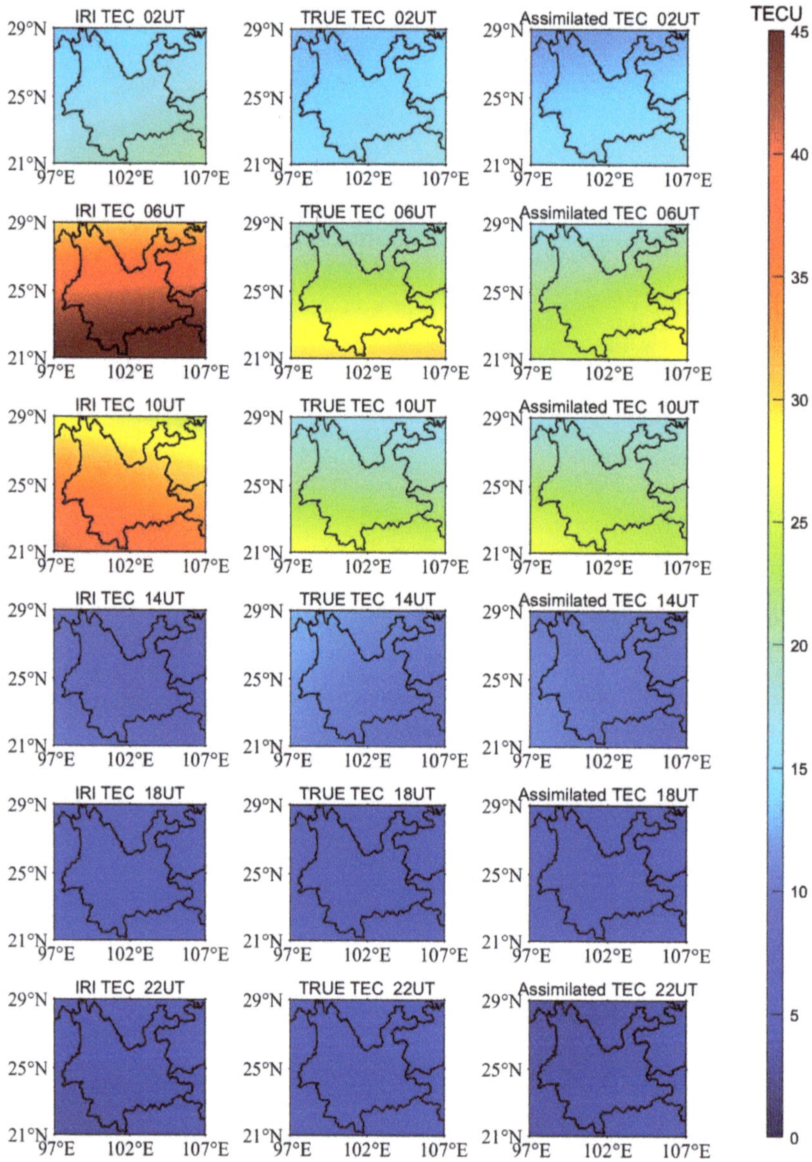

Figure 8. Distribution of ionospheric TEC in Yunnan and surrounding regions throughout the day on day 132, 2017.

4.2. Under Ionospheric Geomagnetic Disturbance Conditions

Figure 9 gives a comparison of the true TEC and the TECs before and after assimilation for the KMIN and XIAG observatories during the 10-day period from day 142 to 151 in 2017; a geomagnetic disturbance occurred on day 148. As can be seen in Figure 9, similar to in Figure 3, the agreement between assimilated TEC and the true value is significantly improved compared to IRI TEC on geomagnetically calm days while the agreement between assimilated TEC and the true value is weakened on the perturbed days 147 to 148 but still improved compared to IRI TEC.

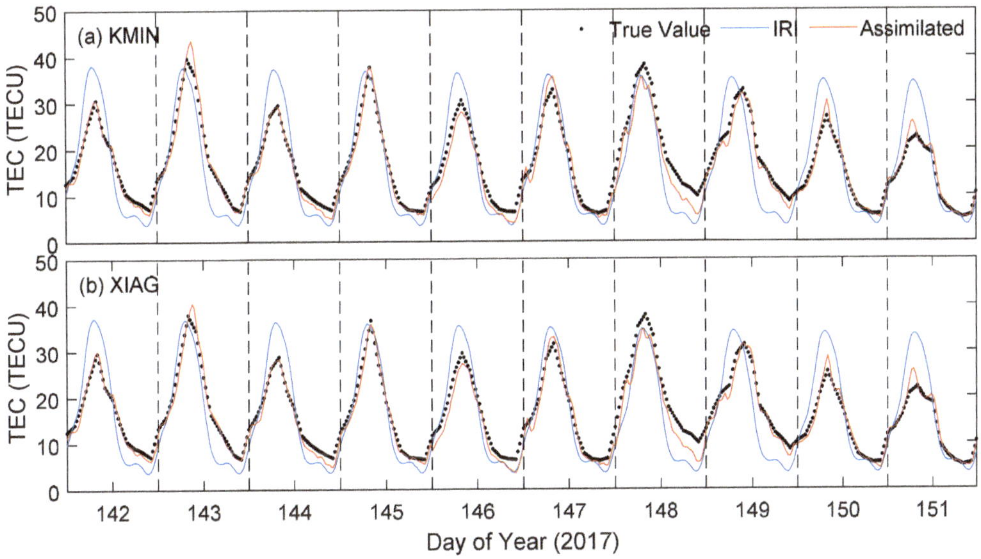

Figure 9. Comparison of true TEC and TECs before and after assimilation for KMIN and XIAG observatories during 10 days from day 142 to day 151 in 2017.

The daily average SKS of the data assimilation and the daily average RMSE of the ionospheric TECs before and after the assimilation are given in Figure 10 for days 142 to 148 during the 10-day period in 2017. It can be seen in Figure 10 that, similar to in Figure 4, the RMSE values of the assimilated TEC during the 10-day period are reduced compared to the pre-assimilation IRI TEC, and the SKS values are all greater than 0, indicating that the assimilated TEC data are improved relative to the background model IRI TEC data. It can also be seen that the SKS of the assimilated TEC on the 148th geomagnetic disturbance day is reduced compared to the calm day, i.e., the difference between the RMSE of the assimilated TEC and the IRI TEC is reduced.

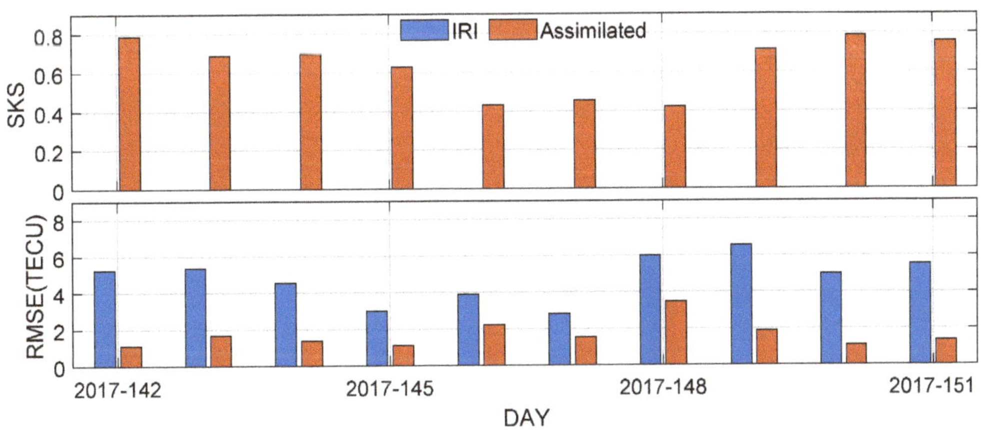

Figure 10. Average daily SKS of assimilated TEC and average daily RMSE of TECs before and after assimilation during 10 days from day 142 to day 151, 2017.

The SKS of the ionospheric TEC assimilated by the data and the RMSE values before and after assimilation over the period of 24 h during the geomagnetic disturbance event day in 2017, day 148, are given in Figure 11. As can be seen in Figure 11, the assimilated SKS is greater than 0 for most moments, and only a few moments show negative values, i.e., the RMSE of the assimilated TEC is greater than that of the IRI TEC, and the assimilation is not as effective as the background data before assimilation. The maximum and minimum values of SKS during this day were 0.85 and −0.21, respectively. The average SKS was 0.42, and the average RMSE values before and after assimilation were 5.99 and 3.46, respectively. From the daily average values of SKS and RMSE, we conclude that the data assimilation still had some improvement effect on the background model on the geomagnetic disturbance day.

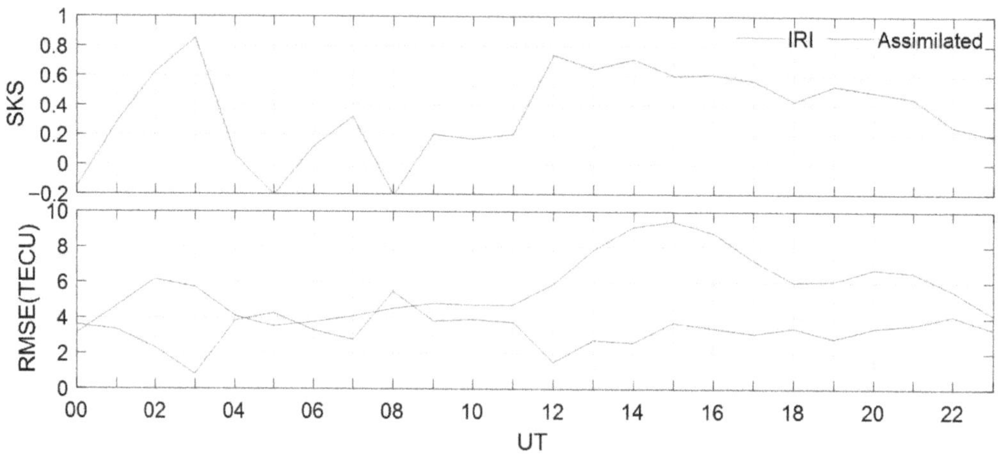

Figure 11. SKS of assimilated TEC and RMSE values of TEC before and after assimilation during 24 h on day 148, 2017.

Figure 12 shows the distribution of IPP TEC, the background model IRI TEC, assimilated TEC, and true TEC for the observed data at 06:00 UT on day 148 in 2017, the geomagnetic disturbance day, for the ionospheric TEC distribution in Yunnan and the surrounding region. As can be seen in Figure 12, the values of the background model IRI TEC are significantly lower compared to the true value, especially between 25°N and 29°N. The assimilated TEC is closer to the real value. This shows that the assimilated TEC is in better agreement with the true value than the IRI TEC on stormy days.

Figure 13 shows the TECs before and after data assimilation compared to the true data for day 148 in 2017, the day of geomagnetic disturbance. The residual histograms of the background model IRI TEC and assimilated TEC compared to the true value are given in Figure 13a,b, respectively, which show that the residual distribution of the assimilated TEC is closer to the unbiased Gaussian distribution compared to the IRI TEC. The mean and RMSE values of the residuals between the IRI TEC and the true value are −5.32 TECU and 5.99 TECU, respectively, while the mean and RMSE values of the residuals between the assimilated TEC and the true value drop to −3.22 TECU and 3.46 TECU, respectively. Figure 13c,d give the scatter plots of TEC before and after assimilation compared with the true value, respectively, and it can be seen that the correlation coefficient with the true value increases from 0.98 to 0.99 before assimilation.

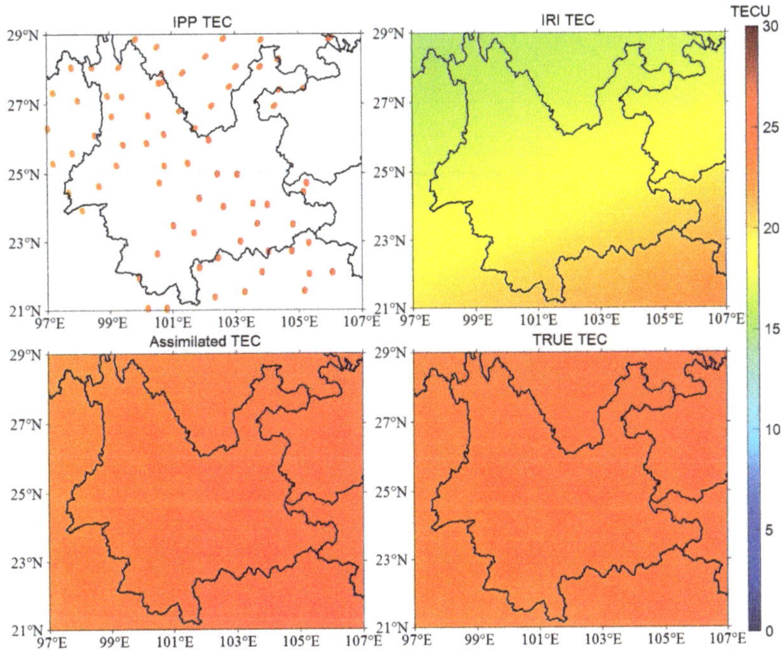

Figure 12. Distribution of ionospheric TEC for IPP, IRI, assimilation, and true values for 148 days in 2017 at 06:00 UT.

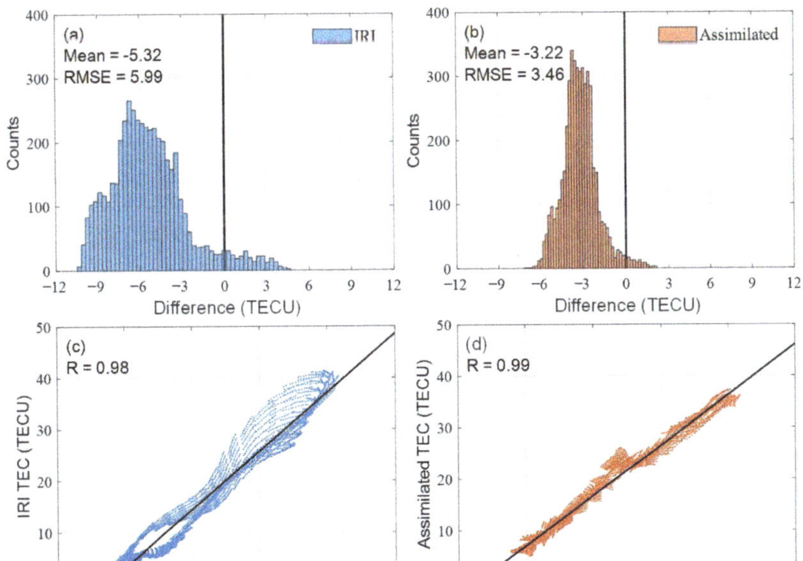

Figure 13. Residual histogram and scatter plot of IRI and assimilated TEC versus true values, respectively, for day 148 in 2017. (**a**,**b**) are difference distribution; (**c**,**d**) are correlation.

Figure 14 shows the background model IRI TEC, true TEC, and assimilated TEC distributions of ionospheric TEC in Yunnan and the surrounding region throughout the day on the geomagnetic disturbance day in 2017, day 148. It can be seen from Figure 14 that the background model gives a larger difference between the ionospheric TEC and the true value compared to the true TEC distribution, especially at 10:00 UT, 14:00 UT, and 18:00 UT, where the IRI TEC is significantly underestimated, while the TEC distribution after assimilation is not very different from the true value but is not as effective as the assimilation distribution on geomagnetically calm days.

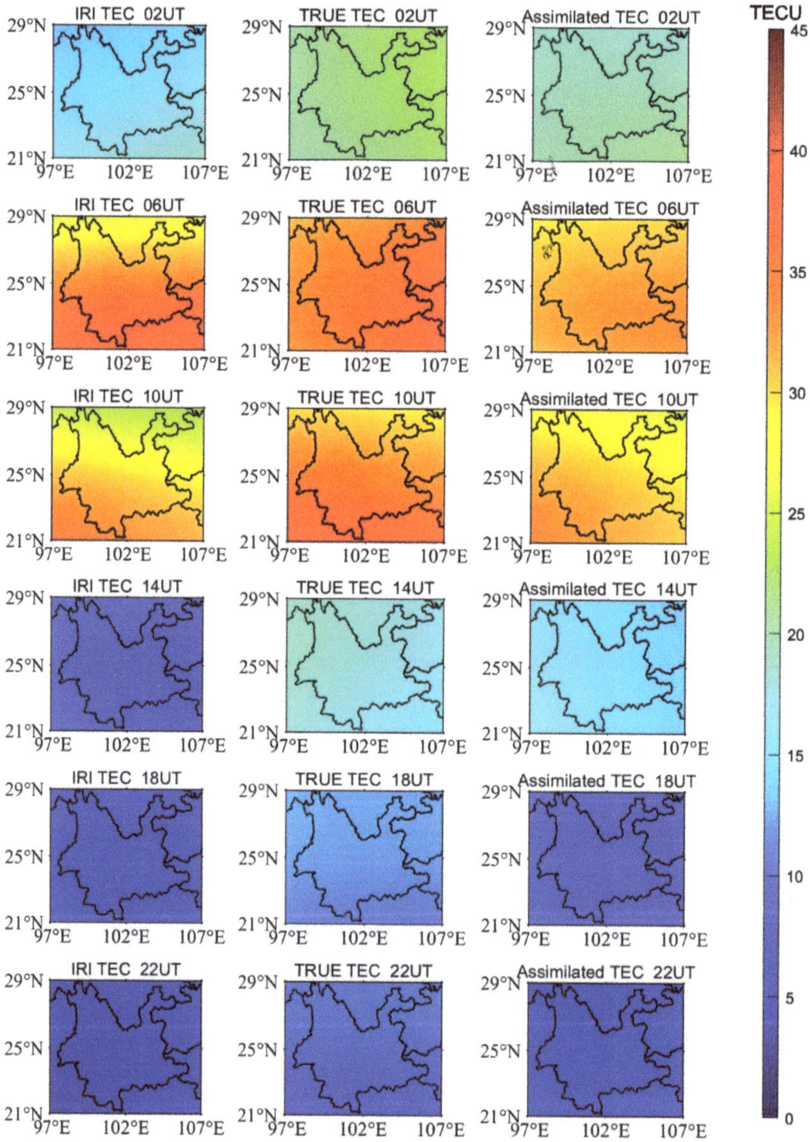

Figure 14. Distribution of ionospheric TEC in Yunnan and surrounding regions on day 148, 2017.

5. Discussion

To make for a better statistical analysis, Figure 15 shows the average monthly RMSE values of TEC before and after assimilation in 2017. The RMSE after assimilation is larger in May and September 2017. The poor performance of these two months is due to the magnetic storms that occurred in both months. In general, the assimilation method performs well on different days, months, and times of the day under different geomagnetic conditions.

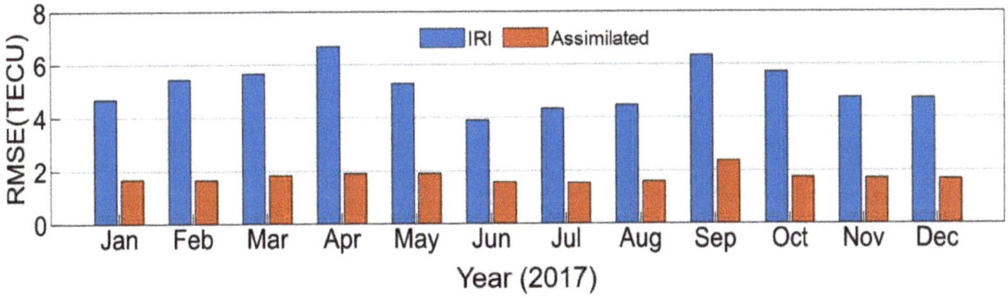

Figure 15. The average monthly RMSE values of TEC before and after assimilation in 2017.

The LETKF is an integrated square root filter that uses a low-rank estimate of the forecast covariance matrix to compute its analysis. This analysis is computed independently, grid point by grid point, by simultaneously absorbing nearby observations. First, the background perturbation matrix in the model and observation space is computed globally. Then, for each grid point, the observation vector mapped to the observation space and the associated background perturbation matrix are restricted to include only observations from the region surrounding the point [26]. Under quiet geomagnetic conditions, the residual distribution before data assimilation is significantly eccentric and deviates from the self-modeled TEC (Figure 7). Under disturbed geomagnetic conditions, the residual distribution before data assimilation is significantly eccentric and deviates from the self-modeled TEC (Figure 13). In Figure 4, the average daily SKS of assimilated TEC is 0.66 and the average daily RMSE values of TEC before and after assimilation during the 10 days from day 125 to day 134 in 2017 are 5.33 TECU and 1.69 TECU, respectively. In Figure 10, the average daily SKS of assimilated TEC is 0.64 and the average daily RMSE values of TEC before and after assimilation during the 10 days from day 142 to day 151 in 2017 are 4.79 TECU and 1.68 TECU, respectively. For the 31 days from day 121 to day 151 in 2017, the average daily SKS of assimilated TEC is 0.55 and the average daily RMSE values of TEC before and after assimilation are 5.03 TECU and 2.13 TECU, respectively. These experimental results show a generally good agreement between the self-modeled TEC and the assimilated TEC. The LETKF model can be effectively used for ionospheric data assimilation.

6. Conclusions

In this paper, the first ionospheric TEC assimilation modeling using the LETKF assimilation method was conducted to establish a regional ionospheric TEC assimilation model for Yunnan using the IRI model as the background model and the observation data from 27 ground-based GNSS stations. The results show that the improvement effect of data assimilation on the IRI background model is more obvious in periods of geomagnetic calm than in periods of geomagnetic disturbance. During the quiet period on day 127 in 2017, the SKS after assimilation is 0.60, the RMSE values before and after assimilation are 5.08 TECU and 2.02 TECU, respectively, and the correlation coefficients after assimilation are 0.94 and 0.99. During the storm period on day 148 in 2017, the SKS after assimilation is 0.42, the RMSE values before and after assimilation are 5.99 TECU and 3.46 TECU, respectively, and the correlation coefficients after assimilation are 0.98 and 0.99. The assimilation method performs well on different days, months, and times of the day under different geomagnetic conditions. The data assimilation can effectively fuse the input observations in both geomagnetic calm and geomagnetic disturbance periods, and the assimilated TEC accuracy is significantly improved compared with the pre-assimilation IRI TEC.

Author Contributions: J.T. and S.Z. conceived and designed the experiments; J.T. and S.Z. performed the experiments; S.Z. wrote the paper; D.Y. analyzed the data; J.T. and X.W. revised the paper. All authors have read and agreed to the published version of the manuscript.

Funding: This research was supported by the National Natural Science Foundation of China (NO. 42261074) and National Natural Science Foundation of China (NO. 41764002).

Data Availability Statement: The IRI-2016 data can be downloaded from the IRI's official website (http://www.irimodel.org (accessed on 10 January 2023)).

Acknowledgments: The authors acknowledge the Crustal Movement Observation Network of China (CMONOC) for providing GNSS data and the IRI Working group for providing IRI-2016 codes. They would also like to thank the anonymous reviewers for their constructive comments in improving this paper.

Conflicts of Interest: The authors declare no conflict of interest.

References

1. Kintner, P.M.; Ledvina, B.M. The ionosphere, radio navigation, and global navigation satellite systems. *Adv. Space Res.* **2005**, *35*, 788–811. [CrossRef]
2. Dubey, S.; Wahi, R.; Gwal, A. Ionospheric effects on GPS positioning. *Adv. Space Res.* **2006**, *38*, 2478–2484. [CrossRef]
3. Lejeune, S.; Wautelet, G.; Warnant, R. Ionospheric effects on relative positioning within a dense GPS network. *GPS Solut.* **2012**, *16*, 105–116. [CrossRef]
4. Hajj, G.A.; Wilson, B.D.; Wang, C.; Pi, X.; Rosen, I.G. Data assimilation of ground GPS total electron content into a physics-based ionospheric model by use of the kalman filter. *Radio Sci.* **2004**, *39*, RS1S05. [CrossRef]
5. Fuller-Rowell, T.; Araujo-Pradere, E.; Minter, C.; Codrescu, M.; Spencer, P.; Robertson, D.; Jacobson, A.R. US-TEC: A new data assimilation product from the space environment center characterizing the ionospheric total electron content using real-time GPS data. *Radio Sci.* **2006**, *41*, RS6003. [CrossRef]
6. Yue, X.; Schreiner, W.S.; Lin, Y.C.; Rocken, C.; Kuo, Y.H.; Zhao, B. Data assimilation retrieval of electron density profiles from radio occultation measurements. *J. Geophys. Res.-Space Phys.* **2011**, *116*, A03317. [CrossRef]
7. Gardner, L.C.; Schunk, R.W.; Scherliess, L.; Sojka, J.J.; Zhu, L. Global assimilation of ionospheric measurements-Gauss Markov model: Improved specifications with multiple data types. *Space Weather* **2014**, *12*, 675–688. [CrossRef]
8. Aa, E.; Huang, W.; Yu, S.; Liu, S.; Shi, L.; Gong, J.; Chen, Y.; Shen, H. A regional ionospheric TEC mapping technique over China and adjacent areas on the basis of data assimilation. *J. Geophys. Res.-Space Phys.* **2015**, *120*, 5049–5061. [CrossRef]
9. Adebiyi, S.J.; Adebesin, B.O.; Ikubanni, S.O.; Joshua, B.W. Performance evaluation of GIM-TEC assimilation of the IRI-Plas model at two equatorial stations in the American sector. *Space Weather* **2017**, *15*, 726–736. [CrossRef]
10. Chen, M.; Liu, L.; Xu, C.; Wang, Y. Improved IRI-2016 model based on BeiDou GEO TEC ingestion across China. *GPS Solut.* **2019**, *24*, 20. [CrossRef]
11. Ssessanga, N.; Kim, Y.H.; Habarulema, J.B.; Kwak, Y.S. On imaging south African regional ionosphere using 4D-var technique. *Space Weather* **2019**, *17*, 1584–1604. [CrossRef]
12. He, J.; Yue, X.; Le, H.; Ren, Z.; Wan, W. Evaluation on the quasi-realistic ionospheric prediction using an ensemble Kalman filter data assimilation algorithm. *Space Weather* **2020**, *18*, e2019SW002410. [CrossRef]
13. He, J.; Yue, X.; Astafyeva, E.; Le, H.; Ren, Z.; Pedatella, N.M.; Ding, F.; Wei, Y. Global gridded ionospheric electron density derivation during 2006–2016 by assimilating COSMIC TEC and its validation. *J. Geophys. Res.-Space Phys.* **2022**, *127*, e2022JA030955. [CrossRef]
14. Bust, G.S.; Garner, T.W.; Gaussiran, T.L. Ionospheric data assimilation three-dimensional (IDA3D): A global, multisensor, electron density specification algorithm. *J. Geophys. Res.-Space Phys.* **2004**, *109*, A11312. [CrossRef]
15. Mengist, C.K.; Ssessanga, N.; Jeong, S.H.; Kim, J.H.; Kim, Y.H.; Kwak, Y.S. Assimilation of multiple data types to a regional ionosphere model with a 3D-var algorithm (IDA40). *Space Weather* **2019**, *17*, 1018–1039. [CrossRef]
16. Mengist, C.K.; Seo, K.H.; Kim, Y.H.; Eswaraiah, S.; Ssessanga, N.; Kwak, Y.S. 3-D regional imaging of ionosphere over Africa through assimilating satellite and ground-based data. *J. Geophys. Res.-Space Phys.* **2023**, *128*, e2022JA030859. [CrossRef]
17. Jeong, S.H.; Kim, Y.H.; Kim, J.H.; Kwak, Y.S. A regional ionospheric assimilation study with GPS and COSMIC measurements using a 3D-Var algorithm (IDA4D). *Adv. Space Res.* **2022**, *69*, 2489–2500. [CrossRef]
18. Schunk, R.W.; Scherliess, L.; Sojka, J.J.; Thompson, D.C.; Anderson, D.N.; Codrescu, M.; Minter, C.; Fuller-Rowell, T.J.; Heelis, R.A.; Hairston, M.; et al. Global assimilation of ionospheric measurements. *Radio Sci.* **2004**, *39*, RS1S02. [CrossRef]

19. Lin, C.Y.; Matsuo, T.; Liu, J.Y.; Lin, C.H.; Huba, J.D.; Tsai, H.F.; Chen, C.Y. Data assimilation of ground-based GPS and radio occultation total electron content for global ionospheric specification. *J. Geophys. Res.-Space Phys.* **2017**, *122*, 10876–10886. [CrossRef]
20. Yue, X.; Wan, W.; Liu, L.; Zheng, F.; Lei, J.; Zhao, B.; Xu, G.; Zhang, S.; Zhu, J. Data assimilation of incoherent scatter radar observation into a one-dimensional midlatitude ionospheric model by applying ensemble Kalman filter. *Radio Sci.* **2007**, *42*, RS6006. [CrossRef]
21. He, J.; Yue, X.; Hu, L.; Wang, J.; Li, M.; Ning, B.; Wan, W.; Xu, J. Observing system impact on ionospheric specification over China using EnKF assimilation. *Space Weather* **2020**, *18*, e2020SW002527. [CrossRef]
22. Kosary, M.; Forootan, E.; Farzaneh, S.; Schumacher, M. A sequential calibration approach based on the ensemble Kalman filter (C-EnKF) for forecasting total electron content (TEC). *J. Geod.* **2022**, *96*, 29. [CrossRef]
23. Tang, J.; Zhang, S.; Huo, X.; Wu, X. Ionospheric assimilation of GNSS TEC into IRI model using a local ensemble Kalman filter. *Remote Sens.* **2022**, *14*, 3267. [CrossRef]
24. Forootan, E.; Kosary, M.; Farzaneh, S.; Schumacher, M. Empirical data assimilation for merging total electron content data with empirical and physical models. *Surv. Geophys.* **2023**. [CrossRef]
25. Hunt, B.R.; Kostelich, E.J.; Szunyogh, I. Efficient data assimilation for spatiotemporal chaos: A local ensemble Kalman filter. *Phys. D* **2007**, *230*, 112–126. [CrossRef]
26. Durazo, J.A.; Kostelich, E.J.; Mahalov, A. Local ensemble transform Kalman filter for ionospheric data assimilation: Observation influence analysis during a geomagnetic storm event. *J. Geophys. Res.-Space Phys.* **2017**, *122*, 9652–9669. [CrossRef]
27. Szunyogh, I.; Kostelich, E.J.; Gyarmati, G.; Patil, D.J.; Hunt, B.R.; Kalnay, E.; Ott, E.; Yorke, J.A. Assessing a local ensemble Kalman filter: Perfect model experiments with the National Centers for Environmental Prediction global model. *Tellus A* **2005**, *57*, 528–545. [CrossRef]
28. Bishop, C.H.; Etherton, B.J.; Majumdar, S.J. Adaptive sampling with the ensemble transform Kalman filter. Part I: Theoretical aspects. *Mon. Weather Rev.* **2001**, *129*, 420–436. [CrossRef]
29. Bishop, C.H.; Whitaker, J.S.; Lei, L. Gain form of the ensemble transform Kalman filter and its relevance to satellite data assimilation with model space ensemble covariance localization. *Mon. Weather Rev.* **2017**, *145*, 4575–4592. [CrossRef]
30. Yuan, Y.; Li, Z.; Wang, N.; Zhang, B.; Li, H.; Li, M.; Huo, X.; Ou, J. Monitoring the ionosphere based on the crustal movement observation network of China. *Geod. Geodyn.* **2015**, *6*, 73–80. [CrossRef]
31. Scherliess, L.; Schunk, R.W.; Sojka, J.J.; Thompson, D.C.; Zhu, L. Utah State university global assimilation of ionospheric measurements Gauss-Markov Kalman filter model of the ionosphere: Model description and validation. *J. Geophys. Res.-Space Phys.* **2006**, *111*, A11315. [CrossRef]

remote sensing MDPI

Article

Comparison of the Forecast Accuracy of Total Electron Content for Bidirectional and Temporal Convolutional Neural Networks in European Region

Artem Kharakhashyan and Olga Maltseva *

Institute for Physics, Southern Federal University, Rostov-on-Don 344090, Russia; aharahashyan@sfedu.ru
* Correspondence: oamaltseva@sfedu.ru; Tel.: +7-9-185772533

Abstract: Machine learning can play a significant role in bringing new insights in GNSS remote sensing for ionosphere monitoring and modeling to service. In this paper, a set of multilayer architectures of neural networks is proposed and considered, including both neural networks based on LSTM and GRU, and temporal convolutional networks. The set of methods included 10 architectures: TCN, modified LSTM-/GRU-based deep networks, including bidirectional ones, and BiTCN. The comparison of TEC forecasting accuracy is performed between individual architectures, as well as their bidirectional modifications, by means of MAE, MAPE, and RMSE estimates. The F10.7, 10 Kp, Np, Vsw, and Dst indices are used as predictors. The results are presented for the reference station Juliusruh, three stations along the meridian 30°E (Murmansk, Moscow, and Nicosia), and three years of different levels of solar activity (2015, 2020, and 2022). The MAE and RMSE values depend on the station latitude, following the solar activity. The conventional LSTM and GRU networks with the proposed modifications and the TCN provide results at the same level of accuracy. The use of bidirectional neural networks significantly improves forecast accuracy for all the architectures and all stations. The best results are provided by the BiTCN architecture, with MAE values less than 0.3 TECU, RMSE less than 0.6 TECU, and MAPE less than 5%.

Keywords: ionosphere; total electron content; forecasting; BiGRU; BiLSTM; BiTCN; GNSS; temporal convolution; space weather; machine learning

1. Introduction

The study of the effects of space weather on the environment is one of the prioritized areas of geophysics [1]. One of the components of the environment is the near outer space of the Earth. Among the parts of this space is the ionosphere, which affects the operation of technological systems such as global navigation satellite systems (GNSS), satellite communications, and other space communications applications [2]. The state of the ionosphere is described by such parameters as critical frequency foF2, maximum height hmF2 of the layer F2, total electron content TEC, and others. The study and prediction of TEC attracts special attention, since its values determine the accuracy of positioning [3,4]. Currently, among the methods of forecasting the ionospheric parameters, methods using neural networks are distinguished as one of the most diverse and undergoing active development [5]. However, since each region of the globe has different properties related to the influence of space weather, it is necessary to choose a method that can provide the most accurate TEC forecast for each region. This paper attempts to select from a variety of methods the most appropriate one for the European region. For this purpose, both the conventional methods', long short-term memory LSTM [6], gated recurrent unit GRU [7], and new ones including bidirectional [8] and temporal convolutional networks TCN [9], architectures are used. From the extensive literature, we selected those papers that include quantitative estimates of forecast accuracy in different regions.

An example of using the LSTM architecture for TEC prediction is a paper which proposed a method with an advance time of 24 h [10]. The authors used CODE (Center for Orbit Determination in Europe) map data with a 1 h step for Beijing station, as well as an index solar radio flux at 10.7 cm (F10.7) of solar activity and geomagnetic activity ap as input parameters. The dataset consists of 1 January 1999 to 1 September 2016. The root mean square error (RMSE) was 3.5 TECU. It is interesting to note that the test years were 2001 and 2015, i.e., years of high solar activity. Sun et al. [11] developed and applied the bidirectional long-term memory (BiLSTM) model to TEC prediction on the same database and using F10.7 and ap indices as in [10]. The results obtained were compared with other methods. The RMSE for BiLSTM was 3.35 TECU; 3.48 TECU was obtained for single-layer LSTM and 3.67 TECU was obtained for double-layer LSTM. A feature that improved the LSTM result is the ability to utilize both the past and future data to make prediction.

Sivakrishna et al. [12] applied the BiLSTM algorithm to one-hour-ahead forecast the Indian regional TEC map. The training data included 261 days (from November 2015 to September 2016). The validation dataset included 24 days in September 2016, and the testing data were 31 days on October 20. The maps were plotted against data from 26 Global Positioning System stations and compared with Artificial Neural Network (ANN) and LSTM model results during both geomagnetic quiet (20 October 2016) and disturbed (14 October 2016) periods. For quiet conditions, the difference between the experimental and predicted values was from 1 TECU up 2.5 TECU for BiLSTM, and from 1 TECU to 3.5 TECU for disturbed conditions and were better than the results for ANN and LSTM. The LSTM model performed better than the ANN model. A comparison between the Indian regional ionospheric forecast of TEC maps with and without solar and geomagnetic indices (F10.7, Kp) as input to BiLSTM during quiet and disturbed periods showed little improvement.

The LSTM architecture has become the reference model against which the results of new approaches are compared, and most often, the results of this comparison are given when they are better than for LSTM. However, the LSTM method continues to be modified. In Chen et al. [13], several architectures, single-step self-prediction model, single-step auxiliary prediction model, multistep self-prediction model, and multistep auxiliary prediction (MSAP) model, were used for TEC prediction based on International GNSS Service (IGS) maps with an advance time of 24 h, 48 h, and 144 h. Their algorithms are presented in the appendix of the paper [13]. The database covered a set of TEC values from January 2011 to December 2019 and indices included the following: the sunspot number R, F10.7, ap, and Dst. In contrast to traditional data processing methods, the cycle consisted of 90 days. Among these, the first 30 days are considered as the training set, the middle 30 days are considered as the validation set, and the last 30 days are considered as the test set.

Comparing the interval within 24 h, the averaged MAE (RMSE) values calculated from the single-step self-prediction model within 48 h increases from 2.721 (3.806) TECU to 3.259 (4.483) TECU. For the MSAP model, these values were 2.116 (3.033) TECU for 24 h and 2.225 (3.175) TECU for 48 h. As the prediction interval increases, the differences become larger: 5.53 (7.427) TECU and 2.485 (3.511) TECU for MSAP and 144 h.

A TEC forecasting model based on deep learning was proposed by Tang et al. [14], which consists of a convolutional neural network (CNN), a long short-term memory (LSTM) neural network, and an attention mechanism. The attention mechanism is added to the pooling layer and the fully connected layer to assign weights to improve the model. The dataset included TEC from 24 GNSS stations of China Network over the 9-year period (from 2010 to 2018) to predict the value of the TEC 24 h ahead. The data from 2010 to 2017 is used as the training set and the data from 2018 is used as the test set. Bz, Kp, Dst, and F10.7 indices are used to characterize solar and geomagnetic activity. The results were compared to the accuracy of models such as NeQuick, LSTM, and CNN-LSTM. The statistical results were obtained from the data of 24 GNSS stations, 24 h a day and 365 days a year in the test set. For all stations, the MAE was 2.6 TECU for the NeQuick, 1.53 TECU for the LSTM, 1.36 TECU for the CNN-LSTM, and 1.17 TECU for the CNN-LSTM-Attention model. The RMSE for the NeQuick was 3.59 TECU, 2.25 TECU for the LSTM, 2.07 TECU for the CNN-LSTM, and 1.87 TECU for the CNN-LSTM-Attention model. The latitudinal and longitudinal dependencies of forecast accuracy were also determined from the data of six stations with similar latitudes and longitudes. RMSE decreases with the increase in latitude near the same longitude, which is about 1 TECU in the mid-latitude region and between 1.6 and 3 TECU in six stations around 30°E. Accuracy estimates depending on the level of disturbance were performed separately for days with $Kp < 3$ and $Kp \geq 3$. For a quiet day on 21 August 2018, the statistics showed RMSE = 4.14 TECU, MAE = 3.16 TECU, and the mean error ME = −0.12 TECU for the NeQuick model. For the LSTM model, statistical estimations have provided the following results: RMSE = 4.29 TECU, MAE = 3.0 TECU, and ME = −1.5 TECU. For the other architecture, CNN-LSTM, the following error values were obtained: RMSE = 4.14 TECU, MAE = 2.98 TECU, and ME = −1.62 TECU. For the proposed CNN-LSTM-Attention architecture, the results were better: RMSE = 3.99 TECU, MAE = 2.81 TECU, and ME = −1.46 TECU. During the perturbed day of 26 August 2018, the magnetic storm led to big changes of TEC, and the accuracy of all methods decreased. For the NeQuick model, RMSE = 4.99 TECU, MAE = 3.67 TECU, and ME = 1.71 TECU. The forecast for the LSTM method provided the following values: RMSE = 5.19 TECU, MAE = 3.76 TECU, and ME = −2.07 TECU. These values decreased to RMSE = 4.43 TECU, MAE = 3.06 TECU, and ME = 0.21 TECU for CNN-LSTM and to RMSE = 4.11 TECU, MAE = 2.81 TECU, and ME = −0.64 TECU for the CNN-LSTM-Attention architecture.

Boulch et al. [4] proposed the ConvRNN method and compared the results with the data presented in the papers [15–17] for specific Chinese stations and on a global scale, with a lead time of 2 and 48 h. In the range 22°N–39°N, an increase in the forecast accuracy with increasing latitude was obtained. On a global scale, the method in [17] provided RMSE = 3.1 TECU and the method in [4] gave 0.89 TECU for a lead time of 48 h and 0.38 TECU for 2 h ahead.

Iluore and Lu [18] analyzed the data from the equatorial station MAL2 (Kenya) and determined that GRU was more accurate than LSTM, Multilayer Perceptron (MLP), GIM_TEC, and the IRI-Plas 2017. The data covered 9 years from 1 January 2010 to 31 December 2018. The data from the year 2010 to 2016 were used for the training, while data from the year 2017 were used for validation and the data in the year 2018 were used to estimate the performance of the models. The GRU unit showed a prediction error of 2.004 TECU while the LSTM, MLP, GIM_TEC, and IRI-Plas 2017 models showed a prediction error of 2.055 TECU, 2.336 TECU, 5.913, and 16.183 TECU, respectively.

Kaselimi et al. [19] proposed a spatiotemporal deep learning architecture, CNN-GRU, combining a convolutional neural network to capture the spatial variability of TEC and a gated recurrent unit for temporal variability modeling. Real TEC measurements were used by determining the slant STEC and its conversion to TEC for six stations in different regions of the globe, with F10.7, solar sunspot number SSN, Kp, ap, and Dst as the input parameters. The TEC values obtained by the proposed method were compared not only with the results of the neural network methods, LSTM [20], BiLSTM [11], and RNN [21], but also with the GIM map CODE and IRI model values. The dataset included data from 2014 to 2018 with the test period being during the second half of 2018. Specific estimates of MAE ranged from 0.7–1.8 TECU for CNN-GRU, 0.9–1.8 TECU for Recurrent Neural Network (RNN), and 0.8–2.2 TECU for LSTM and BiLSTM.

Unfortunately, many papers do not provide Mean Absolute Percentage Error (MAPE) values, while a comparison of MAE and RMSE does not always give the full picture. As shown in Table III in Kaselimi et al. [19], different methods can give the best results for different stations. As for the dependence of the results on latitude, they were better for the middle-latitude station Graz than for the high-latitude station Tixi. This paper also compares the processing time in minutes and the number of the required trainable parameters per method. The most efficient was the RNN architecture, which requires 23 min (for 200 training epochs), while the heaviest is the BiLSTM model which requires

280 min (for the same 200 epochs). The proposed CNN-GRU architecture requires 67 min for its training.

However, other approaches are also possible. Using the advantages of neural networks combined with the Global Ionosphere Map GIM, an empirical model to predict TEC, 24 h in advance at the global scale was developed in the work of Cesaroni et al. [22]. A nonlinear autoregressive neural network with eXternal input NARX was used as a neural network. The Polytechnic University of Catalonia's (UPC) products for single point forecasting were used as GIM. To extend the forecasting at a global scale, a NeQuick2 model adapted to an effective sunspot number R12eff was used. TEC and Kp data for 11 years (2005–2015) divided into training (70%), validation (15%), and testing (15%) datasets with a time resolution of 2 h were used. Examples for four points and two quiet days (30 May 2016 and 21 August 2016) yielded an RMSE of 1.5 TECU, 0.32 TECU, 2.3 TECU, and 3.0 TECU. Accumulated statistics showed that the RMSE was 3.5 TECU for the period June 2017 to May 2018. For the disturbed period of 7–11 September 2017, the RMSE = 3.8 TECU. The RMSE for six per disturbed periods of different intensities (from 1 to −142 nT) was in the range 3.4–5.1 TECU and showed no dependence on Dst. This 24 h empirical approach was implemented on the Ionosphere Prediction Service (IPS), a prototype platform to support different classes of GNSS users.

The work of Natras et al. [23] provides an overview of most of the methods used for TEC forecasting: ANN, LSTM, LSTM-CNN, Encoder-Decoder LSTM, Extended ED-LSTME, NARX, conditional Generative Adversial Network cGAN, and others, and gives statistical characteristics of these methods. In particular, for 1 h forecasting and up to 1-day and 2-day forecasting, the RMSE for a 1 h TEC forecast in low latitudes ranges from 2 to 5 TECU for different learning algorithms and different levels of solar activity. For the mid-latitude 1 h VTEC forecast, the RMSE is about 1.5 TECU. For a 1-day TEC forecast, the RMSE was 4 TECU in high solar activity and 2 TECU in low solar activity. However, noting the large complexity of using deep learning methods as a drawback of these methods, the authors used three algorithms: Decision Tree and ensemble learning of Random Forest, Adaptive Boosting (AdaBoost), and eXtreme Gradient Boosting (XGBoost) for 1 h and 24 h VTEC forecasts. According to the authors, the advantages of these methods are that they are simple, non-parametric, fast to optimize, computationally efficient, and can be used on a limited dataset. The dataset for training and cross-validation included the period January 2015–December 2016. The test dataset included the period January–December 2017. Test results of six methods are given for three latitudes 70°N, 40°N, and 10°N (longitude not shown) and two periods: year 2017 and 7–10 September 2017. The results for the method showing the highest accuracy (Random Forest) are as follows. For 1 h ahead and latitude 70°N, the RMSE was 0.54 TECU, for 40°N the RMSE was 0.92 TECU, and for 10°N the RMSE was 1.2 TECU. For 24 h ahead and a latitude of 70°N, the RMSE was 1.06 TECU, for 40°N, the RMSE was 1.86 TECU, and for 10°N, the RMSE was 2.2 TECU. For the disturbed period 7–10 September 2017 and 1 h ahead, the RMSE were 0.73 TECU, 1.31 TECU, and 1.29 TECU for 70°N, 40°N, and 10°N, respectively. For 24 h ahead, these values were 1.77 TECU, 3.95 TECU, and 3.95 TECU.

Thus, an analysis of the literature data shows that there is a certain sequence of using neural network methods LSTM to GRU to bidirectional to TCN, which allows gradually increasing the accuracy of the TEC forecast. However, the results can be highly dependent on the combination of architectures, region, and space weather conditions.

In a previous work [24], the following results for single hidden layer LSTM and gated recurrent unit neural networks were obtained for the Juliusruh station in 2015: MAE = 1.5 TECU, RMSE = 1.9 TECU, MAPE = 17% for GRU, MAE = 1.39 TECU, RMSE = 1.85 TECU, and MAPE = 14% for LSTM.

In this paper, a set of multilayer neural network architectures, including both LSTM and GRU neural networks and temporal convolutional networks, are proposed and considered. The main aim of this paper is to complement them with bidirectional architectures and to determine the method that provides the highest prediction accuracy in the European region.

In Section 2, experimental data is described and the behavior of the basic indices of solar and geomagnetic activity and the total electron content is illustrated for the chosen years. Additionally, Section 2 presents the neural networks-based forecast methods. The results of the proposed methods are given in Section 3. Section 4 contains a discussion of the results. Section 5 provides the conclusions.

2. Materials and Methods

In this section, information about Materials is given in Section 2.1. Information about Methods is presented in Section 2.2, where a detailed description, including diagrams and formulas, for each method is provided in the corresponding partitions.

2.1. Experimental Data and TEC Behavior

The comparison of the forecasting accuracy of TEC using recurrent and convolutional neural networks, including conventional and bidirectional implementations, was carried out for 3 selected years with significantly different geomagnetic conditions and solar activity levels. The values of global JPL GIM-TEC maps were calculated from IONEX files with a time step of 2 h (https://urs.earthdata.nasa.gov (accessed on 10 February 2023)) for Juliusruh (54.6°N, 14.6°E), Murmansk (69°N, 33°E), Moscow (55.6°N, 37.2°E), and Nicosia (35.1°N, 33.2°E) for three years 2015, 2020, and 2022. The data on the indices of solar and geomagnetic activity, Dst, F10.7, proton density Np, planetary 10 Kp index, and speed of solar wind Vsw, was taken from SPDF OMNIWeb Service (http://omniweb.gsfc.nasa.gov/form/dx1.html (accessed on 10 February 2023)).

The state of the ionosphere and the dynamics of the total electron content depend on the level of solar and geomagnetic activity. The behavior of the basic indices influencing the TEC changes is given in Figure 1 for the years chosen for the analysis: 2015 (near to a maximum of 24 solar cycle), 2020 (a minimum of 24 cycle and transition to 25 cycle), and 2022 (an ascending branch of 25 cycle).

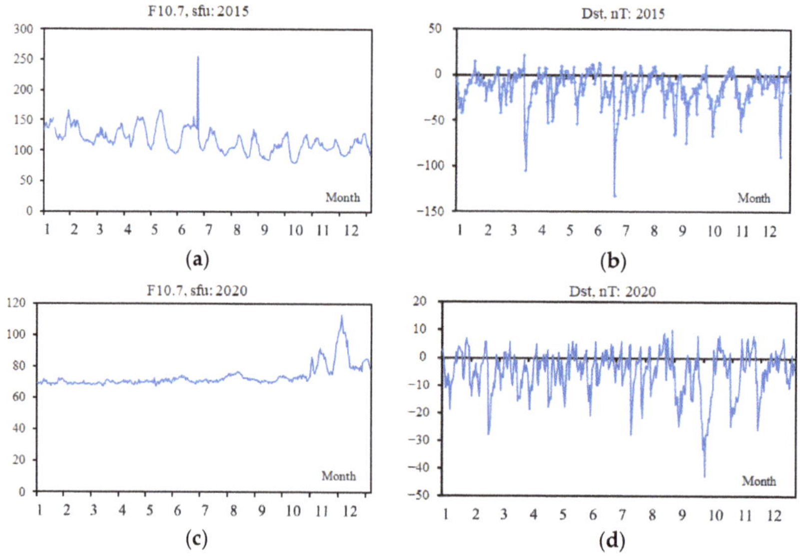

Figure 1. *Cont.*

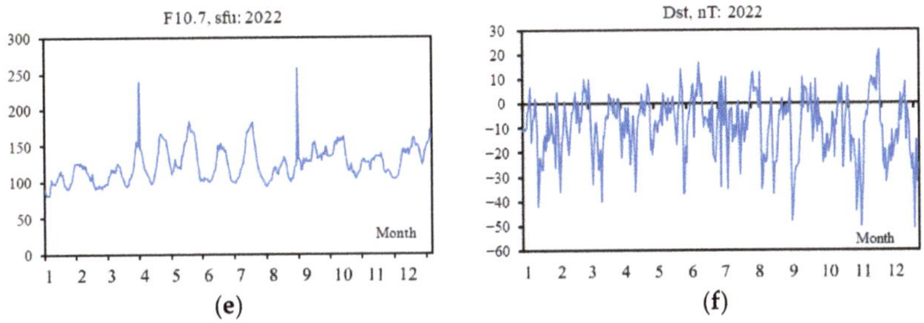

Figure 1. Average daily values of F10.7 and Dst indices for different years: (**a**,**b**) 2015; (**c**,**d**) 2020; (**e**,**f**) 2022.

There is a clear downward trend in F10.7 in 2015, a persistence in 2020 with a rather sharp increase at the end of the year, and an increase in 2022. The maximum disturbance was observed in 2015 and the minimum in 2020.

Figure 2 shows the median values, TEC(med), for the Juliusruh station (left panels) and stations along the 30°E meridian (right panels).

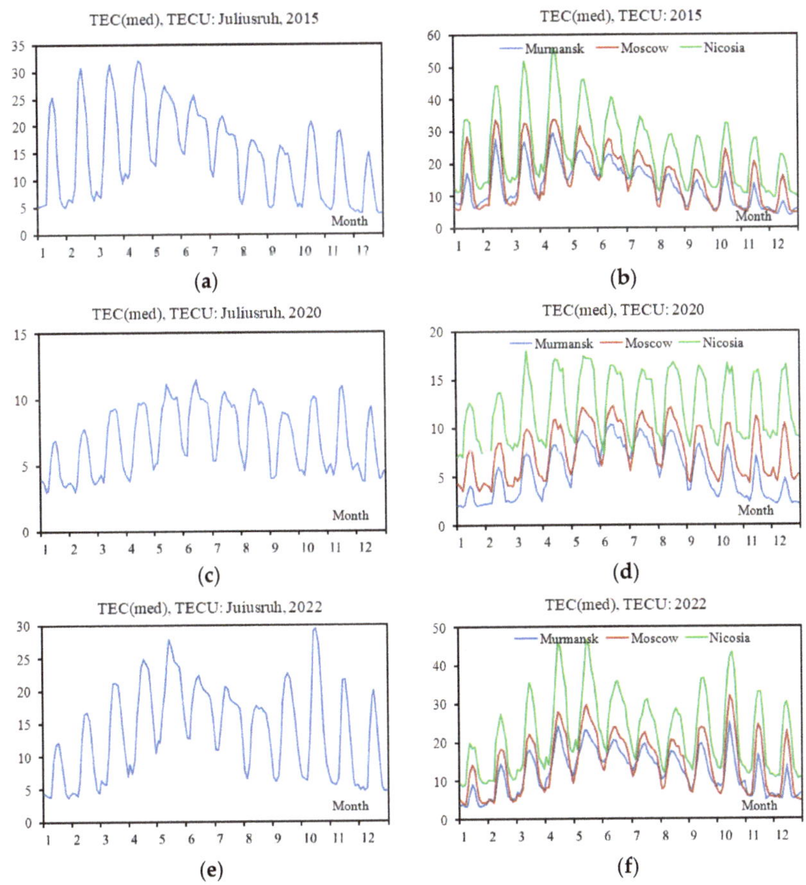

Figure 2. Variations of medians, TEC(med), for the reference station, Juliusruh, and stations along the meridian 30°N: (**a**,**b**) 2015; (**c**,**d**) 2020; (**e**,**f**) 2022.

The year 2015 is on the descending branch of the 24th cycle, which determines higher TEC values in the first half of the year compared to the second half. The year 2020 falls on the minimum of solar activity and is characterized by a more uniform seasonal course and 2–3 times lower maximum values compared to the year 2015. The year 2022 is on the rising branch of cycle 25, and the seasonal dependence differs from the first two cases both in the dynamics of change and in the maximum values of TEC. For stations along the meridian 30°E, this trend is preserved with the corresponding ratio between the maximum values, and this will be reflected in the MAE and RMSE values for different forecast architectures. The dynamics of TEC(med) for the Juliusruh station shows trends determined by the course of the F10.7 index. A comparison of the dynamics of TEC(med) along the meridian shows that for any considered level of solar activity, TEC predominantly increases with decreasing latitude. As noted above, many papers show that taking indices into account allows an increase of the accuracy of TEC prediction. Calculation of the correlation coefficients of TEC with these indices showed that in this case, the most significant were the Dst and solar wind speed Vsw indices. The correlation coefficients are shown in Figure 3.

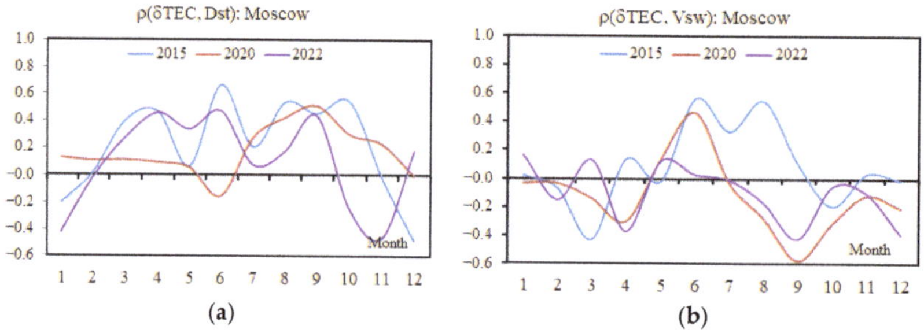

Figure 3. Behavior of TEC correlation coefficients with indices which make the main contribution to TEC change for the Juliusruh station: (**a**) Dst, (**b**) Vsw.

The most commonly used indices are F10.7 and Dst. In this work, indices F10.7, Dst, 10 Kp, Np, and Vsw were used separately and together. It can be seen that these two factors can influence in phase or in antiphase. Figures 4 and 5 show that the influence of Dst, which led to a negative response, was stronger than that of Vsw for June 2015.

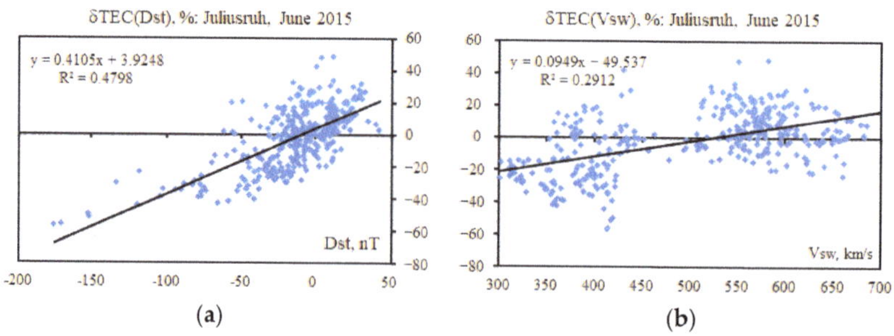

Figure 4. The scatter plot of changes in the relative deviations of TEC from the median (δTEC) with indices: (**a**) Dst, (**b**) Vsw. Regression equations and squares of correlation coefficients are shown inside the figures.

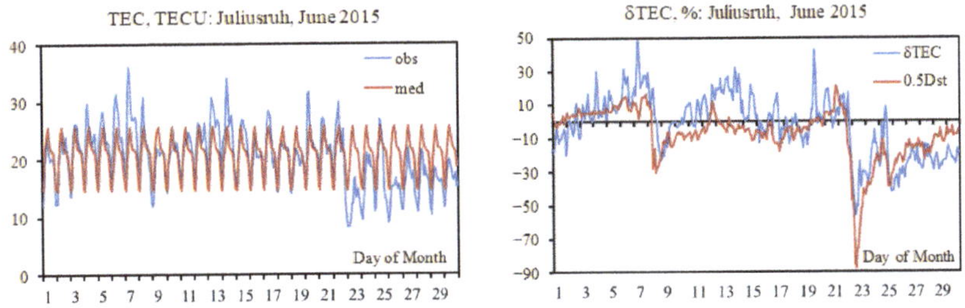

(a) (b)

Figure 5. Illustration of negative disturbances during periods of large negative values of the Dst index: (**a**) influence on behavior of TEC, (**b**) response δTEC.

2.2. Neural Networks Architectures for TEC Forecasting

In this paper, a set of architectures of multilayer neural networks based on GRU, LSTM, and temporal convolution are proposed and considered. The proposed non-bidirectional architectures are first considered in comparison with each other and are then subjected to further modification using bidirectional processing. After that, a comparison is made of both bidirectional architectures with each other and with the original options. This section describes these modifications in detail.

2.2.1. Data Preprocessing

For each year and each station, the data preparation procedure was as follows. Datasets for each year considered consisted of samples with a time step of 2 h for both TEC values and solar and geomagnetic activity indices. For indices that are provided only on a daily basis, daily values were used in 2 h increments throughout the respective day.

Each dataset was split into training, validation, and test subsets. The first 40% of the samples from yearly datasets were used as training data, the next 10% were used for validation, and the last 50% of the samples corresponding to the second half of the year were used for testing. TEC samples are organized into subsequences formed by the applied sliding window with a width equal to 12 samples. All the architectures in this paper are considered for two sets of indices of solar and geomagnetic activity. One set included only Np, 10 Kp, Dst, and F10.7; the second set included Np, 10 Kp, Dst, F10.7, and Vsw. One of the tasks of this work was to determine the possibility of improving the accuracy of TEC values' prediction by including Vsw in the training set due to a high correlation with TEC in this case (Figure 3). The indices are organized into individual sequence features, but no sliding window is applied. Each index array is shifted so the TEC value at time step t was predicted using the index at time step $t-1$. Thus, the TEC value at the time step t is predicted using 12 previous TEC values and a single previous value of the respective included indices. Data arrays are truncated appropriately to maintain the same array lengths for all features.

2.2.2. Recurrent Neural Networks

Among the variety of different types of recurrent neural networks, neural networks based on the employment of LSTM and GRU cells are most widely used in solving problems of forecasting and modeling time series. Long short-term memory neural networks were developed to solve the long-term dependency problem and make it possible for neural networks to effectively learn information over long periods of time. This architecture was originally introduced by Hochreiter and Schmidhuber [6] and became one of the most widespread architectures that has found an application in almost every area. LSTM networks store the information about learned temporal dependencies between each element of the input sequences during its operation in the internal memory state cell. The main advantage of LSTM is the ability to handle longer time sequences compared to a conventional recurrent neural network, which faces the vanishing gradient problem.

The set of equations defining the LSTM cell is given as follows:

$$i_t = \sigma(w_i \cdot [h_{t-1}, x_t] + b_i), \tag{1}$$

$$f_t = \sigma\left(w_f \cdot [h_{t-1}, x_t] + b_f\right), \tag{2}$$

$$o_t = \sigma(w_o \cdot [h_{t-1}, x_t] + b_o), \tag{3}$$

$$\widetilde{C}_t = tanh(w_C \cdot [h_{t-1}, x_t] + b_C), \tag{4}$$

$$C_t = f_t \odot C_{t-1} + i_t \odot \widetilde{C}_t, \tag{5}$$

$$h_t = o_t \odot tanh(C_t), \tag{6}$$

where i_t represents the input gate, f_t is the forget gate, o_t is the output gate, σ and $tanh$ are the sigmoid and hyperbolic tangent activation functions, respectively, C_t is the internal memory state, $\widetilde{C}_t$ represents the cell state candidate, C_{t-1} is the state at a timestamp $t-1$, h_{t-1} is the output of the previous LSTM cell at a timestamp $t-1$, h_t is the output, x_t is the input, w_C, w_i, w_f, and w_o are weights for the respective gates, and b_C, b_i, b_f, and b_o are biases. The $\odot$ symbol denotes the Hadamard product.

An alternative solution to the problem of learning long-term dependencies is gated recurrent unit-based neural networks, originally introduced by Cho et al. [7]. The gated recurrent unit is a simplified modification of the LSTM cell that does not store the internal state of the cell but instead uses an update gate and a reset gate to control the progress of the data transformation.

Data processing using GRU cells is determined by the following equations:

$$z_t = \sigma(w_z \cdot [h_{t-1}, x_t] + b_z) \tag{7}$$

$$r_t = \sigma(w_r \cdot [h_{t-1}, x_t] + b_r) \tag{8}$$

$$\widetilde{h}_t = tanh(w_h \cdot [r_t \odot h_{t-1}, x_t] + b_h) \tag{9}$$

$$h_t = (1 - z_t) \odot h_{t-1} + z_t \odot \widetilde{h}_t \tag{10}$$

where z_t represents the update gate, r_t is the reset gate, $\widetilde{h}_t$ is the output candidate, w_z, w_r, and w_h are weights for the respective gates, and b_z, b_r, and b_h are biases.

The structures of individual cells that make up the neural networks of the corresponding type are shown in Figure 6.

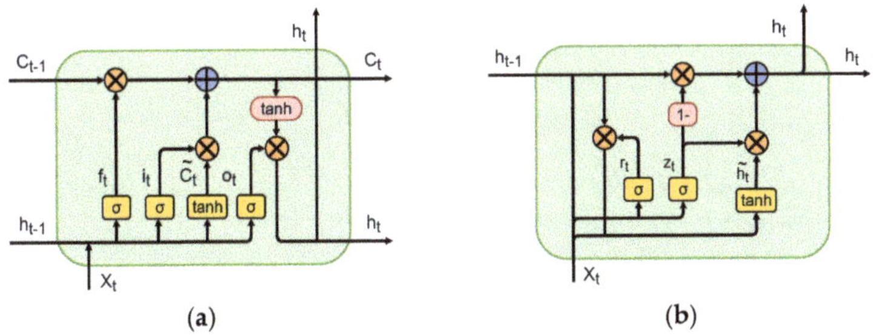

Figure 6. Schematic diagrams of the internal structure of cells: (**a**) LSTM, (**b**) GRU.

2.2.3. Temporal Convolutional Neural Networks

Convolutional neural networks have become widespread in solving image processing problems and have also been successfully applied to time series modeling. It has been shown that convolutional neural networks are able to give results comparable to recurrent neural networks or even surpass them. At the same time, they have more flexibility in meta-parameters management and also have a higher performance in terms of computation time.

The temporal convolutional network consists of a set of dilated causal convolution layers that are processing time steps of each input sequence.

The common approach for 1 d temporal convolution implementation is using a stack of consecutive residual blocks including several dilated convolutional layers and an arbitrary selection of utility, normalization, and activation layers, supplemented by an optional convolution skip connection.

Assuming that d is the dilation factor, x is the input sequence, s is the element of the input sequence, f is the applied filter, k is the filter size, and $s - d{\cdot}i$ represents the dilation in the backward direction, the 1D dilated convolution can be represented as follows:

$$F(s) = (x *_d f)(s) = \sum_{i=0}^{k-1} f(i){\cdot}x_{s-d{\cdot}i}. \quad (11)$$

Depending on the dilation factor, this equation can be reduced either to the usual convolution when $d = 1$ or can be used to increase the receptive field of a neural network by introducing different time steps between filter passes.

2.2.4. Bidirectional Neural Networks

Bidirectional recurrent neural networks are a modification of conventional recurrent neural networks designed to improve their learning capabilities. This architecture is widely used in tasks where the input data has a hard-to-determine or complex time dependence as well as in solving context-sensitive problems or when establishing a connection in a series of consequential events.

While conventional LSTM and GRU neural networks are processing data in one direction, learning the temporal dependencies according to the input sequence order, their learning capabilities can be further enhanced by introducing the bidirectional architecture. Bidirectional RNN consists of cells that are arranged into two layers, one of which processes the original sample in the forward direction and the second one, in the backward direction. The outputs of each of the intermediate layers are concatenated to form a combined output. This allows the network to determine the temporal dependencies in the input information from both the past and the future during the learning process, rather than simply relying on inputs corresponding to the current time step or previous steps. A similar data processing technique can be applied to convolutional neural networks. The schematic diagram of data processing in a bidirectional recurrent neural network is shown in Figure 7.

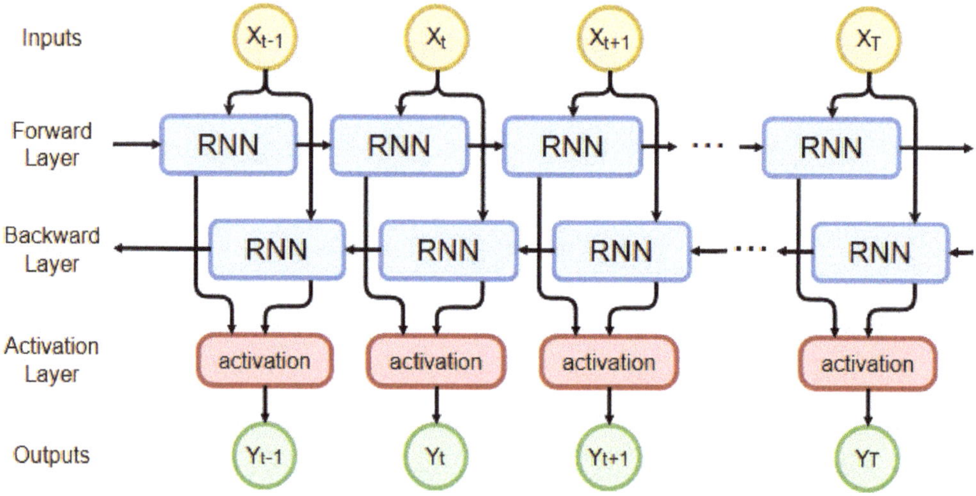

Figure 7. Schematic diagram of the bidirectional recurrent neural network.

2.2.5. The Proposed Deep Neural Networks for TEC Forecasting

The architectures considered in the work can be divided into two main categories—recurrent neural networks and temporal convolutional neural networks. Each of these categories, in turn, is further subdivided into conventional and bidirectional. Hereinafter, the following notion will be employed: the prefix "Bi" denotes bidirectional architecture; for recurrent neural networks, the postfix "RFF" or "FRF" is used to indicate the recurrent layer position in the layer stack, defining the line of architectures. The postfix of the abbreviation "RFF" denotes a recurrent layer preceding two fully connected layers, and the ending "FRF" corresponds to a recurrent layer enclosed between two fully connected layers. Activation functions and normalization layers are not noted in the abbreviations. The "LSTM" or "GRU" sub-prefix specifies whether the long short-term memory or gated recurrent unit layer is used as a recurrent layer; temporal convolutional networks are marked using the "TCN" postfix.

During the first stage of this research, the development of RFF and FRF multilayer architectures was undertaken. The architectures of the RFF line include the recurrent layer preceding two fully connected layers, while the FRF-line architectures consist of the recurrent layer enclosed between the two fully connected layers. The schematic diagrams of the proposed recurrent architectures are shown in Figure 8. The value of N determines the number of cells in each respective layer. The recurrent layer included in the architecture differs in each individual case and is defined either as LSTM, or as GRU, or as BiLSTM, or as BiGRU. These layers were followed by a parametric rectified linear unit (PReLU) activation function with the slope parameter specified using parentheses. PreLU multiplies the negative input values by the given parameter. For RFF architectures, the batch normalization layer was included to perform interlayer input normalization across mini-batches and to stabilize the learning process. The same approach applied to the FRF architecture provided no positive feedback, thus the batch normalization layer is not included.

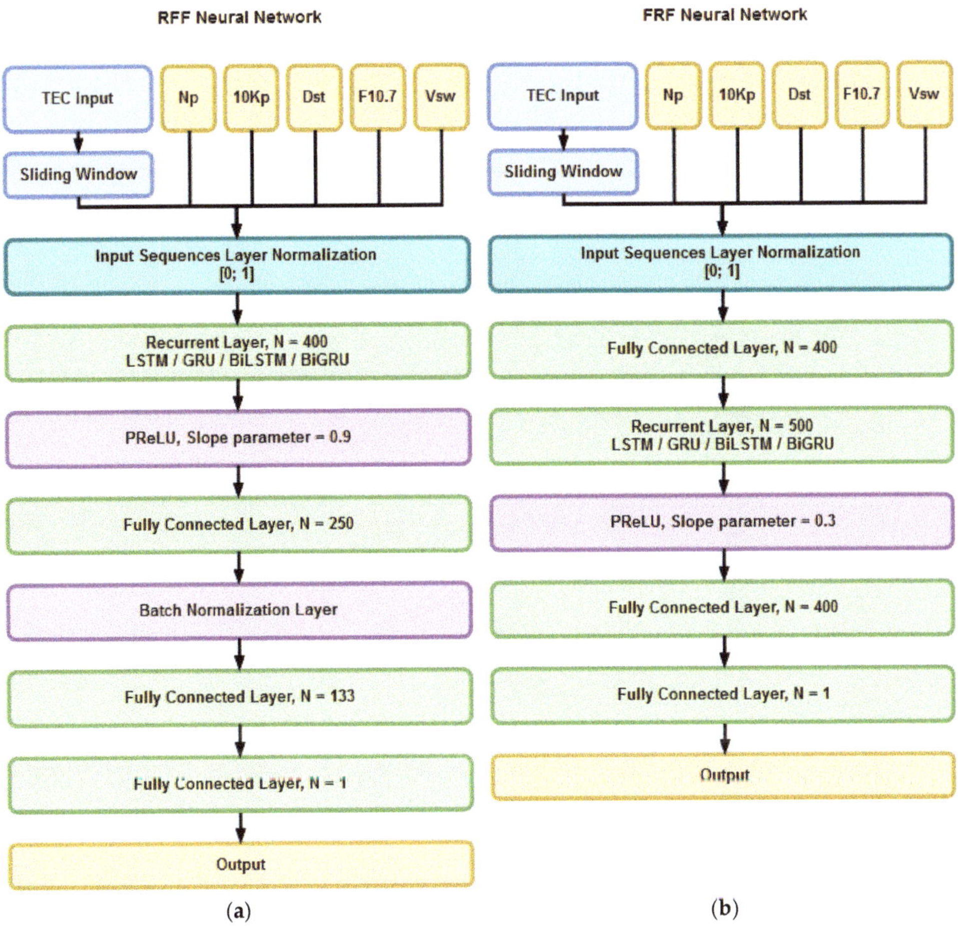

Figure 8. Schematic diagrams of the proposed multilayer recurrent neural network architectures: (**a**) RFF, (**b**) FRF. The selected recurrent layer and indices vary in each specific case.

Weight initialization in fully connected layers is performed using the Glorot initializer [25]. Input and recurrent weights for recurrent layers were initialized using an orthogonal matrix decomposition initializer [26]. For recurrent layers, the tanh activation function is used for state activation and sigmoid is used for gate activation, and each recurrent layer is followed by the PreLU activation function. The schematic diagram of the proposed temporal convolutional network is presented in Figure 9.

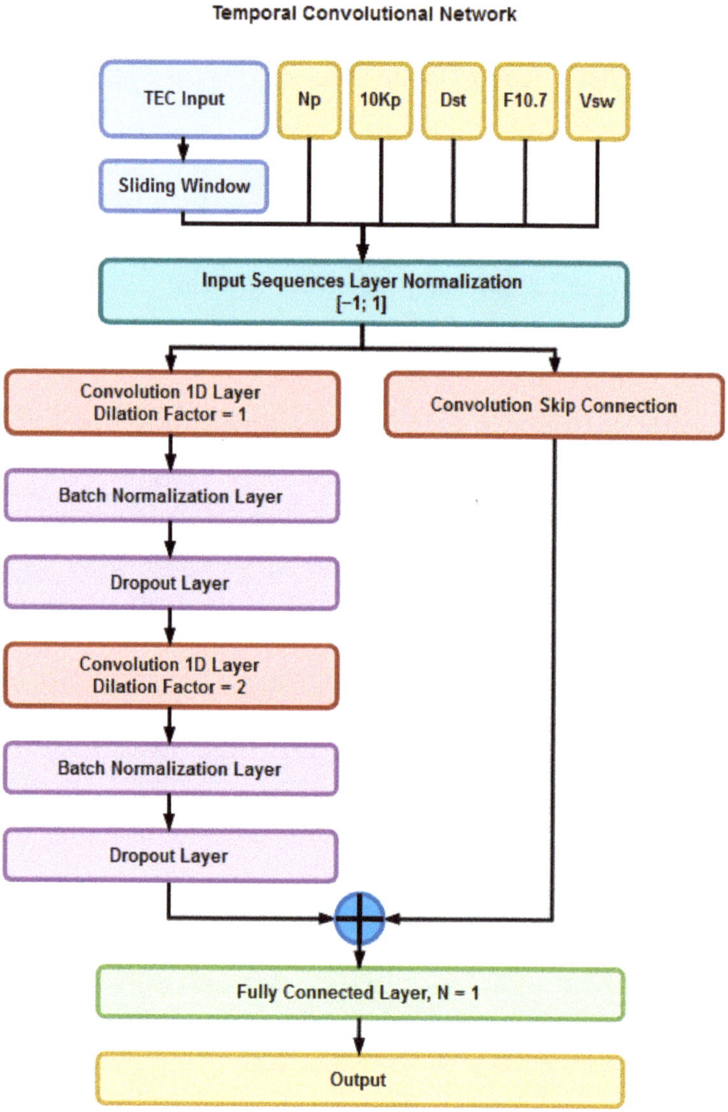

Figure 9. Schematic diagram of the proposed temporal convolutional network.

For temporal convolutional networks, the number of filters was equal to 256 with a filter kernel size of 2. Two dropout layers with a dropout factor of 0.005 were used in the main branch. The outputs are connected using addition. Per-channel symmetric rescaling was used for input data normalization so that each feature was in the range of $[-1; 1]$. For conventional temporal convolutional networks, the optional convolution skip connection was used. For the bidirectional temporal convolutional network, the convolution skip layer was connected in parallel to the layers forming the bidirectional part. The schematic diagram of the bidirectional temporal convolutional network is presented in Figure 10.

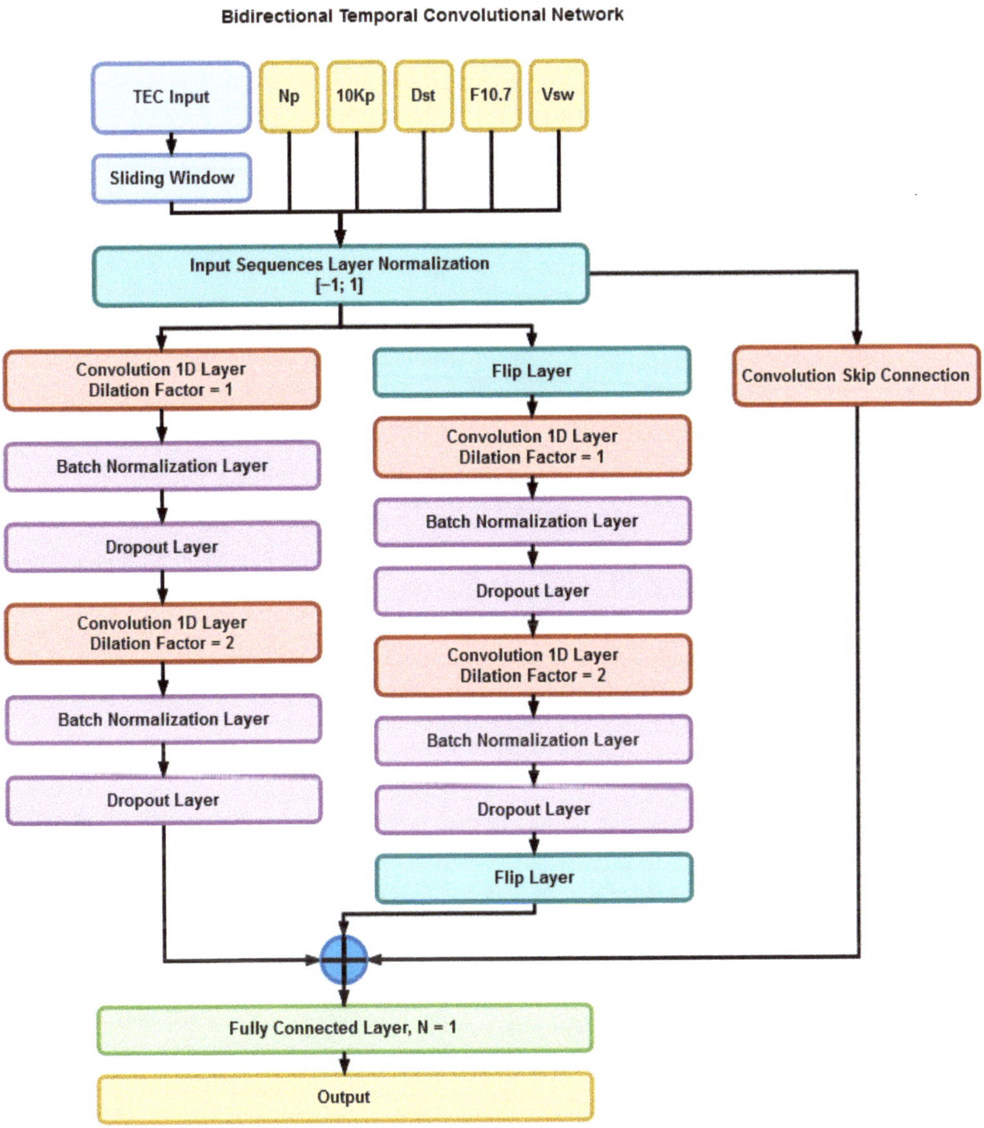

Figure 10. Schematic diagram of the proposed bidirectional temporal convolutional network.

The training was conducted using MATLAB environment and Deep Learning and Parallel Computing Toolboxes. In all the considered cases, both for bidirectional and conventional architectures, the Adam optimizer with the following parameters was used during the training:

GradientDecayFactor: 0.9000;
Squared GradientDecay Factor: 0.9990;
Epsilon: 1×10^{-8};
InitialLearnRate: 0.0030;
Learn RateSchedule: 'none';
LearnRateDropFactor: 0.1000;

LearnRateDropPeriod: 10;
L2Regularization: 0.0001;
GradientThresholdMethod: 'l2norm';
GradientThreshold: Inf;
MaxEpochs: 100;
MiniBatch Size: 1;
ValidationFrequency: 20;
ValidationPatience: Inf;
Shuffle: 'every-epoch';
ExecutionEnvironment: 'gpu';
Sequence Length: 'longest';
SequencePaddingValue: 0;
SequencePaddingDirection: 'left';
ResetInputNormalization: 0.

The mean squared error (MSE) was chosen as a loss function intended for training neural networks, minimizing the difference between real and predicted values during the learning process:

$$loss = \frac{1}{n}\sum_{i=1}^{n}(\hat{y}_i - y_i)^2, \tag{12}$$

here and below, $\hat{y}_i$ stands for the predicted values and y_i stands for the observed values, and n is the number of samples.

2.2.6. Statistical Metrics Adopted in the Validation Process

To compare the accuracy of forecasting TEC values using different architectures, a set of statistical metrics was adopted, including Mean Absolute Error, Mean Absolute Percentage Error, root mean square error, and Pearson's correlation coefficient.

The computation is performed using the following equations:

$$MAE = \frac{1}{n}\sum_{i=1}^{n}|\hat{y}_i - y_i|, \tag{13}$$

$$MAPE = \frac{1}{n}\sum_{i=1}^{n}\left|\frac{\hat{y}_i - y_i}{y_i}\right| \times 100\%, \tag{14}$$

$$RMSE = \sqrt{\sum_{i=1}^{n}\frac{(\hat{y}_i - y_i)^2}{n}}, \tag{15}$$

$$\rho = \frac{cov(y, \hat{y})}{\sigma_y \cdot \sigma_{\hat{y}}}, \tag{16}$$

where n is the number of samples, $cov(y, \hat{y})$ is the covariance between the predicted and observed values, σ_y is the standard deviation of the observed values, and $\sigma_{\hat{y}}$ is the standard deviation of the predicted values.

3. Results

The statistical results on accuracy of TEC forecasting for a set of 10 considered architectures and the two sets of indices (including and excluding Vsw) are presented in Table 1.

Table 1. Forecasting results for Juliusruh station in 2015.

	Architecture and Indices	MAE (TECU)	MAPE (%)	RMSE (TECU)	ρ
1	LSTM-FRF(10 Kp, Np, F10.7, Dst)	1.141	11.915	1.574	0.967
2	LSTM-FRF(10 Kp, Np, F10.7, Dst, Vsw)	1.150	12.156	1.594	0.967
3	LSTM-RFF(10 Kp, Np, F10.7, Dst)	1.187	12.309	1.623	0.969
4	LSTM-RFF(10 Kp, Np, F10.7, Dst, Vsw)	1.207	12.301	1.669	0.967
5	GRU-FRF(10 Kp, Np, F10.7, Dst)	1.163	11.261	1.586	0.967
6	GRU-FRF(10 Kp, Np, F10.7, Dst, Vsw)	1.273	13.430	1.681	0.965
7	GRU-RFF(10 Kp, Np, F10.7, Dst)	1.152	11.890	1.569	0.971
8	GRU-RFF(10 Kp, Np, F10.7, Dst, Vsw)	1.199	12.521	1.634	0.968
9	TCN(10 Kp, Np, F10.7, Dst)	1.047	11.118	1.426	0.973
10	TCN(10 Kp, Np, F10.7, Dst, Vsw)	1.078	11.649	1.453	0.973
11	BiLSTM-FRF(10 Kp, Np, F10.7, Dst)	1.480	14.371	1.943	0.960
12	BiLSTM-FRF(10 Kp, Np, F10.7, Dst, Vsw)	1.308	14.423	1.686	0.966
13	BiLSTM-RFF(10 Kp, Np, F10.7, Dst)	0.583	6.420	0.797	0.992
14	BiLSTM-RFF(10 Kp, Np, F10.7, Dst, Vsw)	0.570	6.441	0.870	0.990
15	BiGRU-FRF(10 Kp, Np, F10.7, Dst)	0.621	6.952	0.800	0.992
16	BiGRU-FRF(10 Kp, Np, F10.7, Dst, Vsw)	0.835	9.497	1.067	0.986
17	BiGRU-RFF(10 Kp, Np, F10.7, Dst)	0.473	5.231	0.740	0.993
18	BiGRU-RFF(10 Kp, Np, F10.7, Dst, Vsw)	0.544	6.115	0.771	0.992
19	BiTCN(10 Kp, Np, F10.7, Dst)	0.191	2.615	0.555	0.996
20	BiTCN(10 Kp, Np, F10.7, Dst, Vsw)	0.161	2.192	0.507	0.997

The metrics considered were MAE, MAPE, RMSE, and ρ, averaged over half a year during single step 2 h ahead TEC forecasting for three years separately and four selected stations. The results for station Juliusruh are presented in Table 1 for 2015.

The results for every station considered are shown in Figure 11 as a dependence of MAE, MAPE, and RMSE on the architecture's variant number. Thus, the results are presented for the following architectures, indicated by numbers: (1) LSTM-FRF, (2) LSTM-RFF, (3) GRU-FRF, (4) GRU-RFF, (5) TCN, (6) BiLSTM-FRF, (7) BiLSTM-RFF, (8) BiGRU-FRF, (9) BiGRU-RFF, and (10) BiTCN. Due to the fact that there is no unambiguous trend in the influence of the Vsw index on the accuracy of the forecast, only the option with the best result (with and without Vsw) was chosen from the two options.

From Figure 11, it can be seen that among the prediction accuracy metrics, three sections can be distinguished with similar characteristics, depending on the architectures used. The first flat area (numbers 1–5) with the highest values for errors refers to traditional LSTM and GRU architectures with FRF and RFF modifications. The second section (numbers 6–8), which includes results for bidirectional networks, has a minor slope and error values are 1.5–2 times smaller compared to the first section. Finally, the third section, which includes BiGRU and BiTCN architectures, provides the lowest error values, i.e., the highest forecasting accuracy. It should be noted that although there is a certain latitude dependence of the statistical characteristics in the first two groups, for the BiTCN architecture, it practically ceases to exist.

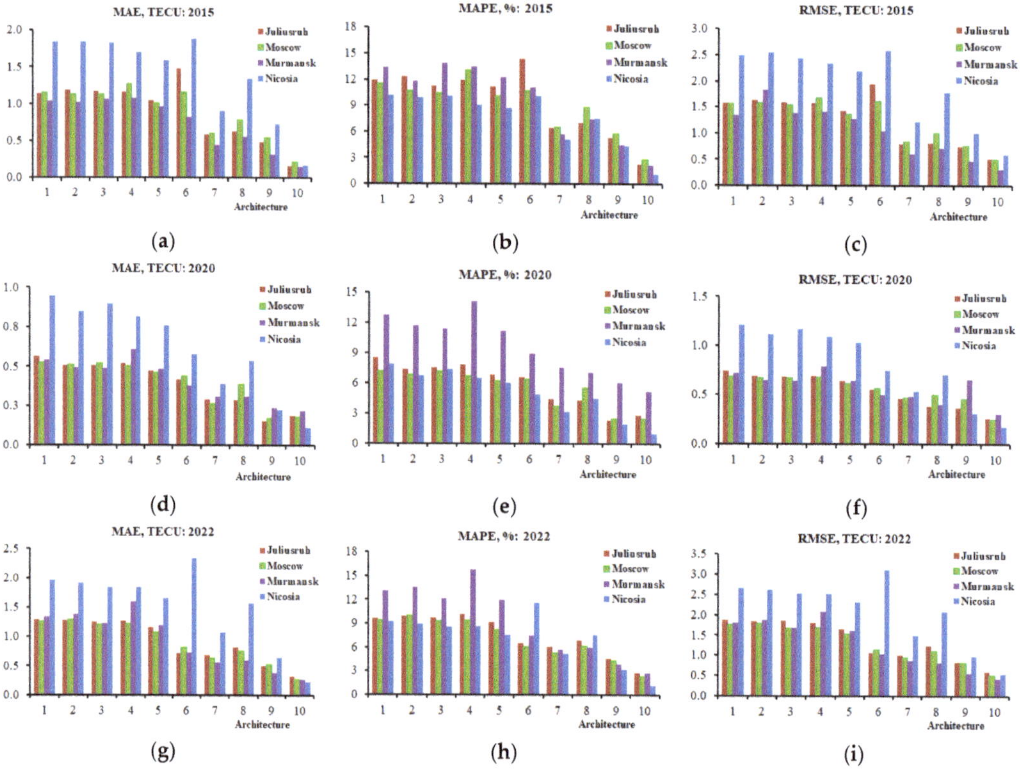

Figure 11. Statistical characteristics of TEC forecast in the European zone depending on a combination of architectures and indices for 2015 (**a**–**c**), 2020 (**d**–**f**), and 2022 (**g**–**i**).

It can be seen that MAPE and RMSE are related to the absolute values of TEC, depending on the level of solar activity. Conventional LSTM and GRU with FRF and RFF modifications and conventional TCN provide results at the same level of accuracy. The usage of bidirectional architecture essentially improves the accuracy of the forecast for almost all architectures and across all stations. It should be noted that during high solar activity, bidirectional architectures of the FRF line (BiLSTM-FRF and BiGRU-FRF) do not improve the accuracy of the forecast for a low-latitude station even compared to conventional methods. The highest accuracy is achieved using BiGRU-RFF and BiTCN architectures: MAPE values are less than 0.6 TECU, RMSE values are less than 0.8 TECU, and MAPE values are less than 7.5%. The best forecasting accuracy is achieved using BiTCN architecture, with less than 0.3 TECU for MAE, less than 0.6 TECU for RMSE, and less than 5% for MAPE.

The forecast errors during the second half of each year for the Moscow station are shown in Figure 12. Each panel for the 2015, 2020, and 2022 years shows the variations of TEC between DoY 185 and 365 along with the Dst index and compares the prediction accuracy for two architectures: LSTM-FRF and BiTCN. All the cases are provided with the Vsw index included in the training data (10 Kp, Np, F10.7, Dst, and Vsw).

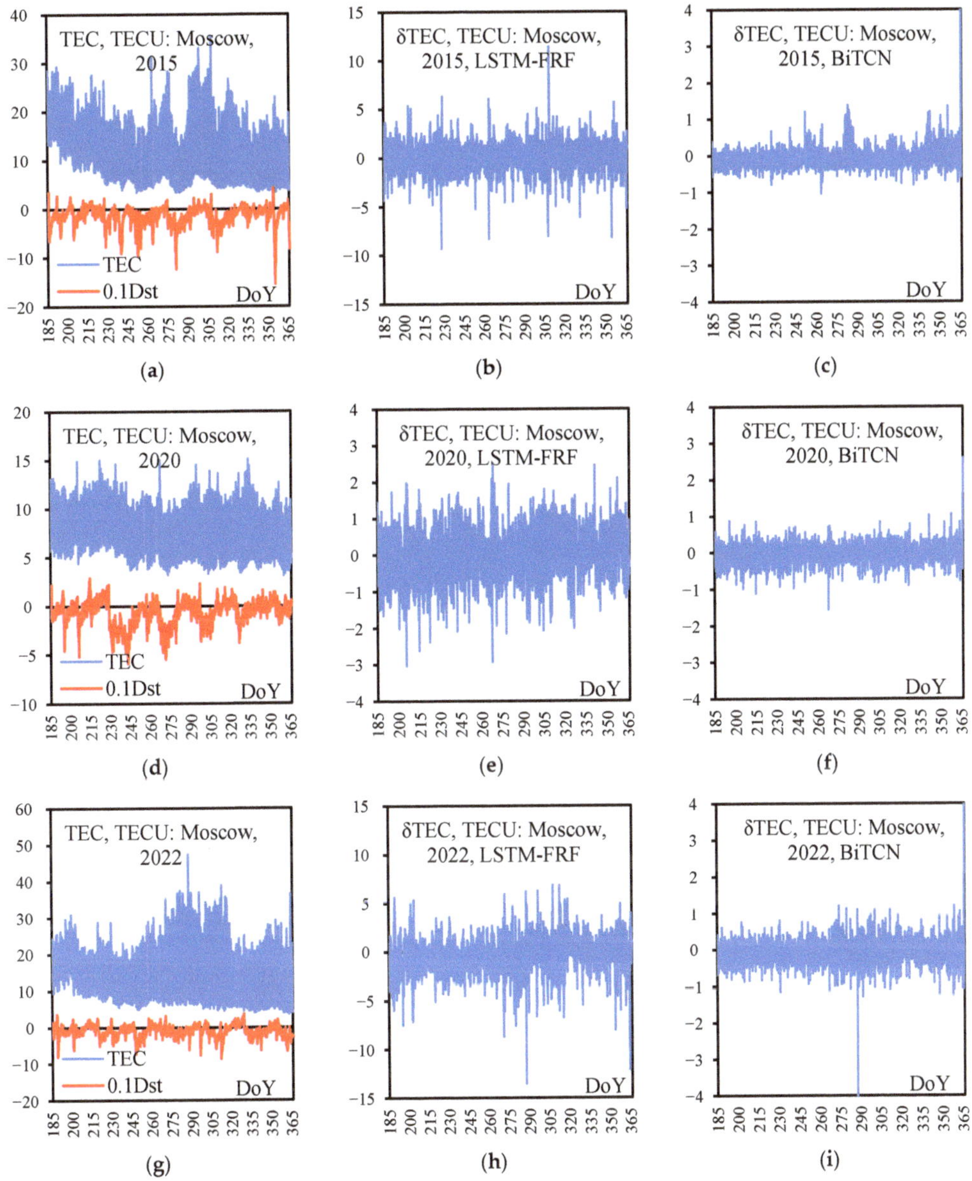

Figure 12. Comparison of the prediction accuracy for two architectures—LSTM-FRF and BiTCN: 2015 (**a**–**c**), 2020 (**d**–**f**), and 2022 (**g**–**i**).

Figure 12 shows the instantaneous TEC and δTEC values for the two selected architectures from among the unidirectional and bidirectional ones, using the Moscow station as an example. Unlike the median values, the TEC values (plots a, d, and g on the left) show the response to disturbed conditions. The average δTEC values for the first and third plots (b,c) are 1.15 TECU and 0.22 TECU in 2015, an improvement of ~5 times can be observed. In 2020 (Figure 12e,f), these values were 0.52 TECU and 0.2 TECU, i.e., an improvement of ~2.5 times, and for the year 2022 (Figure 12h,i), higher values of 1.32 TECU and 0.27 TECU were obtained and an improvement of ~5 times was achieved. Regardless of the year, the accuracy for BiTCN is 2–5 times higher than for LSTM-FRF: the error is in the range of ±1 TECU. The δTEC outlier in the last panel refers to 14 October 2022, when a positive disturbance occurred during the main phase of the magnetic storm (minimum Dst = −62 nT) due to a sharp increase in Vsw, which was minimal and equal to 300 km/s in UT0 on 14 October and gradually increased to 600 km/s in UT0 on 16 October. It is interesting to note that during the main phase of the second magnetic storm (minimum Dst = 76 nT), there was also a positive deviation δTEC, but of less intensity, since the velocity Vsw was constant, 355 km/s.

4. Discussion

In this paper, a set of architectures of multilayer neural networks based on GRU, LSTM, and temporal convolution is proposed and considered. The implementation of the RFF and FRF architectures improves the results compared to [24]. The further modification of those architectures using bidirectional processing led to a 2-fold improvement in the forecast accuracy. The greatest improvement in accuracy of the forecast was achieved with the employment of a bidirectional temporal convolutional network architecture, though its initial variant provides results at a level comparable to traditional methods. Its bidirectional enhancement allows one to increase the accuracy of the forecast by 5–10 times. The observed dependence of the results on solar activity, a combination of indices, and the latitude shows that to determine the best method in each region, it is necessary to examine not only a set of architectures but also indices, as evidenced by the example with the Vsw index.

It is important to note that the number of cells in bidirectional layers is two times higher than in unidirectional ones, but at the same time, a twofold increase in the number of cells in unidirectional networks did not lead to an increase in prediction accuracy but, on the contrary, reduced it. This result is due to overfitting. The bidirectional neural network includes two separate backpropagation passes: one is for the forward direction of inputs, and the second one is for the reverse direction. During the forward pass, the forward layer processes the input sequence in the original input direction and makes predictions for the output sequence. The backward layer processes the input sequence in reverse order during the backward pass and predicts the output sequence. When both passes are completed, their predictions are compared to the target output sequences and the error is propagated back to each respective layer through the network to update the weights. The forward and backward layer weights are updated based on the errors computed during the forward and backward passes, respectively. This process allows you to optimize the weights on a double amount of data—the original inputs and reverse inputs. For each of the layers of a bidirectional network, optimization is carried out on its own subspace of parameters.

As for the comparison with the results of other works, as noted in the introduction, in many works, only MAE and RSME are mentioned as statistical characteristics, but MAPE is not indicated. The results obtained show that this is not sufficient. This is especially evident from the results for the Nicosia station, which has the highest TEC values. This can be demonstrated by the example of [23], where RMSE ~0.5 TECU were obtained without specifying MAPEs. In this work, even smaller RMSE values were obtained regardless of latitude, indicating that the capabilities of the LSTM, GRU, and TCN methods for TEC prediction have not been exhausted.

Apart from the differences in accuracy, it is important to note that significant discrepancies between the architectures in terms of computational costs are also observed. For bidirectional architectures, the computational costs are significantly higher than for unidirectional ones. In particular, among unidirectional architectures, for the FRF line of architectures, the training time was about 11 s, for the RFF line, the training time was about 9 s, and for the TCN, it was 2 s on average. The difference between the LSTM and GRU variants did not exceed 2 s on average, and the GRU-based architectures were converging at a higher rate. With the transition to bidirectional architectures, the training time for FRF and RFF architectures increased by more than 2 times and amounted to 27 and 22 s on average. For BiTCN, the training time increased to 4 s. The training was performed using a single RTX4080 and AMD Ryzen 9 7900X.

It is important to note that the training was carried out on a relatively small amount of data that did not contain complete information about the annual behavior of TEC and the indices used; nevertheless, the proposed architectures of bidirectional networks completed the forecasting task with consistently high accuracy.

5. Conclusions

This work was devoted to finding a combination of the latest neural network architectures, which, under specific conditions of TEC variations over three years, according to data from four European stations can provide the most accurate forecast with an advance of 2 h. Of the three selected years, 2015 belonged to the phase of solar activity decline in cycle 24, 2020 had minimal activity during the transition from one cycle to another, and 2022 came to the phase of solar activity increase in cycle 25. According to the annual mean values of TEC and Dst, the year 2022 had higher solar activity than 2015 but lower geomagnetic activity. It is shown that the maximum values of TEC change in accordance with the change in solar activity, which leads to the same trend of changes in the statistical characteristics of the forecast, i.e., the level of solar activity played a greater role than the geomagnetic activity.

The use of solar and geomagnetic activity indices is known to affect the accuracy of the TEC forecast; therefore, when choosing the most effective index, estimates were made for a wide set. However, unlike the results presented in [24], a significant contribution of the Vsw index was observed and preliminary calculations were carried out for all architectures for the two sets of parameters, including and excluding the Vsw index as an input parameter. The results for these cases differed markedly; therefore, out of 20 options, only 10 architectures with the best statistical characteristics (with or without Vsw) were selected for presentation. It is significant that among the 10 presented architectures, there are groups of them that provide the same accuracy. So, the traditional LSTM and GRU methods gave estimates [24] at a level corresponding to the literature data. Modifications of the FRF and RFF architecture lines led to a minor increase in accuracy. The use of bidirectional models led to an additional increase in accuracy by a factor of 1.5–2 as compared to these methods. The next group, including BiGRU and BiTCN architectures, provided the lowest errors, i.e., the highest forecasting accuracy. The error estimates for the BiTCN architecture were the best, with less than 0.3 TECU for MAE, less than 0.6 TECU for RMSE, and less than 5% for MAPE.

Thus, it was found that the combined use of bidirectional and TCN architectures significantly increases the accuracy of the forecast and levels out the latitudinal dependence. It is planned to apply this approach to 24 h ahead forecasting.

Author Contributions: Conceptualization, O.M.; methodology, A.K. and O.M.; software, A.K.; validation, O.M.; formal analysis, A.K. and O.M.; investigation, A.K. and O.M.; resources, A.K.; data curation, O.M.; writing—original draft preparation, A.K. and O.M.; writing—review and editing, A.K. and O.M. All authors have read and agreed to the published version of the manuscript.

Funding: This research was financially supported by the Ministry of Science and Higher Education of the Russian Federation (State task in the field of scientific activity 2023).

Data Availability Statement: The OMNI data (F10.7, Np, 10 Kp, Dst, and Vsw indices) were obtained from the GSFC/SPDF OMNIWeb interface at https://omniweb.gsfc.nasa.gov (accessed on 22 January 2022). The data on JPL, GIM, and TEC were obtained from IONEX files located at https://urs.earthdata.nasa.gov (accessed on 10 February 2023).

Acknowledgments: The TEC samples data was obtained from https://urs.earthdata.nasa.gov (accessed on 10 February 2023). Data on solar and geomagnetic activity was taken from SPDF OMNI Web Service http://omniweb.gsfc.nasa.gov/form/dx1.html (accessed on 10 February 2023).

Conflicts of Interest: The authors declare no conflict of interest.

References

1. McGranaghan, R.M.; Camporeale, E.; Georgoulis, M.; Anastasiadis, A. Space Weather research in the Digital Age and across the full data lifecycle: Introduction to the Topical Issue. *J. Space Weather Space Clim.* **2021**, *11*, 50. [CrossRef]
2. Goodman, J. Operational communication systems and relationships to the ionosphere and space weather. *Adv. Space Res.* **2005**, *36*, 2241–2252. [CrossRef]
3. Xie, T.; Dai, Z.; Zhu, X.; Chen, B.; Ran, C. LSTM-based short-term ionospheric TEC forecast model and positioning accuracy analysis. *GPS Solut.* **2023**, *27*, 66. [CrossRef]
4. Boulch, A.; Cherrier, N.; Castaings, T. Ionospheric activity prediction using convolutional recurrent neural networks. *arXiv* **2018**, arXiv:1810.1327312.
5. Yu, S.; Ma, J. Deep learning for geophysics: Current and future trends. *Rev. Geophys.* **2021**, *59*, e2021RG000742. [CrossRef]
6. Hochreiter, S.; Schmidhuber, J. Long Short-term Memory. *Neural Comput.* **1997**, *9*, 1735–1780. [CrossRef] [PubMed]
7. Cho, K.; van Merrienboer, B.; Gulcehre, C.; Bahdanau, D.; Bougares, F.; Schwenk, H.; Bengio, Y. Learning Phrase Representations using RNN Encoder-Decoder for Statistical Machine Translation. *arXiv* **2014**, arXiv:1406.1078v3.
8. Lei, D.; Liu, H.; Le, H.; Huang, J.; Yuan, J.; Li, L.; Wang, Y. Ionospheric TEC Prediction Base on Attentional BiGRU. *Atmosphere* **2022**, *13*, 1039. [CrossRef]
9. Lea, C.; Flynn, M.D.; Vidal, R.; Reiter, A.; Hager, G.D. Temporal Convolutional Networks for Action Segmentation and Detection. In Proceedings of the IEEE Conference on Computer Vision and Pattern Recognition, Honolulu, HI, USA, 21–26 July 2017; pp. 1003–1012. [CrossRef]
10. Sun, W.; Xu, L.; Huang, X.; Zhang, W.; Yuan, T.; Chen, Z.; Yan, Y. Forecasting of ionospheric vertical total electron content (TEC) using LSTM networks. In Proceedings of the 2017 International Conference on Machine Learning and Cybernetics, Ningbo, China, 9–12 July 2017.
11. Sun, W.; Xu, L.; Huang, X.; Zhang, W.; Yuan, T.; Yan, Y. Bidirectional LSTM for ionospheric vertical total electron content (TEC) forecasting. In Proceedings of the IEEE Visual Communications and Image Processing (VCIP), St. Petersburg, FL, USA, 10–13 December 2017; pp. 1–4.
12. Sivakrishna, K.; Venkata Ratnam, D.; Sivavaraprasad, G. A Bidirectional Deep-Learning Algorithm to Forecast Regional Ionospheric TEC Maps. *IEEE J. Sel. Top. Appl. Earth Obs. Remote Sens.* **2022**, *15*, 4531–4543. [CrossRef]
13. Chen, Z.; Liao, W.; Li, H.; Wang, J.; Deng, X.; Hong, S. Prediction of global ionospheric TEC based on deep learning. *Space Weather* **2022**, *20*, e2021SW002854. [CrossRef]
14. Tang, J.; Li, Y.; Ding, M.; Liu, H.; Yang, D.; Wu, X. An Ionospheric TEC Forecasting Model Based on a CNN-LSTM-Attention Mechanism Neural Network. *Remote Sens.* **2022**, *14*, 2433. [CrossRef]
15. Chunli, D.; Jinsong, P. Modeling and prediction of TEC in China region for satellite navigation. In Proceedings of the 15th Asia-Pacific Conference on Communications, Shanghai, China, 8–10 October 2009; pp. 310–313.
16. Huang, Z.; Yuan, H. Ionospheric single-station TEC short-term forecast using RBF neural network. *Radio Sci.* **2014**, *49*, 283–292.
17. Niu, R.; Guo, C.; Zhang, Y.; He, L.; Mao, Y. Study of ionospheric TEC short-term forecast model based on combination method. In Proceedings of the 2014 12th International Conference on Signal Processing (ICSP), Hangzhou, China, 19–23 October 2014; pp. 2426–2430.
18. Iluore, K.; Lu, J. Long short-term memory and gated recurrent neural networks to predict the ionospheric vertical total electron content. *Adv. Space Res.* **2022**, *70*, 652–665. [CrossRef]
19. Kaselimi, M.; Voulodimos, A.; Doulamis, N.; Doulamis, A.; Delikaraoglou, D. Deep Recurrent Neural Networks for Ionospheric Variations Estimation Using GNSS Measurements. *IEEE Trans. Geosci. Remote Sens.* **2021**, *60*, 1–15. [CrossRef]
20. Kaselimi, M.; Voulodimos, A.; Doulamis, N.; Doulamis, A.; Delikaraoglou, D. A causal long short-term memory sequence to sequence model for TEC prediction using GNSS observations. *Remote Sens.* **2020**, *12*, 1354. [CrossRef]
21. Tang, R.; Zeng, F.; Chen, Z.; Wang, J.-S.; Huang, C.-M.; Wu, Z. The comparison of predicting storm-time ionospheric TEC by three methods: ARIMA, LSTM, and Seq2Seq. *Atmosphere* **2020**, *11*, 316. [CrossRef]
22. Cesaroni, C.; Spogli, L.; Aragon-Angel, A.; Fiocca, M.; Dear, V.; De Franceschi, G.; Romano, V. Neural network based model for global Total Electron Content forecasting. *J. Space Weather Space Clim.* **2020**, *10*, 11. [CrossRef]
23. Natras, R.; Soja, B.; Schmidt, M. Ensemble Machine Learning of Random Forest, AdaBoost and XGBoost for Vertical Total Electron Content Forecasting. *Remote Sens.* **2022**, *14*, 3547. [CrossRef]
24. Kharakhashyan, A.; Maltseva, O.; Glebova, G. Forecasting the total electron content TEC of the ionosphere using space weather parameters. In Proceedings of the 2021 IEEE International Conference on Wireless for Space and Extreme Environments (WiSEE), Cleveland, OH, USA, 12–14 October 2021; pp. 31–36. [CrossRef]
25. Glorot, X.; Bengio, Y. Understanding the Difficulty of Training Deep Feedforward Neural Networks. In Proceedings of the Thirteenth International Conference on Artificial Intelligence and Statistics, AISTATS, Sardinia, Italy, 13–15 May 2010; pp. 249–356.
26. Saxe, A.M.; McClelland, J.L.; Ganguli, S. Exact solutions to the nonlinear dynamics of learning in deep linear neural networks. *arXiv* **2013**, arXiv:1312.6120.

remote sensing

MDPI

Technical Note

On Some Challenges for National and Global Space Weather Services

Maria A. Sergeeva [1,2,*], Juan Americo Gonzalez-Esparza [1], Victor Jose Gatica-Acevedo [1], Luis Xavier Gonzalez [1,2], Pedro Corona-Romero [1,2], Ernesto Aguilar-Rodriguez [1], Angela Melgarejo-Morales [1], Isaac David Orrala-Legorreta [1], Julio Cesar Mejia-Ambriz [1,2] and Jose Juan Gonzalez-Aviles [1,3]

1 SCiESMEX, LANCE, Instituto de Geofisica, Unidad Michoacan, Universidad Nacional Autonoma de Mexico, Morelia 58089, Michoacan, Mexico
2 CONAHCYT, Instituto de Geofisica, Unidad Michoacan, Universidad Nacional Autonoma de Mexico, Morelia 58089, Michoacan, Mexico
3 Escuela Nacional de Estudios Superiores, Unidad Morelia, Universidad Nacional Autónoma de México, Morelia 58190, Michoacan, Mexico
* Correspondence: maria.a.sergeeva@gmail.com

Abstract: Space Weather (SW) hazards are discussed in terms of the operation of national SW services and global SW centers for the International Civil Aviation Organization (ICAO). The definition of threshold values of monitored parameters which are used to identify moderate and severe SW events is one of the critical problems. Due to the lack of both physical data on severe events and user feedback, we tried to approach the problem statistically. In particular, we pursued the answer to the question about what intensity of ionospheric storms and flare effects should be reported by national and global SW entities to their users. We also discussed the possible role of an active region on the Sun, and the cosmic rays' issues that may be helpful regarding SW operational work. The presented considerations are based on examples of the ionosphere state assessment for the low-latitude American sector with a focus on the Mexican region. This work attempts to argue the possible approaches to resolve the tasks that the SW national services and global centers face.

Keywords: Space Weather Service; thresholds; ionospheric storm; flare; TEC; active region

1. Introduction

There are many scientific works dedicated to Solar–Terrestrial relationships, e.g., Refs. [1–13] and many others. These studies are critical to deepening our knowledge of the physics of natural processes in the near-Earth space. The concept of Space Weather includes not just a combination of physical phenomena (specifically the conditions on the Sun, in the solar wind and interplanetary medium, the magnetosphere, atmosphere, thermosphere and ionosphere), but also the harmful effects of these phenomena on modern technology. Indeed, Space Weather (SW) event studies include the degradation and failure of technological system operations. In other words, SW studies are the branch of applied research in heliophysics, Solar–Terrestrial physics and Earth's science.

International organizations like the World Meteorological Organization (WMO), the International Space Weather Service (ISES), the United Nations Office for the Peaceful Uses of Outer Space (UNCOPUOS) and for Disaster Risk Reduction (UNDRR), the International Civil Aviation Organization (ICAO), and the Committee on Space Research (COSPAR), promote collaboration to share data to monitor and study Space Weather events [14–18]. A severe Space Weather event is a global hazard, but with particular regional effects depending on the geomagnetic latitude and characteristics of the region. For that reason, it is necessary to have local regional measurements.

The organizations responsible for the provision of the official SW information to state authorities are the national SW services. A national SW service provides the interested users with SW-related products that may include the now-casting and short-term forecasting (if possible) of SW conditions, early warnings, alerts, notifications and other messages on SW conditions, analysis of particular events on demand, etc. In most cases, an SW service operation implies 24/7 continuous quasi-real-time monitoring and diagnostics of current conditions. Usually, both automatic features and operator work are combined. Further, an SW service is expected to estimate the potential severity of an SW event when it occurs, responding to the following questions. What is happening or has just happened? Can it constitute a danger and, if so, to what applications? Which region can be affected? What can reasonably be expected in the next few hours–days? There are more assigned tasks, such as the operation of scientific instruments, data collection, processing and classification, data exchange with other services, standardization and automatization, coordination of information, etc. For more information on the SW services, the reader is referred to the official ISES web page (www.spaceweather.org, accessed on 1 September 2023). Some details of SW service tasks are discussed in [8,19–28]. It may be noted here that the preparedness for a severe SW event is one of the most important designated problems. When should an SW event be reported to authorities? What can happen in the worst-case scenario? Governmental agencies, operators and decision-makers in the vulnerable industry sectors are the primary users of national SW service products. Interdisciplinary working groups are created to develop national strategic plans for the case of a severe event. If a national SW service (endorsed by a national government) meets the specific requirements, it may become a Regional Warning Center (RWC) of ISES. The latter coordinates the work of the national services on SW hazards on a global scale. At the moment of this article's writing, 22 RWCs of different nations and the European Space Agency belong to ISES. Please see a brief review on the international collaboration in the field in [8] and the references therein.

One of the main SW-information users is the aviation sector represented by the ICAO bodies. Veritably, the risks posed to a flight's safety by a significant SW activity include the degraded performance of the aircraft's avionics, radio communication systems, navigation means and the potential overdosed radiation for the crew and the passengers [9,29,30]. For example, the case of the storms in October–November 2003 may be recalled, when several flights were rerouted to avoid the increased radiation levels and degraded communication [10,31,32]. Other economy sectors such as GNSS navigation precision, power transmission lines, low-orbit satellite operations, space launches, human work in space and many others are also vulnerable to SW events [8,29,31,33–35]. To provide SW information to ICAO worldwide in real-time, a 24/7 international operational service was launched in 2019 [30,36]. The responsibilities within this service are rotated among the assigned SW centers every two weeks. At present, the following SW global centers (SWXC) participate in this rotation [37,38]: the ACFJ consortium, consisting of the national services of Australia, Canada, France and Japan [39]; the PECASUS consortium, consisting of the national services of Austria, Belgium, Cyprus, Finland, Germany, Italy, Netherlands, Poland and United Kingdom [40]; the Space Weather Prediction Center of the USA [41,42]; and a consortium consisting of the national services of the People's Republic of China and the Russian Federation (CRC) [43,44]. More details can be found in the ICAO's official documentation (https://www.icao.int/SAM/Documents, accessed on 13 March 2023).

In most cases, the occurrence of significant SW disturbances is monitored automatically. This is done using threshold values of SW parameters introduced in the monitoring systems. The automatically issued messages should be verified by a service/center operator who takes the final decision (both in the national services and SWXC) on issuing a message (advisory/warning/alert) through the assigned communication channels to their users. The approach should be complex: all the SW conditions should be diagnosed (occurrence of flares, coronal mass ejections, etc.), especially if only one of the monitored parameters exceeded its threshold value. Obviously, even with a systematic approach to the issue, there is space for the subjectivity of the operator on duty. This is probably not a bad issue as there is a lack of statistics on significant SW events. The professional experience, training and scientific knowledge, if there is a possibility to consult with researchers, are beneficial in the decision taking and may be more effective than an automatic control. At the same time, it is possible that different decisions may be taken from the same borderline parameter values. This is due to the possibility of different opinions on some edge cases and the fact that each particular national service or SWXC has its own algorithm of actions.

This work discusses some challenges that could be encountered in the operation of the national SW services and SWXC. One of the challenges that the operational community faces is the thresholds for the monitored SW parameters which mark moderate and severe events. The point in question may be divided in two. For national services, there is a task of regional thresholds' definition, since it is known that the response to an SW event depends on the coordinates of the observation point. The thresholds used by the SWXC are defined by the ICAO's *Manual on Space Weather Information in Support of International Air Navigation* [30], based as much as possible on the impacts to systems. As it is stated in [30], "the data does not exist for a one-to-one correspondence between system degradation and Space Weather intensity, so educated estimates are necessary in some cases" and "periodic updates" on the issue are necessary. Indeed, there is a lack of statistics on significant events and their effects. Moreover, though there are notable advances, the scientific knowledge of the solar–terrestrial relationships still needs to be improved. Nonetheless, we should begin with something because specific instructions must be given to the automatic algorithms and the service/center operators. Therefore, we suggest departing from simple considerations. Suppose it is not possible to define the thresholds physically, the question may be posed on how often a service/center considers it possible to alert its users/authorities with the messages on SW. Though this approach may seem too simplistic, it makes sense because people can pay no attention to frequent alerts. The aim of this work is to illustrate the application of this approach for the impacts on the ionosphere to be reported. The paper is organized as follows. Section 2 discusses the intensity thresholds for ionospheric storms caused by geomagnetic disturbances whichshould be reported by the national and global SW entities to their users. In Section 3, the answer on what intensity of solar flares to be reported is pursued. The possible role of an active region of the Sun, and the cosmic rays' issues are discussed in Sections 4 and 5, respectively, as they may be helpful regarding the SW's operational work. Final remarks are given in the Conclusions section.

2. Ionospheric Storms to Report

Usually, within the SW context, two types of ionospheric disturbances with the potential to affect GNSS operation are considered to be monitored and reported by the SW services/centers: ionospheric storms and ionospheric scintillation events. The latter are not considered in this study due to their high regional specificity and because this type of disturbance deserves a separate discussion. The ionospheric storms are viewed in terms of vertical Total Electron Content (TEC) [45,46] whose values can exceed a certain level. At first glance, and from a scientific point of view, TEC may seem a rather simplistic approach considering only one parameter. Nonetheless, it is logical. First, TEC is one of the most widely used ionospheric parameters due to the extensive coverage of GNSS receivers worldwide and the continuous development of global and regional GNSS networks. Furthermore, its value variations are directly correlated to the GNSS positioning error and other SW adverse effects related to the changes in the ionosphere [45]. Usually, significant TEC deviations are associated with geomagnetic storms [47–49], though TEC enhancements due to other phenomena (usually of shorter duration, such as intense solar flares, particle precipitations, etc.) are not discarded. The TEC behavior is influenced by atmospheric chemistry, transport due to the combination of directions of the electric E and magnetic field B over the equator (ExB plasma drift), neutral winds, ambipolar diffusion and other physical processes. Please note that, from the GNSS point of view, only TEC enhancements can represent danger (due to the increase in positioning error) [50–52]. The electron concentration decreases that are dangerous for high-frequency (HF) communications are estimated using the ionospheric sounding data and are not considered in this paper.

The TEC thresholds suggested by the ICAO documents [30] are global and not specified for different regions. This means that, probably, these thresholds will never be exceeded at high latitudes due to the generally smaller electron concentrations in the high-latitude ionosphere. This is confirmed by the results of [53] (see Table 1 of this work). Namely, the threshold values suggested in [30] for moderate and severe TEC disturbances are 125 and 175 TECU, respectively. For further assumptions in this work, we use the examples of low-latitudes in the American sector, with a particular focus on Mexico. It is worth noting that the national services may apply regionally adapted scales. For instance, the Space Weather Service Mexico (SCiESMEX) currently applies the following regional thresholds for the ionospheric storm detection over Mexico: TEC $\geq$ 125 TECU (moderate), TEC $\geq$ 175 TECU (significant) and TEC $\geq$ 250 TECU (severe).

Table 1. TEC responses to Dst < −200 nT at two observation points. The values that exceeded the SGXC threshold for a moderate storm are marked with a red color.

		MEXICO				BRAZIL			
Dstmin, nT	**Geomagnetic Storm**	**>100 [a] TECU**	**TECmax, TECU**	**TEC Response**	**Near Regular**	**>100 [a] TECU**	**TECmax, TECU**	**TEC Response**	**Near Regular**
−205	4–6 May 1998	no data	no data	no data	no data	no data	no data	no data	no data
−207	25–26 September 1998		68.3	↑↓			85	-	v
−237	22–24 October 1999		80.2	↑↓	v	+	109.1	↓	v
−292	6–8 April 2000	+	165.3	↑		+	143.7	↑	v
−300	15–17 July 2000	+	116.1	↑↓			83.5	↑	
−234	12–13 August 2000		78.2	↑↓			87	↑	
−201	17–19 September 2000	(+)	96.1	↑			91.3	↑	v
−284	31–1 March 2001	+	123.4	↑	v	+	147.5	↑	
−271	11–13 April 2001	+	143.4	↑↓		+	122.7	↓	v
−292	6–8 November 2001	+	105.3	↑	v	+	122.8	↓	v
−221	24–26 November 2001	(+)	95.9	↑↓	v	+	106.3	↓	v
−383	28–3 October 2003	+	208.4	↑		+	99.2	↑↓	
−422	20–23 November 2003	+	107.8	↑			81.6	↑↓	v
−374	6–12 November 2004		53.8	↑	v		64.4	↓	
−247	15–16 May 2005		46.1	↓	v		61.2	↑	
−225	17–19 March 2015		71.5	↑↓	v		93.6	↑↓	

[a] The column is filled assuming that the accuracy of GIM TEC is about 2–5 TECU. The sign (+) is used when the value was close to 100 TECU. The sign ↑ stands for an increase and ↓ stands for a decrease of TEC. v means that TEC values during and after the storm were close to or defined by their regular variations.

Ionospheric storms are mostly caused by geomagnetic storms [30]. Their possible impacts are discussed in many works such as [8,9,11–13,54,55] and the references therein. Different geomagnetic storm classifications exist, e.g., [56–58]. Usually, a storm is considered intense if characterized by Dst-index [59,60] value lower than −100 nT. According to [57], Dst < −100 nT means a strong geomagnetic storm, Dst < −200 nT means a severe geomagnetic storm and Dst < −350 nT means a great geomagnetic storm. The authors of [61] showed that TEC over Mexico can show ionospheric disturbance beginning with Dst-index being lower than −30 nT, and that the Dst value has no direct correlation to the intensity of the ionospheric storm. For this study, we considered TEC behavior during and after the storms characterized by Dst < −200 nT.

In general, the highest TEC values are observed at low and equatorial latitudes. That is why our focus is on them. We analyzed TEC response at two observation points: Mexico (geographic lat 20°N, lon 100°W) and Brazil (geographic lat 20°S, lon 55°W), representing low and equatorial geomagnetic latitudes, respectively. The Mexican RWC—SCiESMEX—performs a continuous regional TEC monitoring since 2015 [62]. Unfortunately, this provides relatively small statistics that do not cover even one solar cycle. Such a small data volume does not allow us to draw even rough conclusions. This was the reason for estimating TEC variations withthe global ionospheric TEC maps (GIM TEC) instead of using the regional GNSS data. As GIM TEC have been available since 1998, our study is based on the statistics from 1998 to the present (February 2023). The given estimates are only of qualitative character as GIM of a low time resolution (2 h) were used. In addition, GIM TEC values bydifferent IGS analysis centers may vary within certain limits [63–65]. For the present study, we assume that the difference between TEC values by different centers is about (2–5) TECU. Note that TEC can show different behavior within even a limited latitudinal/longitudinal sector due to local factors. The last means that one TEC value (for a particular observation point) does not represent an entire country or sector. However, GIM TEC (by JPL) allowed us to make at least a rough assessment of the regional ionospheric responses to the storms. The results are given in Table 1.

GIM TEC data was available for 15 out of 16 geomagnetic storms. Only positive ionospheric responses were of interest due to the objectives of the study. TEC absolute values that reached or exceeded 100 TECU were revealed for nine cases in Mexico and seven (not the same) cases in Brazil. Some of the rather low maximum TEC values (TECmax) in Table 1 are explained by the backgroundregular TEC variations at its observation point (low diurnal values due to the season and the solar cycle branch). Even a high percentage of deviation from a low initial value means no alarm. On the other hand, it is noteworthy that according to GIM data, some of the intense geomagnetic storms caused little or no effect on TEC. This may be due to the interpolation applied in GIM TEC. It also may be related to the fact that TEC is an integral parameter (the changes of electron content at different heights offset each other to some extent), or it may be attributed to some other (even physical) cause. However, from an operational point of view, this is not very important. As for the high values, the analysis of GIM TEC statistics showed that, sometimes, high values originate from the regular variations of TEC in a particular region of the Earth. Figure 1 provides examples of two independent events registered at two different observation points. The TEC enhancement during the disturbance is shown in the upper panel. The high regular TEC values in the diurnal variation are observed before the disturbance, in the lower panel.

The fact of the possibility of high regular values emphasizes the significance of regular (systematic) variation knowledge on a regional scale. The majority of scientific articles is focused on the responses to moderate and intense SW events. At the same time, for operational purposes, regular parameter variations are very important. Further, the following question arises. If the regular regional trends include the periods of high values of the parameter, does this mean that the parameter's thresholds should be increased for the considered region?

The first decision would be to answer in the affirmative and report only the drastic deviations from the regular (reference) values. This would help avoid sending users too many messages/advisories. This might be an option for the national SW services because they possess updated information on the regular trends for their regions due to the continuous monitoring and diagnostics of conditions. They can even rapidly consult these trends with the national research entities. Eventually, the decision to release a warning message would be taken, in each case, individually. At the same time, this is not an option for an SWXC on duty, as there is no "global information" of this kind to the best of our knowledge. By "global information" we mean the information on all the regionalregular (systematic) trends, which are different in different zones. Moreover, theregular trends are constantly changing, making it more difficult to rapidly take an individual decision on a global scale.

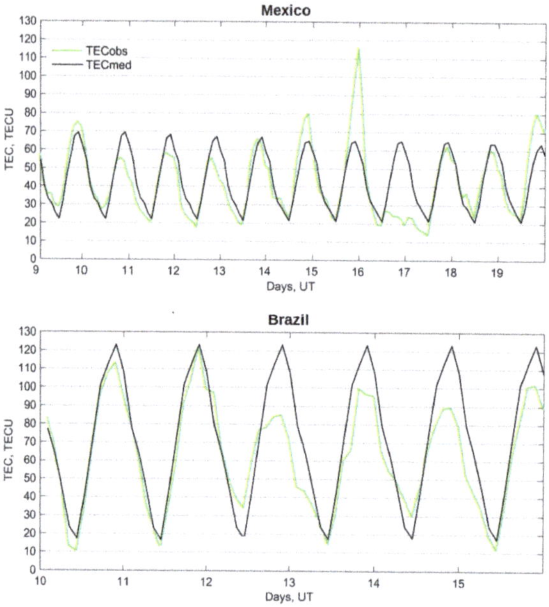

Figure 1. The pronounced TEC response to the storm of 15–17 July 2000 (**upper panel**), and the high TEC reference values before and during the storm of 11–13 April 2001 (**lower panel**). TECobs stands for the observed value, and TECmed stands for the 27-day running median representing regular variations (reference values).

The problem may be approached from another point of view. The services should answer the question of whom their messages are directed to. The highregular values may be reported as often as they occur, but the messages should be sent only to those users for whom this information is critical. For example, the output data of a single-frequency GNSS receiver is affected by any high TEC value, whether it is a result of some SW disturbance or a temporal manifestation of a regular TEC trend in some region. Consequently, a user of this receiver may consider each of many messages/advisories reporting high values. At the same time, the users of more complex systems which make no use of such messages/advisories should not receive them.

This approach brings us to an essential aspect of an SW service operation—the definition of the parameter thresholds, individually, for each of the users of the SW information (this is not the case of SWXC, as its only user is the ICAO). This concerns not only TEC values, which are only the example for this study, but the thresholds for all the monitored SW parameters. One of the problems is the absence of feedback from users from different sectors of the national economy. It would be beneficial if the users provided information on which SW disturbance affected which technological system and to what extent. It might be achieved by comparing the lists of detected technological problems and the time-moments of SW disturbances of different kind. However, this task is not as easy as it may seem. First, the significant SW events (especially the severe events) occur rarely, which gives only a few cases to analyze. Second, the information is probably not stored by users indefinitely. Considering that the majority of the significant SW events occurred before the solar cycle 24, the impacts recorded by users may be absent. Finally, even if all the records were available, it would not be easy to define the parameter thresholds well due to the two circumstances. (a) The overall statistics would still be little as the technology affected by

SW was designed recently in terms of the time scales of severe SW event occurrences. (b) Technology is constantly being improved, meaning that what was vulnerable ten years ago may not be affected now.

Nevertheless, the statistics should be collected and the protocols for sharing data on the outages should be designed at least on a national scale, as there is no approach other than the collaborative efforts between the SW services and the industry.

Let us illustrate the application of the considered statistical approach to threshold definition. While the TEC statistics are not very large and are only qualitative, they allow us to make at least a crude estimate. All the TEC values from Table 1 were divided into ranges (limited by two boundary values) and semi-ranges (limited by one boundary value) by their intensity (Figure 2). According to GIM data, the threshold for a severe disturbance (175 TECU) was not exceeded in Brazil and was exceeded once in Mexico. Using the plots in Figure 2, the TEC thresholds may be defined from the number of messages considered appropriate for a national SW service to issue during the considered period. For example, if issue four messages in Mexico, the threshold of TEC $\geq$ 125 TECU may be applied. The inverse problem solution provides us with a possible number of advisories issued for this region by SWXC. For example, if report only TEC $\geq$ 125 TECU, then two messages would be issued in Brazil during the considered period.

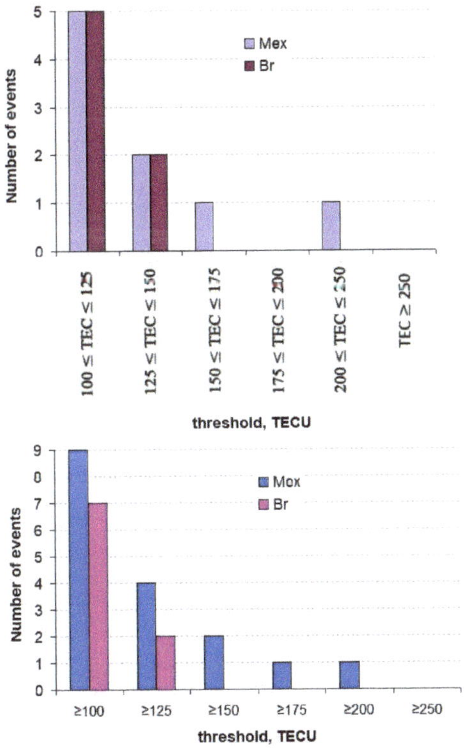

Figure 2. Maximal TEC values registered during and after the geomagnetic storms characterized by Dstmin < −200 nT (assuming that the accuracy of GIM TEC is about 2–5 TECU), grouped by ranges (**upper panel**) and semi-ranges (**lower panel**).

It is known that the regional TEC value (similar to GIM TEC by different analysis centers) may differ when obtained with different calculation methods, e.g., Ref. [66]. Therefore, applying different TEC calibration techniques compounds the problem of thresholds choice for this parameter on a global scale. In a real-case scenario, it could occur that one national SW service or SWXC would issue a message/advisory on TEC exceeding an accepted threshold while others would not. This is a consistency issue. In this case, the advisory issuance will depend on which SWXC is on duty today. Is this a severe problem? The higher the thresholds, the lower the possible uncertainty. Considering that the TEC threshold suggested by the ICAO for a severe event is 175 TECU, the ambiguity due to the TEC calculation method does not seem very important. At the same time, considering the results in Table 1, the threshold for moderate events (125 TECU) may be affected.

3. The Role of the X-Class Intensity of the Flare

An intense flare affects sky-wave radio propagation and can cause a total or partial shortwave fade-out. From a physical point of view, it was proven that the flare intensity defined in terms of the X-ray class might not be a measure of the effects on the ionosphere and radio propagation conditions without applying other criteria such as the solar zenith angle, active region position on the Sun, etc. (e.g., [67–70]). Without going into detail, we note that, from the operational point of view of the national SW services, the observer has not yet many alternative parameters to rapidly evaluate the flare intensity (within seconds–minutes). Moreover, when speaking about automatic messages, the decision is made quite immediately. In regard to the SWXC advisories, they are issued on a global scale, which makes the situation even more complex.

Considering the above, let us depart from the simple considerations, as in the previous section. An answer is needed to the question of how many times an SW service may address the authorities or its users during some period. The ICAO document [30] suggests the X1-class as a threshold for the moderate effects and the X10-class for the severe flare effects. X1 seems a reasonable lowest level for the Mexican region [71]. Therefore, only the X-class flares are considered further. Figure 3 illustrates a not surprising dependence of the X-class flare occurrence on solar activity: more flares occur during high solar activity periods. Here, and further in the figures, the X-class number values are given just in numbers (without a logarithmic scale applied for the X-ray flux in Watts/m^2) for illustrative purposes.

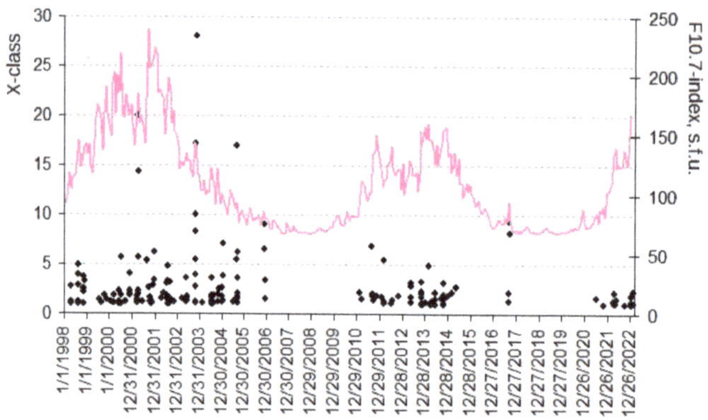

Figure 3. X-class flares occurred between January 1998 and February 2023, during approximately two solar cycles. The black points mark the moments of X-flares whose intensity is given in the left Y-axis. The rose curve stands for the F10.7-index series (right Y-axis).

All the flares from Figure 3 were divided into ranges and semi-ranges according to their intensity. Only the daytime flares are hazardous. From the SWXC's point of view, all the flares whose radiation is emitted towards the Earth's orbit should be considered because the SWXC perform monitoring on a planetary scale, in which case there are always sunlit regions. From a national SW service's perspective, only the local daytime events are of interest. In particular, in the case of the SCiESMEX (Mexico), the focus is on the flares that occurred during the daytime local hours (LT) for the North-American sector. For this study, we considered a flare to be a daytime flare if it occurred between 07 and 21 LT

Central Mexican Time. These limits were chosen taking into account the fact that Mexico administratively has four time zones. Figure 4 shows the statistical results.

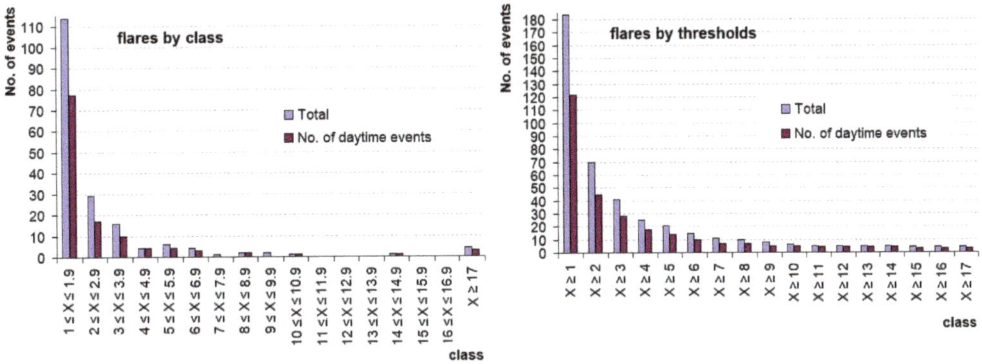

Figure 4. The number of flares of each X-ray class (**left**) and the number of flares exceeding a certain intensity level characterized by the X-ray class (**right**). Daytime events are for Mexico. The covered period is the same as in Figure 3.

Similarly to the previous section, a decision is needed on the reasonable number of messages/advisories that may be issued by an SWXC and by a national SW service. Let us consider our examples. The vast majority of events (114 in total and 77 Mexican daytime) were below the level of X2 (Figure 4 left). If moderate flare events (from X1-class by ICAO thresholds) were reported each time, then 114 events would be reported globally and 77 events would be reported in Mexico. Is this a lot or a little for a 25–26 year period? On average, 4–5 events would be reported each year globally and about three events would be reported in Mexico. At the same time, it should be recalled that the events would not be distributed evenly in time (see Figure 3). In addition, there would be issued messages/advisories related to other SW hazardous events (besides the flares). The concern is that there might be a situation where many messages are released, but the technological systems would not be affected each time. This would lead to the ignoring of future messages. Let us provide an example. Suppose it is appropriate to address national authorities/users once in 2–3 years, which in Figure 4 (right) would correspond to 10 events in ~25 years. In that case, the threshold for Mexico may be defined as the flare intensity higher than X6-class. The threshold of X10-classwould mean a potential message once in approximately five years. Each operational SW entity may individually decide on the number of messages, according to its tasks and experience.

There are only four administrative time zones in Mexico. What if a country covers a larger longitudinal sector, like, for example, Canada (six time zones) or Russia (eleven time zones)? Considering only severe events, the national and the global criteria would probably be the same. This is because (a) severe events are relatively rare and (b) flare effects cover a large longitudinal sector (with the strongest effects over the noon region and the less intense effects in the morning and evening sectors).

Considering all the above, the threshold of X10 for the severe flares suggested in the ICAO documents seems reasonable. According to Figure 4 (right), it would be reported only six times since 1998.

4. Role of Active Region Responsible for the X-Class Flare Event

Let us estimate the probability of a significant flare event considering the activity of the active region (AR) on the Sun responsible for the flare. It should be recalled that, each time the same AR appears on the solar disk after being invisible due to the solar rotation, it is assigned a new number. Therefore, here, we only deal with the AR lifetime limited approximately by ~13 days. The SW services usually base their monitoring on the data of

GOES satellites at the Earth's orbit [72]; therefore, the events on the far side of the Sun are not taken into account.

We considered the flare events that occurred between 1998 and February 2023. All the ARs that produced X-class flares during this period were divided into two groups: those that\produced only one flare and those that produced multiple (more than one) flares.

In regard to the first group, the vast majority of the ARs responsible only for one flare (46 of 58) produced the events of X-class $1 \leq X \leq 1.9$. At that, seven ARs of the first group were responsible for flares $2 \leq X \leq 2.9$ and two ARs for $3 \leq X \leq 3.9$. The flares of the classes $4 \leq X \leq 4.9$, $5 \leq X \leq 5.9$ and $6 \leq X \leq 6.9$ were produced only once. The ARs of the first group did not produce the flares with higher intensities. The impression is that, in the case of the ARs responsible only for one flare, there is a flare intensity limit. Statistically, it seems unlikely that such an AR would produce a flare of X > 10 class that is the threshold accepted in the ICAO's official documents for severe shortwave fade-outs. However, in practice, the difficulty is that the observer never knows if more flares will be produced after the first flare. We tried to associate the lowest intensity of the produced flares with the number of flares produced by the same AR, but, as it is shown below, the lowest intensity always begins with the X1 class.

Indeed, let us consider the ARs of the second group. The majority of ARs (17 of 38) responsible for multiple flares produced two flares whose intensity mainly varied between the classes X1 and X3. There is one case of X5.4 and one case of X6.2. Seven ARs responsible for three flares provoked mainly the intensities characterized by X1–X3 classes, and there was one case of X4.8 and one case of X5.7. Another seven ARs were responsible for four flares. Their intensities laid mostly between classes X1 and X4. There were also two flares of X9 and one of X20 class. The last facts call attention. Four ARs of the second group were responsible for five flares whose intensities were mostly between X1 and X5, and there was one case of X14.4 class. Only one AR produced six flares. Their intensities were between X1 and X4 classes (which probably implies the lack of events for statistics). Only one AR produced seven flares with the intensities of X1, X5, X8, X10, X17 and X28 classes. Likewise, only one AR produced ten flares. Their intensities included classes X1–2 (six events), X3, X5, X6 and X17.

Apparently, the more flares produced by the same AR, the greater the probability of a severe flare event. Based on the available statistics (presented above), the AR already responsible for three flares of X-class seems promising in terms of a significant (X > 10) for an SW service flare event (Figure 5). We would like to repeat that this is only a statistical result for the chosen period, without considering the physics of the processes within the particular AR.

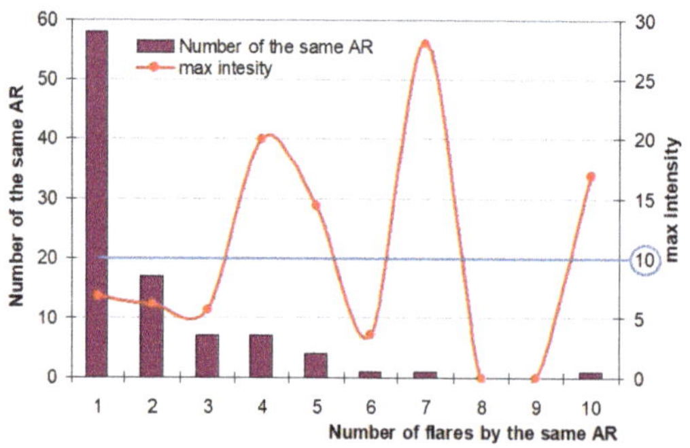

Figure 5. The role of an AR's activity during the time that it is visible on the solar disk. "Number of the same AR" stands for the number of the particular ARs, each of which is responsible for multiple flares.

5. Cosmic Rays Perspective

Cosmic ray phenomena [73,74] are among the SW events that are considered in [30] because they can cause an elevated radiation. To estimate the effective radiation (danger for technological systems and humans in flights), the national SW services may use not only the geosynchronous satellite data and models, but also their own ground-based instruments (neutron monitors). Let us consider the regional cosmic rays' statistics in regard to the SW effects to be reported by the SW national services. We continue to use the examples for the particular region of Mexico. For this low-latitude region, the particles accelerated by shock waves are not sufficiently intense to cause a significant (from an operational point of view) particle event. That is why only the events caused by flares and magnetic storms are considered below.

First, the Solar Energetic Particle (SEP) events that form part of SW hazards are contemplated. In the Mexican latitudes, these are caused by solar flares [75–77]. The SCiESMEX considers a SEP event's presence if the particle flux intensity exceeds its average value during the last three hours. Only two events were registered in Mexico during the considered period: one with a flux exceeding 8.5 and another exceeding 2.89 standard deviations on the 7 September 2005 (X17-flare) and on the 4 November 2003 (X28-flare), respectively. Even with only two cases, it is evident that the X-ray class of the flare may not be an appropriate parameter for the assessment of possible SEP effects. In addition, it is noteworthy that no effect was registered after the daytime X20-flare. First, the mentioned facts argue for the impossibility of estimating the potential SEP effects based only on the flare's X-class intensity, without the measurements of cosmic rays. Second, this means the necessity to know the physics of the processes, which indicates the direction for future research work.

In regard to cosmic rays' response to geomagnetic storms, the SCiESMEX considers an event to be significant if the cosmic rays' flux increases its regular value of diurnal variation by at least three standard deviations. Two such events were registered over Mexico since 1998: on the 20 November 2003 (flux enhancements by 7.2 standard deviations) and on the 6 April 2000 (7.6 standard deviations). The mentioned geomagnetic storms were characterized by the Dst minimum values of −422 nT and −291 nT, respectively. Even with the limited statistics of only two events, the Dst value does not seem promising to evaluate the potential intensity of a particle event. Considering that other intense geomagnetic storms (see Table 1) did not result in the cosmic rays events over Mexico, Dst may not be used even to identify the possibility of the eventoccurrence. More research is needed to provide the SW services with more parameters that may be involved in the diagnostics.

The question on how regional these effects are may be posed. Again, we depart from simple considerations. Let us consider an event of particle precipitation that represents no danger for Mexico because its territory is at low latitudes. However, operations within the aviation sector are interconnected globally. For instance, the popular flight routes between Europe and Mexico usually lie through the mid- and high-latitudes. The latter could be affected drastically by particle precipitations due to the geomagnetic field lines' configuration. In practice, this means that, even with a weak event registered over Mexico, the operation in the Mexican aviation sector may be affected due to the effects in other regions. One of the possible approaches to particle precipitation risk estimation suggested for the national SW services would be to act as an SWXC, assessing radiation risks globally.

6. Conclusions

Without statistics on extreme events and feedback from industry users, the SW services and centers must have an approach to address the challenges they face. The operational nature of their work demands immediate action, rather than waiting for an ideal approach in the future. It is crucial to implement feasible solutions now to mitigate potential risks and provide valuable information to users. The frequency of SW disturbances' occurrence can be a helpful reference. The intensity of SW events represented by parameter thresholds must be taken into account as, from an operational standpoint, there are no other criteria for decision-making processes. The definition of the parameter thresholds used for the detection of moderate and severe SW events is one of the challenges to which the national SW services and SWXC stand up. It is difficult to define these thresholds physically due to the small statistics of the registered significant SW events, the relative recency of the majority of vulnerable technological systems and their constant improvement, and other discussed causes. The possibility of the statistical approach based on SW events' occurrence was shown to solve this task. The question on how frequently the SW messages/advisories should be issued should be answered.

(1) The answers for the SGXC and the national SW services may differ as the national services may have more than one user. Moreover, there is no obvious response on whether the regional SW thresholds should coincide with the global SW thresholds for the same user. For example, at present, the TEC thresholds used by the SCiESMEX are similar, but are not the same as those used by the SWXC.
(2) It is important that the warning messages by the national SW services are user-targeted and that the parameter thresholds for each user are chosen individually (communications, national electricity companies, etc.). The only user of the SW information provided by the SWXC is the ICAO.
(3) The application of the suggested statistical approach was illustrated for the low-latitude American sector (primarily the Mexican region). In particular, the intensity of the ionospheric storms (TEC threshold) and the intensity of the flares (X-class threshold) that should be reported were under analysis.
    - In regard to ionospheric storms, the regular (systematic) variation knowledge on a regional scale is important, as high TEC values exceeding thresholds can be not only the product of disturbance, but also a part of a regular trend. The absolute TEC value's dependence on the calculation method is probably not very important for the detection of severe event (175 TECU threshold suggested by the ICAO). Still, it may play some role in a moderate event's detection (125 TECU), especially during periods of high, regular TEC values due to TEC's systematic variations at a particular observation point. The higher the thresholds, the lower the uncertainty about exceeding them.
    - For the example of Mexico, the suggested approach results in the severe flare's threshold by the ICAO ($X \geq 10$), if the possible number of issued messages is considered to be once in approximately five years. In addition, it was shown that, statistically, for the last two solar cycles, the more flares produced by the same AR, the greater the probability of a severe flare event. The AR already responsible for three flares of X-class seems promising in terms of a significant ($X \geq 10$) for an SW service flare event.
    - The flare intensity defined in terms of the X-ray class of the flares is insufficient to estimate either the effects on HF communications or the SEP event's occurrence. There also needs to have more than the Dst-index to estimate the possibility of a cosmic rays' event caused by a geomagnetic storm.
    - It is suggested that the particle precipitation risk estimation by the national SW services is considered not regionally, but as a global event. In other words, in the case of cosmic rays, it is suggested that a national SW service assesses radiation risks globally, as the SWXC.

Finally, we emphasize the importance of the statistics collected on physical events and technology outages, as well as the close collaboration and data sharing between the researchers, operational SW services and industry sectors. This will provide more ideas for approaching the current SW challenges.

Author Contributions: Conceptualization, M.A.S. and J.A.G.-E.; methodology, M.A.S. and J.A.G.-E.; data acquisition, V.J.G.-A., E.A.-R. and A.M.-M.; data processing and validation, I.D.O.-L., J.C.M.-A. and J.J.G.-A.; data analysis, M.A.S., L.X.G. and P.C.-R.; visualization, V.J.G.-A. and A.M.-M.; resources,

J.A.G.-E.; writing—original draft preparation, M.A.S. and J.A.G.-E. All authors have read and agreed to the published version of the manuscript.

Funding: LANCE acknowledges partial support from the CONAHCyT-AEM, Grant 2017-01-292684, CONAHCyT LN-315829 and PAPIIT IN116023. L.X. Gonzalez was supported by the CONAHCyT-AEM Grant AEM-2018-01-A3-S-63804. E. Aguilar-Rodriguez was supported by the DGAPA/PAPIIT project IN103821. P. Corona-Romero is grateful for the Investigadores por México-CONAHCYT (Catedras CONAHCYT) project 1045 Space Weather Service, financed by the Consejo Nacional de Humanidades Ciencia y Tecnología (CONAHCYT). J.J. Gonzalez-Aviles acknowledges to CONAHCyT 319216 project "Modalidad: Paradigmas y Controversias de la Ciencia 2022". V.J. Gatica-Acevedo and A. Melgarejo-Morales express their gratitude to the CONAHCyT.

Data Availability Statement: The OMNI data (Dst and F10.7 indices) were obtained from the GSFC/SPDF OMNIWeb interface at https://omniweb.gsfc.nasa.gov (accessed on 25 December 2022). The authors also thank the NOAA's National Centers for Environmental Information and the NOAA's Space Weather Prediction Center for the opportunity to use the GOES satellite data available at https://data.ngdc.noaa.gov/platforms/solar-space-observing-satellites/goes/ (accessed on 25 December 2022) and at www.swpc.noaa.gov/products/goes-x-ray-flux (accessed on 25 December 2022), correspondingly. The GIM TEC data provided by IGS were obtained from https://cddis.nasa.gov/archive/gnss/products/ionex (accessed on 13 March 2023).

Acknowledgments: The authors thank the NOAA's National Centers for Environmental Information for the opportunity to download the GOES satellite data. The authors express their gratitude to the services of the IGS for the opportunity of using the IONEX data via the Internet. The authors thank the Editor and the anonymous reviewers for their valuable comments.

Conflicts of Interest: The authors declare no conflict of interest.

References

1. Buonsanto, M.J. Ionospheric storms—A review. *Space Sci. Rev.* **1999**, *88*, 563–601. [CrossRef]
2. Blagoveshchensky, D.V.; MacDougall, J.W.; Piatkova, A.V. Ionospheric effects preceding the October 2003 Halloween storm. *J. Atmos. Sol. Terr. Phys.* **2006**, *68*, 821–831. [CrossRef]
3. Kutiev, I.; Tsagouri, I.; Perrone, L.; Pancheva, D.; Mukhtarov, P.; Mikhailov, A.; Lastovicka, J.; Jakowski, N.; Buresova, D.; Blanch, E.; et al. Solar activity impact on the Earth's upper atmosphere. *J. Space Weather Space Clim.* **2013**, *3*, A06. [CrossRef]
4. Laštovička, J. Lower ionosphere response to external forcing: A brief review. *Adv. Space Res.* **2009**, *43*, 1–14. [CrossRef]
5. Karan, D.K.; Eastes, R.W.; Daniell, R.E.; Martinis, C.R.; McClintock, W.E. GOLD mission's observation about the geomagnetic storm effects on the nighttime equatorial ionization anomaly (EIA) and equatorial plasma bubbles (EPB) during a solar minimum equinox. *Space Weather* **2023**, *21*, e2022SW003321. [CrossRef]
6. Huang, C.-S.; Zhang, Y.; Wang, W.; Lin, D.; Wu, Q. Behaviors of ionospheric topside ion density, ion temperature, and electron temperature during the 20 November 2003 superstorm. *J. Geophys. Res. Space. Phys.* **2022**, *127*, e2022JA030468. [CrossRef]
7. Yu, T.; Wang, W.; Ren, Z.; Yue, J.; Yue, X.; He, M. Middle-low latitude neutral composition and temperature responses to the 20 and 21 November 2003 superstorm from GUVI dayside Limb measurements. *J. Geophys. Res. Space Phys.* **2021**, *126*, e2020JA028427. [CrossRef]
8. Sergeeva, M.A. Space Weather General Concepts. In *Space Weather Impact on GNSS Performance*; Springer: Cham, Switzerland, 2022; pp. 89–150. [CrossRef]
9. Cannon, P. *Extreme Space Weather: Impacts on Engineered Systems and Infrastructure*; Royal Academy of Engineering, Prince Philip House: London, UK, 2013; ISBN 1-903496-95-0. Available online: https://www.raeng.org.uk/spaceweather (accessed on 13 March 2023).
10. National Research Council. *Severe Space Weather Events—Understanding Societal and Economic Impacts, Workshop Report*; National Academies Press: Washington, DC, USA, 2009; ISBN 13: 978-0-309-13811-6; Available online: https://nap.nationalacademies.org/read/12643/chapter/1 (accessed on 13 March 2023).
11. Moldwin, M. The technological impacts of space storms. In *An Introduction to Space Weather*; Cambridge University Press: Cambridge, UK, 2008; pp. 79–94. [CrossRef]
12. Baker, D.; Bodeau, J.M.; Cameron, E.; Fennell, J.F.; Fisher, G.M.; Forbes, K.F.; Kintner, P.L.; Leffler, L.G.; Lewis, W.S.; Reagan, J.B.; et al. *Severe Space Weather Events—Understanding Societal and Economic Impacts*; The National Academy Press: Washington, DC, USA, 2008.
13. Bothmer, V.; Daglis, I.A. *Space Weather—Physics and Effects*; Springer: Berlin/Heidelberg, Germany, 2007.
14. Mann, I.R.; Di Pippo, S.; Opgenoorth, H.J.; Kuznetsova, M.; Kendall, D.J. International collaboration within the United Nations Committee on the Peaceful Uses of Outer Space: Framework for international space weather services (2018–2030). *Space Weather* **2018**, *16*, 428–433. [CrossRef]

15. UNDRR. Hazard Definition and Classification Review (Technical Report). 2020. Available online: https://www.undrr.org/quick/12955 (accessed on 20 September 2023).
16. UNDRR. Hazard Information Profiles (HIPs). 2020. Available online: https://www.undrr.org/quick/66872 (accessed on 20 September 2023).
17. WMO. 2023. Available online: https://community.wmo.int/en/activity-areas/wmo-space-programme-wsp/space-weather-introduction (accessed on 20 September 2023).
18. Schrijver, C.J.; Kauristie, K.; Aylward, A.D.; Denardini, C.M.; Gibson, S.E.; Glover, A.; Gopalswamy, N.; Grande, M.; Hapgood, M.; Heynderickx, D.; et al. Understanding space weather to shield society: A global road map for 2015–2025 commissioned by COSPAR and ILWS. *Adv. Space Res.* **2015**, *15*, 2745–2807. [CrossRef]
19. Simpson, S. Europe to develop new space weather services. *Space Weather* **2003**, *1-2*, 1–2. [CrossRef]
20. Belehaki, A.; Watermann, J.; Lilensten, J.; Glover, A.; Hapgood, M.; Messerotti, M.; Van der Linden, R.; Lundstedt, H. Renewed support dawns in Europe: An action to develop space weather products and services. *Space Weather* **2009**, 7. [CrossRef]
21. Onsager, T. Advancing space weather services through international coordination. *Space Weather* **2012**, 10. [CrossRef]
22. Kotzé, P.B.; Cilliers, P.J.; Sutcliffe, P.R. The role of SANSA's geomagnetic observation network in space weather monitoring: A review. *Space Weather* **2015**, *13*, 656–664. [CrossRef]
23. Lanabere, V.; Dasso, S.; Gulisano, A.M.; López, V.E.; Niemelä-Celeda, A.E. Space weather service activities and initiatives at LAMP (Argentinean Space Weather Laboratory group). *Adv. Space Res.* **2020**, *65*, 2223–2234. [CrossRef]
24. Lam, H.-L. From Early Exploration to Space Weather Forecasts: Canada's Geomagnetic Odyssey. *Space Weather* **2011**, *9*. [CrossRef]
25. Ahn, B.-H. South Korea's renewed focus on Space Weather. *Space Weather* **2011**, *9*. [CrossRef]
26. Nagatsuma, T. New Ages of Operational Space Weather Forecast in Japan. *Space Weather* **2013**, *11*, 207–210. [CrossRef]
27. Wilkinson, P. Space Weather Studies in Australia. *Space Weather* **2009**, *7*. [CrossRef]
28. Ishii, M.; Shiota, D.; Tao, C.; Ebihara, Y.; Fujiwara, H.; Ishii, T.; Ichimoto, K.; Kataoka, R.; Koga, K.; Kubo, Y.; et al. Space weather benchmarks on Japanese society. *Earth Planets Space* **2021**, *73*, 108. [CrossRef]
29. Jones, J.B.L.; Bentley, R.D.; Hunter, R.; Iles, R.H.A.; Taylor, G.C.; Thomas, D.J. Space weather and commercial airlines. *Adv. Space Res.* **2005**, *36*, 2258–2267. [CrossRef]
30. Doc 10100. *Manual on Space Weather Information in Support of International Air Navigation*, 1st ed.; ICAO Publications: Montreal, QB, Canada, 2019; ISBN 978-92-9258-784-0.
31. Geomagnetic Storms. *As a Contribution to the OECD Project—Future Global Shocks*; Centra Technology, Inc.: Burlington, MA, USA, 2011; 69p. Available online: http://www.oecd.org/gov/risk/46891645.pdf (accessed on 13 March 2023).
32. Service Assessment. *Intense Space Weather Storms 19 October–7 November 2003*; NOAA, National Weather Service: Silver Spring, MD, USA, 2004; 50p. Available online: http://www.weather.gov/media/publications/assessments/SWstorms_assessment.pdf (accessed on 13 March 2023).
33. Koskinen, H.; Tanskanen, E.; Pirjola, A.E.R.; Pulkkinen, A.; Dyer, C.; Rodgers, D.; Cannon, P.; Mandeville, J.-C.; Boscher, D. Space Weather Effects Catalogue, 2001, ESWS-FMI-RP-0001 (2.2). Available online: http://www.spaceweather.org/ISES/swxeff/1.pdf (accessed on 13 March 2023).
34. Oughton, E.J. The Economic Impact of Critical National Infrastructure Failure Due to SPACE weather. In *Oxford Research Encyclopedia of Natural Hazard Science*; Oxford University Press: New York, NY, USA, 2018. [CrossRef]
35. Eastwood, J.P.; Biffis, E.; Hapgood, M.A.; Green, L.; Bisi, M.M.; Bentley, R.D.; Wicks, R.; McKinnell, L.-A.; Gibbs, M.; Burnett, C. The economic impact of space weather: Where do we stand? *Risk Anal.* **2017**, *37*, 206–218. [CrossRef]
36. *Annex 3: Meteorological Service for International Air Navigation*, 20th ed.; ICAO Publications: Montreal, QB, Canada, 2018.
37. Dissemination of Space Weather Advisories. ICAO Publications. 2019. Available online: https://www.icao.int/airnavigation/METP/Panel%20Documents/SWXA_Additional%20Information.pdf (accessed on 13 March 2023).
38. *The Establishment of China/Russian Federation Consortium for ICAO Designated Space Weather Center*; Assembly—40th Session; ICAO Publications: Montreal, QB, Canada, 2019. Available online: https://www.icao.int/Meetings/a40/Documents/WP/wp_336_en.pdf (accessed on 13 March 2023).
39. ACFJ Services for Aviation. Available online: https://www.swpc.noaa.gov/sites/default/files/images/u59/05%20David%20Boteler%20Official.pdf (accessed on 13 March 2023).
40. Kauristie, K.; Andries, J.; Beck, P.; Berdermann, J.; Berghmans, D.; Cesaroni, C.; De Donder, E.; de Patoul, J.; Dierckxsens, M.; Doornbos, E.; et al. Space weather services for civil aviation—Challenges and solutions. *Remote Sens.* **2021**, *13*, 3685. [CrossRef]
41. Bain, H.M.; Steenburgh, R.A.; Onsager, T.G.; Stitely, E.M. A summary of national oceanic and atmospheric administration space weather prediction center proton event forecast performance and skill. *Space Weather* **2021**, *19*, e2020SW002670. [CrossRef]
42. Crown, M.D. Validation of the NOAA space weather prediction center's solar flare forecasting look-up table and forecaster-issued probabilities. *Space Weather* **2012**, *10*, 760. [CrossRef]
43. Kholodkov, K.; Aleshin, I.; Arakelov, A.; Burov, V.; Vasiliev, A.; Ivanov, S. Sino-Russian space weather effort for global air navigation safety. *Russ. J. Earth Sci.* **2021**, *21*, 77. [CrossRef]
44. Aleshin, I.M.; Arakelov, A.S.; Bruevich, E.A.; Burov, V.A.; Ivanov, S.D.; Ochelkov, Y.P.; Repin, A.Y.; Kholodkov, K.I. Methods for monitoring strong space weather disturbances to support international air navigation. *Russ. Meteorol. Hydrol.* **2021**, *46*, 205–211. [CrossRef]

45. Davies, K.; Hartmann, G.K. Studying the ionosphere with the global positioning system. *Radio. Sci.* **1997**, *199732*, 1695–1703. [CrossRef]
46. Afraimovich, E.L.; Perevalova, N.P. *GPS-Monitoring of Earth Upper Atmosphere*; Russian Academy of Sciences, Siberian Branch: Irkutsk, Russia, 2006; p. 460, ISBN 5-98277-033-7.
47. Blagoveshchensky, D.V.; Maltseva, O.A.; Sergeeva, M.A. Impact of magnetic storms on the global TEC distribution. *Ann. Geophys.* **2018**, *36*, 1071. [CrossRef]
48. Mendillo, M. Storms in the ionosphere: Patterns and processes for total electron content. *Rev. Geophys.* **2006**, *44*, 1114. [CrossRef]
49. Tsagouri, I.; Koutroumbas, K.; Elias, P. A new short-term forecasting model for the total electron content storm time disturbances. *J. Space Weather Space Clim.* **2018**, *8*, A33. [CrossRef]
50. Klobuchar, J.A. Ionospheric time-delay algorithm for single-frequency GPS users. *IEEE Trans. Aerosp. Electron. Syst.* **1987**, *3*, 325–331. [CrossRef]
51. Prieto-Cerdeira, R.; Orús-Pérez, R.; Breeuwer, E.; Lucas-Rodriguez, R.; Falcone, M. Performance of the Galileo single-frequency ionospheric correction during in-orbit validation. *GPS World* **2014**, *25*, 53–58.
52. Yuan, Y.; Wang, N.; Li, Z.; Huo, X. The BeiDou global broadcast ionospheric delay correction model (BDGIM) and its preliminary performance evaluation results. *Navigation* **2019**, *66*, 55–69. [CrossRef]
53. Nikitina, L.; Fiori, R.A.D.; Ghoddousi-Fard, R.; Waddington, G.H. Statistical analysis of large and extreme global ionospheric total electron content. *J. Atm. Sol-Ter. Phys.* **2022**, *229*, 105841. [CrossRef]
54. Goodman, J.M. Operational communication systems and relationships to the ionosphere and Space Weather. *Adv. Space Res.* **2005**, *36*, 2241–2252. [CrossRef]
55. Demyanov, V.; Yasyukevich, Y. Space weather: Risk factors for Global Navigation Satellite Systems. *Sol-Terr. Phys.* **2021**, *7*, 28–47. [CrossRef]
56. Gonzalez, W.D.; Joselyn, J.A.; Kamide, D.; Kroehl, H.W.; Rostoker, G.; Tsurutani, B.T.; Vasyliunas, V.M. What is a geomagnetic storm? *J. Geophys. Res.* **1994**, *99*, 5771–5792. [CrossRef]
57. Loewe, C.A.; Prölss, G.W. Classification and mean behavior of magnetic storms. *J. Geophys. Res.* **1997**, *102*, 14209–14213. [CrossRef]
58. Poppe, B.; Onsager, T.G.; Murtagh, W. Halloween storms star in space weather week 2004 meeting. *Space Weather* **2004**, *2*. [CrossRef]
59. Sugiura, M.; Kamei, T. Equatorial Dst Index 1957–1986. In *IAGA Bulletin*; IUGG: Paris, France, 1991; p. 40.
60. Sugiura, M. *Hourly Values of Equatorial Dst for the IGY, Annual International Geophysics Year*; Pergamon Press: Oxford, UK, 1964; Volume 35, p. 9.
61. Sergeeva, M.A.; Maltseva, O.A.; Gonzalez-Esparza, J.A.; De la Luz, V.; Corona-Romero, P. Features of TEC behaviour over the low-latitude North-American region during the period of medium solar activity. *Adv. Space Res.* **2017**, *60*, 1594–1605. [CrossRef]
62. Gonzalez-Esparza, J.A.; De la Luz, V.; Corona Romero, P.; Mejia-Ambriz, J.C.; Gonzalez, L.X.; Sergeeva, M.A.; Romero-Hernandez, E.; Aguilar-Rodriguez, E. Mexican space weather service (SCIESMEX). *Space Weather* **2017**, *15*, 3–11. [CrossRef]
63. Li, W.; Wang, K.; Yuan, K. Performance and Consistency of Final Global Ionospheric Maps from Different IGS Analysis Centers. *Remote Sens.* **2023**, *15*, 1010. [CrossRef]
64. Chen, P.; Liu, H.; Ma, Y.; Zheng, N. Accuracy and consistency of different global ionospheric maps released by IGS ionosphere associate analysis centers. *Adv. Space Res.* **2020**, *65*, 163–174. [CrossRef]
65. Hernández-Pajares, M.; Juan, J.M.; Sanz, J.; Orus, R.; Garcia-Rigo, A.; Feltens, J.; Komjathy, A.; Schaer, S.C.; Krankowski, A. The IGS VTEC maps: A reliable source of ionospheric information since 1998. *J. Geod.* **2009**, *83*, 263–275. [CrossRef]
66. Arikan, F.; Erol, C.B.; Arıkan, O. Regularized estimation of vertical total electron content from Global Positioning System data. *J. Geophys. Res.* **2003**, *108*, 1469. [CrossRef]
67. Syrovatskiy, S.V.; Yasyukevich, Y.V.; Vesnin, A.M.; Edemskiy, I.K.; Voeykov, S.V.; Zhivetiev, I.V. The effect of solar flares on the ionosphere of the Earth during 24th cycle of solar activity. *Mem. Fac. Phys. Mosc. Univ.* **2018**, *4*.
68. Dmitriev, A.V.; Yeh, H.-C.; Chao, J.-K.; Veselovsky, I.S.; Su, S.-Y.; Fu, C.C. Top-side ionosphere response to extreme solar events. *Ann. Geophys.* **2006**, *24*, 1469–1477. [CrossRef]
69. Kunitsyn, V.E.; Nazarenko, M.O.; Nesterov, I.A.; Padokhin, A.M. Solar Flare Forcing on Ionization of Upper Atmosphere. Comparative Study of Several Major X_Class Events of 23rd and 24th Solar Cycles. *Phys. Earth Atmos. Hydrosphere. Mosc. Univ. Phys. Bull.* **2015**, *70*, 312–318. [CrossRef]
70. Le, H.; Liu, L.; Chen, Y.; Wan, W. Statistical analysis of ionospheric responses to solar flares in the solar cycle 23. *J. Geophys. Res. Space* **2013**, *118*, 576–582. [CrossRef]
71. Sergeeva, M.A.; Maltseva, O.A.; Vesnin, A.M.; Blagoveshchensky, D.V.; Gatica-Acevedo, V.J.; Gonzalez-Esparza, J.A.; Chernov, A.G.; Orrala-Legorreta, I.D.; Melgarejo-Morales, A.; Gonzalez, L.X.; et al. Solar flare effects observed over Mexico during March. *Remote Sens.* **2023**, *15*, 397. [CrossRef]
72. Machol, J.L.; Eparvier, F.G.; Viereck, R.A.; Woodraska, D.L.; Snow, M.; Thiemann, E.; Woods, T.N.; Mc Clintock, W.E.; Mueller, S.; Eden, T.D.; et al. GOES-R Series Solar X-ray and Ultraviolet Irradiance. In *GOES-R Series*; Elsevier: Amsterdam, The Netherlands, 2020; pp. 233–242.
73. Gaisser, T.; Engel, R.; Resconi, E. Cosmic rays. In *Cosmic Rays and Particle Physics*; Cambridge University Press: Cambridge, UK, 2016; pp. 1–11. [CrossRef]

74. Valdés-Galicia, J.F.; González, L.X. Solar modulation of low energy galactic cosmic rays in the near-earth space environment. *Adv. Space Res.* **2016**, *57*, 1294–1306. [CrossRef]
75. Sako, T.; Watanabe, K.; Muraki, Y.; Matsubara, Y.; Tsujihara, H.; Yamashita, M.; Sakai, T.; Shibata, S.; Valdés-Galicia, J.F.; González, L.X.; et al. Long-lived Solar Neutron Emission in Comparison with Electron-produced Radiation in the 2005 September 7 Solar Flare. *Astrophys. J.* **2006**, *651*, L69–L72. [CrossRef]
76. González, L.X.; Valdés-Galicia, J.F.; Sánchez, F.; Muraki, Y.; Sako, T.; Watanabe, K.; Matsubara, Y.; Nagai, Y.; Shibata, S.; Sakai, T.; et al. Re-evaluation of the neutron emission from the solar flare of 7 September 2005, detected by the Solar Neutron Telescope at Sierra Negra. *Astrophys. J.* **2015**, *814*, 136. [CrossRef]
77. Muraki, Y.; López, D.; Koga, K.; Kakimoto, F.; Goka, T.; González, L.X.; Masuda, S.; Matsumoto, H.; Mastubara, Y.; Miranda, P.; et al. Simultaneous Observation of Solar Neutrons from the ISS and High Mountain Observatories in Association with a Flare on 8 July 2014. *Sol. Phys.* **2016**, *291*, 1241–1265. [CrossRef]

remote sensing

MDPI

Technical Note

Topside Ionospheric Tomography Exclusively Based on LEO POD GPS Carrier Phases: Application to Autonomous LEO DCB Estimation

Manuel Hernández-Pajares [1,*], Germán Olivares-Pulido [1], M. Mainul Hoque [2], Fabricio S. Prol [2], Liangliang Yuan [2], Riccardo Notarpietro [3] and Victoria Graffigna [1]

1 UPC-IonSAT/IEEC-UPC, Mod.C3 Campus Nord UPC, C/Jordi Girona 1-3, E-08034 Barcelona, Spain
2 German Aerospace Center (DLR), 17235 Neustrelitz, Germany
3 EUMETSAT, 64295 Darmstadt, Germany
* Correspondence: manuel.hernandez@upc.edu; Tel.: +34-93-4016019

Abstract: This paper presents a novel technique to estimate DCBs from GPS transmitters and receivers on-board Low Earth Orbit (LEO) satellites. The technique consists of obtaining the DCBs as residuals from the difference between the ionospheric combination of the code and the associated ionospheric delay. The ionospheric delay is computed with TOMION, a background-model-free ionospheric tomographic technique based on dual-frequency GPS carrier phase data only, and solved with a Kalman filter. Thus, DCBs are also estimated epoch-wise from the LEO Precise Orbit Determination (POD) GPS receiver as a secondary product. The results for GPS satellite DCBs, obtained exclusively from the three MetOp LEO POD GPS receivers over four consecutive weeks, are in full agreement (i.e., at the level of a few tenths of ns) with those reported independently with other techniques from hundreds of ground-based receivers exclusively, by JPL and CODE analysis centers.

Keywords: ionospheric tomography; Global Navigation Satellite Systems; Low Earth Orbiting satellites; differential code biases

1. Introduction

The inter-frequency differential code bias (DCB) is usually one of the most important error sources that affect the accuracy of slant total electron content (TEC) estimation from ground- as well as space-based measurements of the Global Navigation Satellite System (GNSS). They can be understood as hardware delays caused by the digital signal processing components, antenna and cables, when signals at the satellite and receiver are encoded. For example, ref. [1] quantified timing delays between the different legacy and modernized GPS and Galileo signals broadcast on L_1 and their dependencies on factors such as user receiver filter bandwidth, filter transfer function and delay locked loop (DLL) correlator spacing; ref. [2] provided a comprehensive overview of pseudorange biases and their dependency on receiver front-end bandwidth and correlator design, and [3] showed that the long-term variations of current DCB products may vary significantly.

The most common DCB calibration methods can provide high accuracy in single-frequency precise point positioning (ref. [4]), it being common to estimate them in ground-based stations (ref. [5]) and to consider GNSS satellite and receiver DCBs as constant during a single day (refs. [6,7]).

Over the last two decades, however, significant research concerning the Global Positioning System (GPS) DCB has been conducted. For example, ref. [8] analyzed the stability of the GPS instrumental biases; ref. [9] found that the day-to-day and annual variation in the estimated GPS DCB is related to the ionospheric variability; and [10] concluded that DCB variability is attributed to the GPS satellite replacements with different satellite types and the zero-mean condition imposed on all satellite DCBs. Furthermore, refs. [11,12]

found intraday variabilities with clear correlations in temperature. Estimate methods for DCBs have also been developed for on-board Low Earth Orbit (LEO) satellites. In this regard, ref. [13] showed that the STEC estimate might be enhanced if the temperature dependency of DCB estimation is considered as well. Ref. [14] estimated the COSMIC GPS DCB receivers in order to perform plasmaspheric observations from LEO satellites with GNSS data. Moreover, ref. [15] developed a method whereby GNSS observations from standalone LEO satellites (i.e., with no ground-based GNSS data) can estimate GPS satellite DCBs.

Various methods for DCB estimation in LEO satellites were proposed in the past decade. For example, least squares-based estimation using a Fengyun-3C satellite was developed by [16]. A zero method was proposed and tested using long-term observations of the Challenging Minisatellite Payload (CHAMP) and Gravity Recovery and Climate Experiment (GRACE) by [17]. GPS P1-P2 DCBs and vertical TEC (VTEC) above orbit height were estimated with JASON-2 and precise orbit determination (POD) observations epoch by epoch by [18]. Ref. [19] estimated the GPS and Beidou satellite DCBs as well as receiver DCB onboard Fengyun-3C using the same method as [18]. Ref. [15] proposed an approach to estimate the GPS satellite DCB based on the ionospheric spherical symmetry assumption for CHAMP and Constellation Observing System for Meteorology, Ionosphere and Climate (COSMIC) observations. It was found that the GPS satellite DCB estimates based on CHAMP were much better than those based on COSMIC. Refs. [20,21] proposed an inequality constrained least squares method to estimate the COSMIC receiver DCBs with LEO satellites.

In this research, we study the capability of performing topside ionospheric tomography based exclusively on dual-frequency carrier phase LEO POD GPS data and validated in terms of the precision and accuracy of the GPS transmitter DCBs. The advantages of using tomography include the consideration of several ionospheric layers, instead of the common use of unrealistic mapping functions to convert VTEC into the slant direction.

A few studies have already developed tomography approaches to obtain topside electron density representations. For example, ref. [22] presented initial results of the sounding of the topside ionosphere and plasmasphere based on GPS measurements from CHAMP. Ref. [23] introduced a new mathematical approach to imaging the electron density distribution in the high regions of the topside ionosphere and the plasmasphere using GPS measurements from LEO satellites. A data assimilative method based on 3D Var for the sounding of the ionosphere and plasmasphere using COSMIC-GPS measurements was introduced in [24]. Using Jason-1 plasmaspheric total electron content (TEC) measurements, ref. [25] developed a tomographic technique for the reconstruction of the plasmaspheric density. COSMIC satellite data have also been used for a tomographic method that estimates the plasmasphere [26]. However, the common approach is to calibrate the DCB unknowns before the tomography inversion. In [27], the DCBs are estimated simultaneously with the electron densities, regarding background values and model respectively, from input GPS ground- and LEO-based (radio occultation) measurements. In this work, the GPS satellite DCB and on-board LEO DCBs are directly retrieved (without background values) right after the 3D electron density distributions are estimated. This estimation is exclusively based on dual-frequency GPS carrier phase measurements gathered from the LEO zenith antennas (i.e., without incorporating radio occultation measurements) and without any electron density background model.

In this regard we processed 1Hz dual-frequency GPS measurements of the POD receivers of the three MetOp LEO satellites, metb (M01), meta (M02) and metc (M03), provided by the European Organization for the Exploitation of Meteorological Satellites (EUMETSAT) from days 173 to 255, 2020.

2. TOMION in a Nutshell

TOMION is a data-driven ionospheric model mostly developed by the first author of this work during more than 25 years (refs. [28,29]), in collaboration with Raul Orus-

Perez and David Roma-Dollase for the kriging interpolation and multi-GNSS extensions (see [30,31]). TOMION allows a tomographic estimation by a Kalman filter of the number density of free ionospheric electrons from GPS carrier phase dual-frequency data only and without any background model. Then, full ionospheric maps are obtained via a kriging-based interpolation [30] for the vertically integrated electron density (the vertical total electron content, VTEC). Recently, TOMION has been modified with the capability of processing multi-GNSS measurements (ref. [31], pp. 19–22).

The tomographic model assumes that the ionospheric electron content is distributed in voxels throughout the ionosphere and thus the ionospheric, also known as (hereinafter, a.k.a.) geometry-free (GF), combination ($L_{GF} \equiv L_I$) of two GNSS carrier phases in length units, L_1 and L_2 with frequencies $f_1 > f_2$, can be expressed through the fundamental equation Equation (1) solved in TOMION,

$$L_{GF,j} \equiv L_1 - L_2 = I + B_{GF,j} = \alpha \cdot S + B_{GF,j} \simeq \alpha \cdot \sum_j N_{e,j} \cdot l_j + B_{GF,j} \tag{1}$$

where $\alpha = 40.3 \left(\frac{1}{f_2^2} - \frac{1}{f_1^2} \right)$ in International System of Units (IS) ($\alpha \simeq 0.105$ m/TECU for L_1 and L_2 GPS frequencies, see, for example, [32]), S represents the Slant Total Electron Content (STEC), the index j runs over the number of illuminated voxels and l_j is the segment of the straight line between satellite and receiver that pierces the jth voxel. N_e is the ionospheric electron density and $B_{GF,j}$ represents the corresponding carrier phase ambiguity, both solved as an approximate random walk and random variable (constant), respectively, in a Kalman filter. The electron density random process is estimated with an update of the covariance matrix, with a process noise of $9 \cdot 10^{-8}$ (meters of $L_1 - L_2$ delay/km)^2/h. This optimal approximation has been established empirically, after 25 years of daily processing for contributing to IGS with the UPC-IonSAT global ionospheric maps (GIMs), among different tomographic experiments. The initial observation covariances are assumed to be diagonal, with an a priori standard deviation per $L_1 - L_2$ observation equation of 0.05 m.

TOMION is the software used in the generation of UPC-IonSAT GIMs of VTEC for IGS, such as the UQRG one, one of the best, RMS-wise, GIMs in IGS (ref. [33]. UQRG GIM has been assessed with other GIMs and it provides RMS values of 2 TECU [34,35]).

The UQRG GIM produced by TOMION is, for instance, able to detect realistic features of the polar ionosphere as well (ref. [36]) and to provide a realistic and sensitive storm index (ref. [37]). The tomography performed by TOMION is able to ingest different geometries and types of input measurements (ref. [38]), in agreement with independent measurements and models (refs. [39,40]).

3. Methodology for DCB Retrieval

The method introduced in this work computes DCBs as the difference between the Geometry-Free combination of the pseudodistance (a.k.a. code or pseudorange) and the associated ionospheric delay estimated with TOMION, a tomographic model without using background electron density and without using GNSS pseudorange measurements (significantly affected by thermal noise and multipath, see, for instance, [32]) as inputs. This section presents the development from the GNSS equations to the final formula that estimates the combined satellite and receiver DCBs in terms of the GF combinations of the code and the calibrated carrier-phase.

The GNSS observables consist of the linear combination of the geometric distance (between satellite and receiver) and additional propagation delay sources. They can be classified into two groups: dispersive and non-dispersive terms. Namely, the link between observables, geometric distance and additional delay terms is summarized in the following expression:

$$\begin{array}{lcl} P^s_{r,i} & = & \rho^s_r + c(d\tau_r - d\tau^s) + T^s_r + I^s_{r,i} + D_{r,i} + D^s_i + \epsilon_i \\ L^s_{r,i} & = & \rho^s_r + c(d\tau_r - d\tau^s) + T^s_r - I^s_{r,i} + \frac{c}{f_i}\phi^s_r + B^s_{r,i} + \xi_i\,, \end{array} \tag{2}$$

where $P^s_{r,i}$ and $L^s_{r,i}$ are the code and carrier-phase (obtained from integration of the Doppler shift) measurements for the ith frequency; r and s stand for the receiver and satellite indexes; ρ^s_r is the geometric distance; $d\tau_r$ and $d\tau^s$ are the receiver and satellite clock errors, respectively; $I^s_{r,i} = \frac{40.3}{f_i^2} \cdot S$, T^s_r and ϕ^s_r are the ionospheric delay, tropospheric delay and phase wind-up, respectively, between receiver and satellite transmitter. Note that only the code hardware delays are explicitly indicated for the receiver and the satellite, $D_{r,i}$ and D^s_i, respectively, whereas the ambiguity term $B^s_{r,i}$ only affects carrier-phase measurements and it also contains the corresponding carrier-phase hardware delays.

The combination of the code from two different frequencies, for example, corresponding to L_1 and L_2 carriers, removes all non-dispersive terms, thus yielding the GF combination:

$$P^s_{r,GF} \equiv P^s_{r,2} - P^s_{r,1} = I^s_{r,GF} + D_{r,GF} + D^s_{GF}\,, \tag{3}$$

where the associated ionospheric delay $I^s_r \equiv I^s_{r,GF} = \alpha \cdot \int_r^s N_e\, dl$ is proportional to the path integral of the ionospheric electron density, N_e, from the receiver r to the satellite s; and $D_{r,GF}$, D^s_{GF} are the GF combinations of, respectively, the receiver and satellite hardware delays, a.k.a. differential code biases (DCBs).

Similarly to the GF combination of the code in Equation (3), the GF combination of the carrier-phase is as follows:

$$L^s_{r,GF} \equiv L^s_{r,1} - L^s_{r,2} = I^s_{r,GF} + c\left(\frac{1}{f_1} - \frac{1}{f_2}\right)\phi^s_r + B^s_{r,GF}\,, \tag{4}$$

where $B^s_{r,GF} \equiv B^s_{1,r} - B^s_{2,r}$ is the GF combination of the ambiguities in both frequencies and ϕ is the phase wind-up angle, with a term impact that is small (a fraction of the difference of wavelengths of both frequencies $\lambda_1 - \lambda_2$) and can be corrected. The replacement of the ionospheric term in Equation (3) using Equation (4) yields, neglecting the code thermal noise and multipath, the following expression:

$$P^s_{r,GF} = L^s_{r,GF} - B^s_{r,GF} - (\lambda_1 - \lambda_2)\phi^s_r + D_{r,GF} + D^s_{GF}\,, \tag{5}$$

and, thereby, after rearranging terms, it leads to the following expression of DCBs in terms of GF combinations of the observables and the ambiguity:

$$D_{r,GF} + D^s_{GF} = P^s_{r,GF} - \tilde{L}^s_{r,GF}\,, \tag{6}$$

where $\tilde{L}^s_{r,GF} \equiv L^s_{r,GF} - \hat{B}^s_{r,GF} - (\lambda_1 - \lambda_2)\phi^s_r$ is the calibrated carrier-phase of the GF combination. It is computed from the measured carrier phase, $L^s_{r,GF}$, minus the estimated value of the corresponding ambiguity, $\hat{B}^s_{r,GF}$ and the small wind-up term as well, $(\lambda_1 - \lambda_2)\phi^s_r$. In other words, it is the estimated GF ionospheric delay, $I^s_{r,GF}$. Consequently, according to Equation (6), the sum of the transmitter and receiver DCBs can be estimated as residuals from the difference between the GF combinations of the calibrated carrier-phase and the code.

However, in order to further proceed and estimate those DCBs, it is necessary to previously estimate the ambiguity term, $B^s_{r,GF}$. It is possible to do so with further modelling of the ionospheric term in Equation (4). A system of equations based on Equation (4) is rank-deficient. However, it is possible to fix the rank by any model that separates the relative geometry between satellite and receiver and the state of the ionosphere [41]. The reason is that the ionospheric term, I_r^s, as opposed to the ambiguity term, $B_{r,GF}^s$, depends on the relative geometry.

For example, the simplest method would be the so-called thin-shell model, whereby the ionospheric delay is computed as follows:

$$I_{r,GF}^s = \alpha S_r^s = \alpha \left[1 - \left(\frac{R_e \cos E_r^s}{R_e + h} \right)^2 \right]^{-1/2} V_{IPP_r^s}, \tag{7}$$

where S is the Slant Total Electron content (STEC) along the path from the transmitter to the receiver, R_e is the average Earth's radius, h is the effective height (e.g., 450 km adopted in IGS, see [33]), where all the ionospheric electron content is assumed located, E_r^s is the elevation angle of the satellite s above the local horizon of receiver r and $V_{IPP_r^s}$ is the Vertical Total Electron Content at the ionospheric pierce point (a.k.a. IPP) where the straight line between the satellite s and the receiver r pierces the thin shell. This model consists of mapping the VTEC along the line of sight, thus computing the corresponding STEC. Note that, as has been discussed above, $STEC$ (S) is also the path integral of the ionospheric electron, N_e, content along the line of sight between the receiver r and the GNSS satellite s, i.e., $S = \int_r^S N_e \, dl$.

Although the simplest, the thin-shell model is also the less accurate one. See, for example, ref. [42] for further improvement of $VTEC$-based models.

Another approach is the estimation of the carrier phase ambiguity $B_{r,GF}^s$; in addition, the ionospheric electron density, N_e, follows the approach summarized in Section 2.

Regardless of whether the model is based on $VTEC$ or electron density estimation, both strategies can estimate the ambiguity term, $B_{r,GF}^s$, thus calibrating the GF carrier-phase, $L_{r,GF}^s$, and, eventually, can lead to the computation of DCBs according to Equation (6).

Altogether, TOMION combines and solves the following system of equations:

$$\begin{aligned} L_{r,GF}^s(t_k) - (\lambda_1 - \lambda_2)\phi_r^s &= \alpha \cdot \textstyle\sum_i N_{e,i}(t_k) \cdot l_i + B_{r,GF}^s(t_k) \\ N_{e,i}(t_k) &= \hat{N}_{e,i}(t_{k-1}) + \delta(t_k) \\ B_{r,GF}^s(t_k) &= \hat{B}_{r,GF}^s(t_{k-1}), \end{aligned} \tag{8}$$

where t_k stand for the kth time-epoch; ϕ_r^s is the wind-up angle; $N_{e,i}$ and $\hat{N}_{e,i}$ are the electron density and its estimate, respectively, at the ith voxel; δ is the system prediction noise; and $B_{r,GF}^s$ and $\hat{B}_{r,GF}^s$ are the ambiguity term and its estimate, respectively. In cases when a cycle-slip occurs, i.e., when a significant jump in $L_{r,GF}^s$ drift rate occurs (a variant of method described in [43]), the Kalman filter starts computing the new ambiguity, after considering the unknown as a white noise random process right during the first epoch after the cycle-slip.

Note that, as a consequence of estimating $B_{r,GF}^s$ on an epoch-wise basis with Equation (8), DCB estimates and their uncertainty can then be estimated epoch-wise as well either in real-time by solving Equation (6), or in post-process by considering the last estimation of the carrier phase ambiguity given by the Kalman filter for each given transmitter–receiver phase continuous arc, the one typically better estimated. This is the way in which UPC-IonSAT provides the DCBs to IGS.

4. Data and Results

The dataset consists of carrier-phase (L1W and L2W) and code (C1W and C2W) 1 Hz measurements from the POD GPS receivers on-board the three MetOp LEO satellites A, B and C, for day 176 and days 224 to 252, 2020, and the corresponding receiver and transmitter precise positions.

A first, qualitative way of comparing the performance of the TOMION model Equation (8) is looking at the consistency of the electron density solutions, considering the different combinations between the sources of GPS data (ground-based and MetOp POD) and different vertical resolutions: two vs four layers, centered on the MetOp orbital height (i.e., one and two layers of them, respectively, right above MetOp satellites).

A second, quantitative way of comparing the performance of TOMION, focusing on four vs two layers, is looking at the impact on the estimated standard deviations of the corresponding receiver and transmitter C2W-C1W DCB estimate, $D_{r,GF} + D^{s}_{GF}$. They are obtained, one per Line-Of-Sight (LOS), after TOMION directly estimates the carrier phase ambiguity at the same time as the electron density. DCBs are expected to be rather stable within hours, or maybe even days.

Therefore, if the number of layers (e.g., two or four) is not relevant, then the DCBs and their uncertainties estimated by each model should be rather the same values (i.e., below the SNR of the GPS signal). In this regard, this section presents an analysis of the estimated DCBs that demonstrates that the number of layers has an impact on the accuracy of the ionospheric sounding.

The stability of the DCBs estimated by TOMION, from MetOpt POD GPS measurements only and over four consecutive weeks was analyzed. The DCB estimation was assessed by comparing with GPS transmitter DCBs computed with different techniques by CODE and JPL IGS IAACs (Ionosphere Associate Analysis Centers) from ground-based measurements only.

It has been previously reported that the results computed by TOMION with ground receivers only might be affected by vertical correlations when the ionospheric grid is designed with more than two layers [29]. This is due to the relative geometry between satellites and ground receivers. Indeed, GNSS measurements from a ground receiver provide high horizontal resolution, but they lack vertical resolution. However, the MetOp POD GPS measurements from the MetOp orbit height (800 km approximately) provide certain vertical resolution, up to two layers above, thereby removing the vertical correlation artifacts in four-layer models.

4.1. Ionospheric Tomography

We compared, similar to what we did in [38], the tomographic solution under different combinations of sources of dual-frequency GPS measurements, mainly 150 ground GPS receivers, providing ambiguous STEC observations from the GPS transmitters (around 20,200 km height) to the ground and GPS measurements gathered from the POD receiver on board MetOp A, B and C.

A representative example can be seen in Figure 1, where the Vertical Upper Electron Content above 780 km (VUEC) provides more consistent results when the MetOp measurements are considered. In the second and third rows, the negative VUEC values at very high latitudes are negligible, compared with the runs with only ground-based GPS measurements, the first row (it has to be considered that no background model or any constraint on electron density positivityis considered). This improvement is still more clear when two layers above MetOp satellites (four layers in total, second column) are considered instead of just one (first column). The tomographic results with MetOp POD GPS-only (central-right plot at Figure 1) are similar in consistency to the ones with ground and MetOp POD GPS data, just showing a reduced range of values due to the smaller set of illuminated voxels (bottom-right plot at the same figure).

4.2. DCBs of MetOp POD GPS Receivers

In Figure 2, the TOMION daily estimations of the GPS $P_2 - P_1$ DCBs of the POD GPS receivers of the three MetOp LEO satellites referring to the average of the GPS transmitter DCBs (taken as data as usual in IGS) are plotted vs time.

On the one hand the TOMION estimation shows stability within the 28-day period, peak-to-peak, better than 1 TECU, i.e., 0.3 ns. This stability is similar or better than the typical stability around 1 ns observed for ground GPS receivers (e.g., see Figure 16 in [33]).

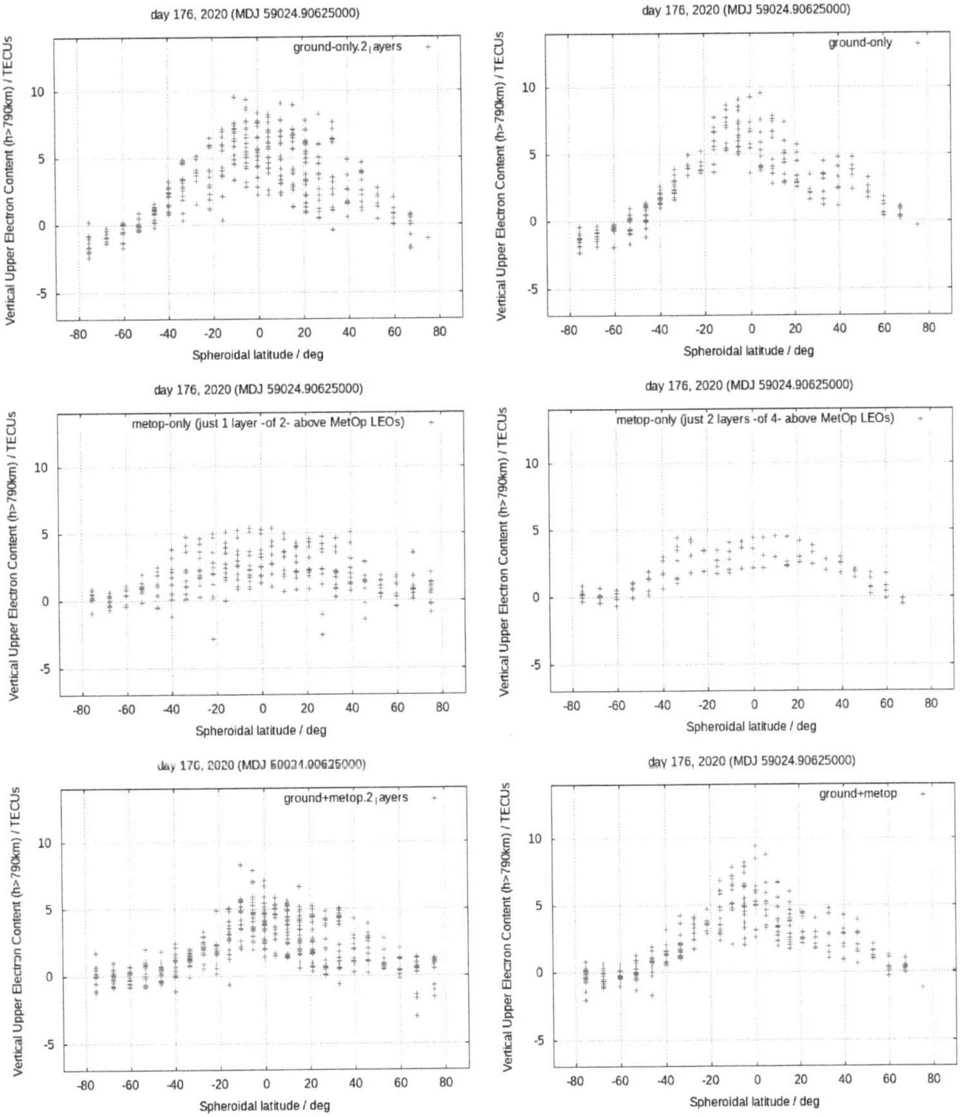

Figure 1. Comparison of the Vertical Upper Electron Content (in TECUs) vs latitude at GPS time 21 h 45 m of day 176 of year 2020, considering: (**left**) 2 layers (1 above MetOp LEO satellites) in the first column and (**right**) 4 layers (2 above MetOp LEO satellites) in the second one. The input measurements consist of ground GPS ground-based measurements only (first row), MetOp POD GPS measurements only (second row) and combined ground GPS and MetOP POD GPS measurements (third row).

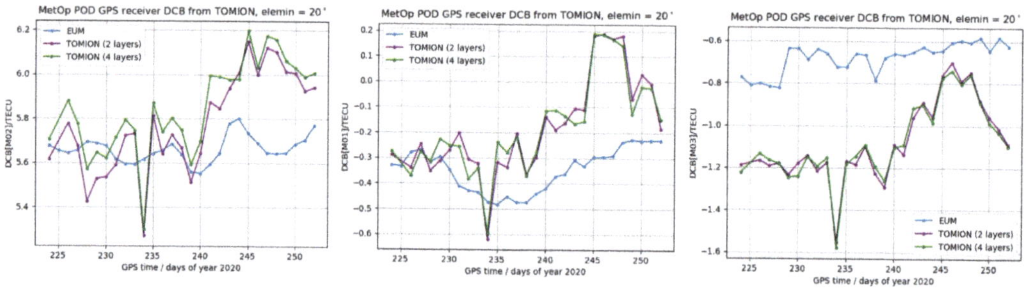

Figure 2. Daily values of the MetOp POD GPS receiver DCBs, in TECUs, from left to right, DCB[meta], DCB[metb], DCB[metc], in TECU, vs. day of year 2020, computed with TOMIONv1 with MetOp GPS data only under 2-layer TOMION run (green points), under 4-layer TOMION model (magenta points) and with the updated EUMETSAT technique (blue).

In detail, the stability of TOMION results with four layers (two above the MetOp LEO satellites) are slightly more stable (at below 0.1 ns level, see left plot in Figure 2) than the results with two layers only, namely it is 1 above. Also note that the tomographic DCB results vary in time. Although we cannot identify the origin of such DCB variability, it has been previously reported that GNSS DCB estimates vary with environmental [11] and/or CPU temperature [44]. Furthermore, if DCB and temperature are correlated as presented in [11] for ground receivers, then it might be possible to model the environmental temperature variability with DCB estimates at LEO orbit heights.

In Figure 2 the new EUMETSAT DCB results, after applying the zero TEC technique as discussed by [44], are also shown, comparing well with the TOMION results. The EUMETSAT approach considers as daily receiver DCB the first quartile of the distribution of the minimum apparent sTEC when they are computed with zenith data over the previous 5 days: only data at high latitudes (above 60 deg north/south), during night local times (from 18:00 to 6:00) and taken above 40 deg elevation, are considered. We can see from Figure 2 that EUMETSAT results seem more stable. This is simply related to the fact that, for estimating the daily DCB for day X, we are taking the first quartile of the minimum sTEC computed for the interval [day X - 4 days; day X]. So, this clearly impacts the stability of the EUMETSAT DCB time series, which are somehow more "smoothed". However, all of them show a variability within 0.3 TECU, without apparent differences between the three Metops.

4.3. External Assessment via Transmitter DCBs

It is also possible to assess the TOMION performance (two and four layers) from MetOp GPS POD data only and without any background model, with a comparison between the corresponding GPS transmitter DCB estimates and those from IGS centers, computed from GPS ground data only and regarding the data given by the sum of transmitter DCBs (daily satellite DCBs can be obtained from Equation (6), see Appendix A for further details).

As can be seen in Figure 3, the GPS transmitter DCBs estimated by TOMION, as a by-product of the tomography run from the MetOp A, B and C dual-frequency carrier phase measurements only, are in good agreement with those from CODE and JPL across days 224 to 252, 2020. They typically agree at a few tenths of ns level and up to 0.5 ns in the worse case, i.e., at the level of the agreement shown between transmitter DCBs provided by different IGS ionospheric analysis centers from ground GPS data only (see, for instance, Figure 7 in [33] and associated comments).

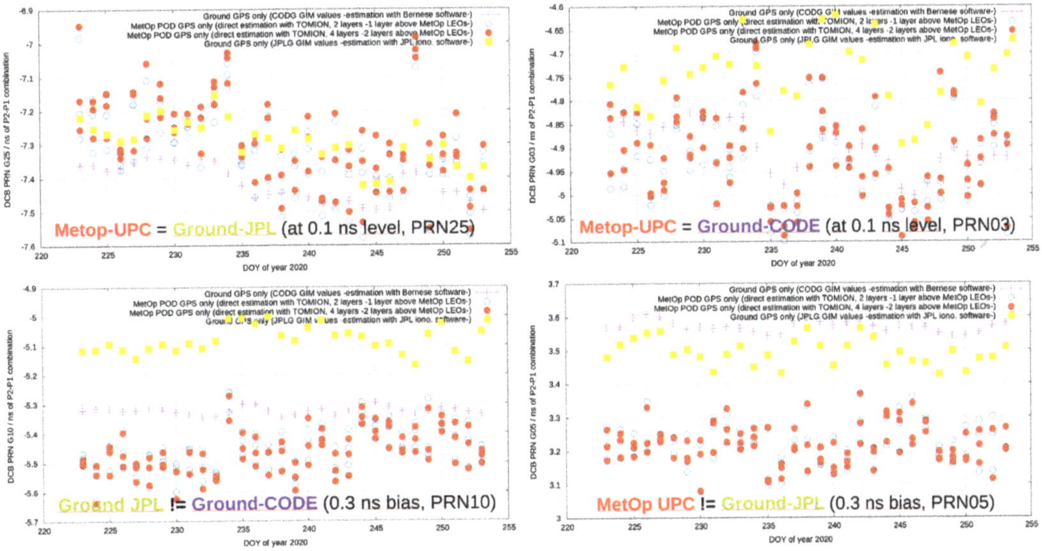

Figure 3. Four typical examples of comparison of GPS transmitter DCBs obtained by TOMION runs based on MetOp LEO data only (2 and 4 layers, i.e., 1 and 2 layers above) vs. ground-based values from IGS centers (CODE and JPL): PRN25 (**top−left**), PRN03 (**top−right**), PRN10 (**bottom−left**) and PRN05 (**bottom−right**).

5. Discussion

In the previous section, we showed that the estimated tomographic model of the LEO topside electron content only fed with raw GPS ionospheric $L_1 - L_2$ measurements and solved without any background model is able to provide directly consistent electron content results. Moreover, as a byproduct, it is able to provide transmitter DCBs comparable to the ones computed from hundreds of GPS ground-based receivers by other analysis centers.

In this section, we focus on the influence of two particular aspects of the modelling which are not usually discussed when topside electron content is estimated and the DCBs of LEO-based GNSS receivers are compared: the influence of changes in the DCB datum and the GPS receiving antenna phase center variation. Moreover the potential influence of the difference between the LEO orbits will also be addressed.

5.1. Influence of Changes in the DCB Data

The estimated DCBs (for GPS in particular) delivered by many ionospheric analysis centers, such as those of IGS (refs. [31,33]) refer to the average of the GPS transmitter values, which is defined as zero. This is the data in the definition of GPS DCBs. As can be seen in the top plots of Figure 4 corresponding to the four analyzed weeks, it might happen that some days some analysis centers (JPL and UPC in this) do not provide the daily DCB values for the complete set of GPS transmitters (top-right plot), but keep the DCB data to the given set of GPS transmitters (top-left plot). However, the effect is small, at or below 0.1 ns at C2W-C1W level, as can be seen, for instance, for GPS10 DCB estimation via JPL in the bottom-left plot of Figure 3 during days 233–243 and 249–250 of 2020, where the JPL DCB datum is based on 30 and not 31 GPS transmitters (top-right plot of Figure 4).

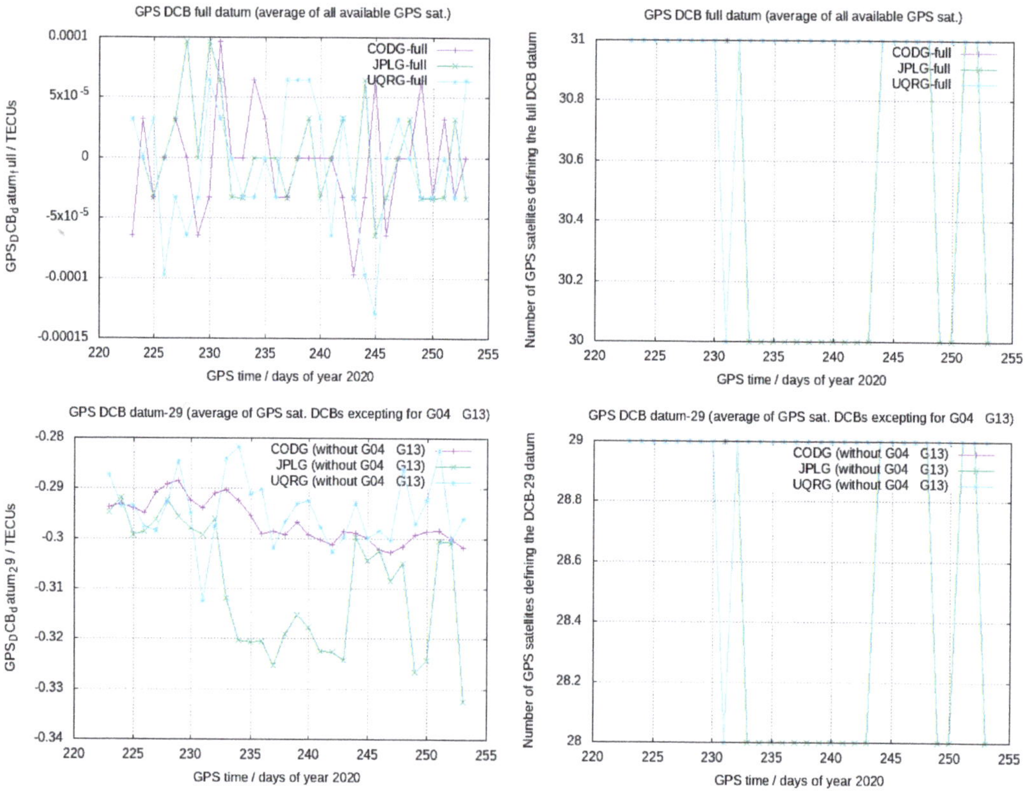

Figure 4. Average of all the available GPS transmitter C2W-C1W DCBs (i.e., GPS data) in the CODE, JPL and UPC-IonSAT GIMs (**top**−**left**); corresponding number of GPS transmitters (**top**−**right**). Similarly, for a slightly different datum of 29 transmitters (DCB average excluding GPS04 and GPS13), the GPS DCB data (**bottom**−**left**) and the corresponding number of GPS transmitters among the 29 ones (**bottom**−**right**).

Similarly, in the estimation of the DCBs based only on MetOp measurements presented in previous sections, two GPS transmitters (G04 and G13) were not available with enough high elevation, from the close orbits of the three LEOs, meta, metb and metb (see Figure 8), and they did not take part in the DCB data of our results, based on the remaining 29 GPS transmitters. Similarly as in the previously noted case of JPL DCBs, the impact is small, at the level of 0.1 ns, i.e., 0.3 TECU approximately. This can be seen in the average of these 29 GPS transmitter DCBs from CODE, JPL and UPC-IonSAT ground-based DCBs, associated with the corresponding GIMs, in the top-bottom plot of Figure 4, where the change of the data for 28 transmitters mostly for JPL (days 233–243 and 249–250) and UPC-IonSAT (day 231) appears in this case as a sort of second-order datum effect. A similar case of loss of one additional GPS transmitter, but during day 234, affects the DCB data in the MetOp-only based solution and explains the jump of -0.3 TECU of the receiver DCB value during this day (see Figure 2).

5.2. Influence of the GPS Receiving Antenna Phase Center Variation

Accurate knowledge of the LEO receiver antenna phase center variation can require dedicated in-flight research, such as the case of the characterization of the zenith GPS antenna of MetOp receivers, meta, metb and metb [45]. Despite the horizontal separation of L_1 and L_2 phase centers being larger than 10 cm, the vertical one is smaller than 1 cm.

This brings us to neglect this correction, assuming that its impact should be very small, because the results in our problem should depend mostly on the close to vertical observations from the LEOs. Indeed, the main correction affects the low elevation measurements (see Figure 5) and the final impact is very low in the tomographic solution (at sub-TECU level, see Figure 6) and in the DCB estimation, typically less or much less than 0.1 ns in C2W-C1W DCB (see representative example for GPS10 in Figure 7).

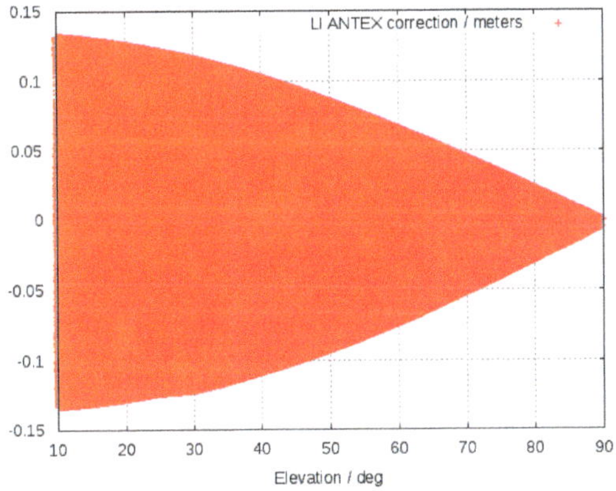

Figure 5. GPS receiving antenna corrections for the MetOp LEOs in ionospheric combination during day 176 of year 2020.

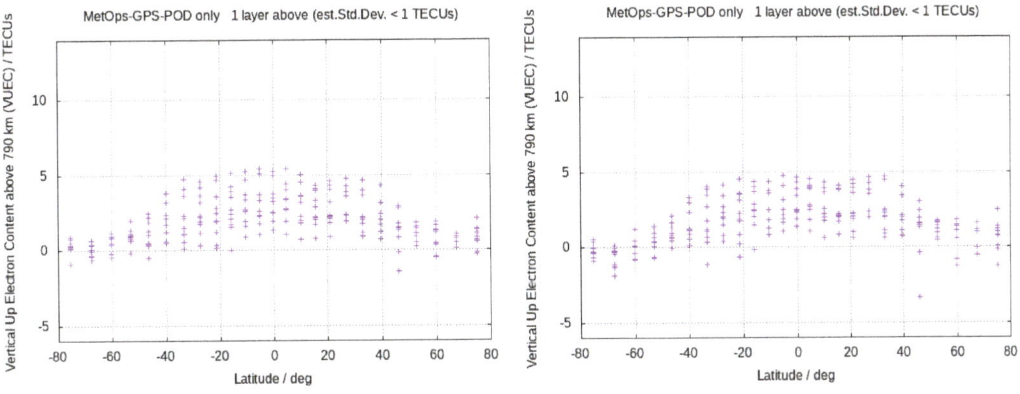

Figure 6. Comparison of the Vertical Upper Electron Content (in TECUs) vs latitude at GPS time 21 h 45 m of day 176 of year 2020, considering 2 layers (1 above MetOp LEO satellites) and MetOp POD GPS measurements only: (**left**) Without receiving antenna corrections and (**right**) with receiving antenna corrections.

5.3. Potential Influence of the Difference between the LEO Orbits

The distribution of the orbits of the three MetOp LEOs, flying at a similar height around 800 km, are identical for metb and metc and similar for meta, but placed at distances of thousands of km between the LEOs (see Figure 8). However, looking at the different estimation of the transmitter DCBs from the measurements separately of meta, metb and metc (Figure 7), or directly comparing the receiver DCBs (Figure 2), it can be seen than there is no evident relationship between the LEO satellite relative spatial location and them (in fact it might seem more similar to the time evolution of meta and metc DCBs, which do not follow the same orbit).

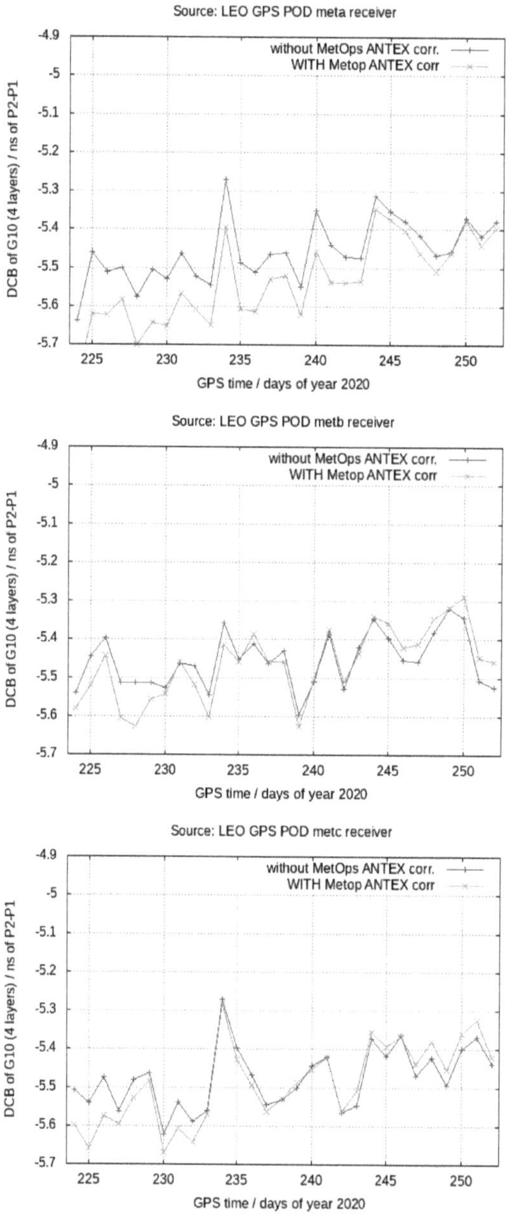

Figure 7. Comparison of the daily GPS10 transmitter DCBs estimated from MetOp zenith measurements only without (magenta) and with (green) receiving antenna corrections, two layers above LEOs and since days 224 to 252 of year 2020, from meta (**top**), metb (**center**) and metc (**bottom**) LEOs.

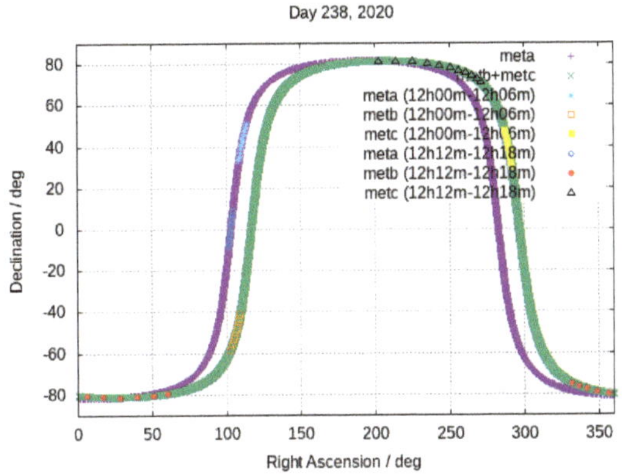

Figure 8. Footprint of the orbits of meta, metb, metc MetOp LEOs, for day 238 of year 2020, showing the displacement between 12 h 00 m–12 h 06 m and 12 h 12 m–12 h 18 m.

6. Conclusions

The characterization of the GPS carrier-phase-based only tomographic estimation of the LEO topside electron content was summarized. It was demonstrated on a single representative day (176, 2020) for different combinations of input data (among ground GPS and MetOp POD GPS) and different degrees of tomographic resolutions (four layers vs two layers). Slightly better results were obtained with ground GPS + MetOp POD GPS data and with MetOp POD GPS only and showing a stability of the MetOp POD GPS receiver DCBs at a few tenths of the TECU level.

The study was extended to 29 consecutive days (days 224 to 252, 2020) showing a consistent result of the TOMION DCBs, obtained in daily independent runs from carrier-phase data only (stability of the DCB differences among MetOp POD GPS receivers at 0.1 TECU level). The associated transmitter DCBs from MetOp-only data are consistent with the IGS ground-based ones at a few tenths of ns in a P2-P1 delay level, as expected for transmitter DCBs

In this way, a novel method for DCB estimation has been introduced. DCBs are estimated as residuals out of the difference between the GF combinations of the code and the calibrated carrier-phase. The carrier-phase has been calibrated by a tomographic model, TOMION, that decorrelates the ionospheric delay and the ambiguity term. The DCBs and their uncertainty are estimated on an epoch basis, thus showing the variability in time of DCBs, which is consistent with previous works from other authors. The method has been assessed with external GPS transmitter DCBs computed by CODE and JPL and their relative differences are within 0.5 ns. The small influence of some modelling aspects which are not frequently considered, such as the change of DCB data and the influence of the LEO receiver zenith antenna phase variations, were also discussed in detail.

Finally, potential future follow-on research could analyze the existence of a correlation between LEO DCBs and temperature. If such a correlation exists for LEO satellites, then it would provide the possibility of using DCBs as a proxy of atmospheric temperature variability.

In conclusion, high accuracy estimation of LEO receiver DCBs provides a method for monitoring the state of GNSS receivers and for a quick and accurate computation of the STEC corresponding to the POD GPS receiver lines-of-sight. This approach has as a main advantage providing a simultaneous estimation of the LEO topside ionospheric electron content distribution and of the transmitter and receiver DCBs—everything in an autonomous way (i.e., only based on raw GPS zenith LEO observations). This is done in a simple manner, where the DCBs are computed as an a posteriori product of the ionospheric tomography (then suitable for multi-GNSS measurements) and without any background model or initial set of DCB values required.

Author Contributions: Conceptualization, M.H.-P. and R.N.; methodology, M.H.-P.; software, M.H.-P. and V.G.; validation, M.H.-P., R.N. and L.Y.; formal analysis, M.H.-P.; investigation, M.H.-P.; resources, R.N.; data curation, R.N., G.O.-P. and M.H.-P.; writing—original draft preparation, G.O.-P. and M.H.-P.; writing—review and editing, M.M.H., F.S.P., L.Y., R.N., V.G., G.O.-P. and M.H.-P.; visualization, M.H.-P.; supervision, R.N., M.M.H., F.S.P. and M.H.-P.; project administration, R.N., M.M.H. and M.H.-P.; funding acquisition, M.M.H. and M.H.-P. All authors have read and agreed to the published version of the manuscript.

Funding: This research was funded by EUMETSAT in the context of the project *Assessment of GRAS Ionospheric Measurements for Ionospheric Model Assimilation* (GIMA, number EUM/CO/21/4600002530/RN).

Data Availability Statement: The external DCB data are openly accessible (https://cddis.nasa.gov/archive/gnss/products/ionex accessed on 5 November 2021) from Crustal Dynamics Data Information System. The input GPS measurements of the MetOp POD GPS receivers and precise MetOp orbits, can be solicited to EUMETSAT if needed.

Acknowledgments: The work has been performed coinciding with the execution of the PITHIA-NRF H2020 project (H2020-INFRAIA-2018-2020101007599).

Conflicts of Interest: The authors declare no conflict of interest.

Appendix A

In order to obtain the transmitter DCB, firstly we average Equation (6) over time, thus yielding the following expression for the daily total DCBs, $\overline{D_r^s}$:

$$\overline{D_r^s} \equiv \overline{D_{r,GF} + D_{GF}^s} = \frac{1}{N}\sum_{i=1}^{i=N} P_{r,GF}^s(t_i) - \tilde{L}_{r,GF}^s(t_i)\,, \tag{A1}$$

where N is the number of epochs and t_i the time at the ith epoch. Secondly, the daily receiver is computed by averaging $\overline{D_r^s}$ over satellites:

$$\hat{D}_r = \frac{1}{M}\sum_{i=1}^{i=M} \overline{D_r^i}\,, \tag{A2}$$

where M is the number of satellites in view by the receiver. Now, notice that $\overline{D_r^i} = \overline{D_r + D^i} = \overline{D_r} + \overline{D^i}$, then:

$$\hat{D}_r = \frac{1}{M}\sum_{i=1}^{i=M} \left[\overline{D_r} + \overline{D^i}\right], \tag{A3}$$

which, after setting the sum of daily satellites DCBs to zero, i.e.,

$$\sum_{i=1}^{i=M} \overline{D^i} = 0\,, \tag{A4}$$

leads to:

$$\hat{D}_r = \frac{1}{M}\sum_{i=1}^{i=M} \overline{D_r}\,, \tag{A5}$$

i.e., the daily satellite-averaged receiver DCBs. Finally, subtraction of $\hat{D}_r$ in Equation (A1) yields the daily satellite DCB, $\hat{D}^s$, as follows:

$$\hat{D}^s = \overline{D_r^s} - \hat{D}_r\,. \tag{A6}$$

References

1. Hegarty, C.J.; Powers, E.D.; Fonville, B. Accounting for timing biases between GPS, modernized GPS and Galileo signals. In Proceedings of the 18th International Technical Meeting of the Satellite Division of The Institute of Navigation (ION GNSS 2005), Long Beach, CA, USA, 13–16 September 2005; pp. 2401–2407.
2. Hauschild, A.; Montenbruck, O. A study on the dependency of GNSS pseudorange biases on correlator spacing. *GPS Solut.* **2016**, *20*, 159–171. [CrossRef]
3. Xiang, Y.; Xu, Z.; Gao, Y.; Yu, W. Understanding long-term variations in GPS differential code biases. *GPS Solut.* **2020**, *24*, 1–11. [CrossRef]
4. Prol, F.d.S.; Camargo, P.d.O.; Monico, J.F.G.; Muella, M.T.d.A.H. Assessment of a TEC calibration procedure by single-frequency PPP. *GPS Solut.* **2018**, *22*, 1–11. [CrossRef]
5. Arikan, F.; Nayir, H.; Sezen, U.; Arikan, O. Estimation of single station interfrequency receiver bias using GPS-TEC. *Radio Sci.* **2008**, *43*, 1–13. [CrossRef]
6. Mannucci, A.; Wilson, B.; Yuan, D.; Ho, C.; Lindqwister, U.; Runge, T. A Global mapping technique for GPS-derived ionospheric total electron content measurements. *Radio Sci.* **1998**, *33*, 565–582. [CrossRef]
7. Sardon, E.; Rius, A.; Zarraoa, N. Estimation of the transmitter and receiver differential biases and the ionospheric total electron content from Global Positioning System observations. *Radio Sci.* **1994**, *29*, 577–586. [CrossRef]
8. Sardón, E.; Zarraoa, N. Estimation of total electron content using GPS data: How stable are the differential satellite and receiver instrumental biases? *Radio Sci.* **1997**, *32*, 1899–1910. [CrossRef]
9. Zhang, D.; Shi, H.; Jin, Y.; Zhang, W.; Hao, Y.; Xiao, Z. The variation of the estimated GPS instrumental bias and its possible connection with ionospheric variability. *Sci. China Technol. Sci.* **2014**, *57*, 67–79. [CrossRef]
10. Zhong, J.; Lei, J.; Dou, X.; Yue, X. Is the long-term variation of the estimated GPS differential code biases associated with ionospheric variability? *GPS Solut.* **2016**, *20*, 313–319. [CrossRef]
11. Zha, J.; Zhang, B.; Yuan, Y.; Li, M. Use of modified carrier-to-code leveling to analyze temperature dependence of multi-GNSS receiver DCB and to retrieve ionospheric TEC. *GPS Solut.* **2019**, *23*, 1–12. [CrossRef]
12. Mi, X.; Zhang, B.; Odolinski, R.; Yuan, Y. On the temperature sensitivity of multi-GNSS intra-and inter-system biases and the impact on RTK positioning. *GPS Solut.* **2020**, *24*, 1–14. [CrossRef]
13. Yue, X.; Schreiner, W.S.; Hunt, D.C.; Rocken, C.; Kuo, Y.H. Quantitative evaluation of the Low Earth Orbit satellite based slant total electron content determination. *Space Weather* **2011**, *9*. [CrossRef]
14. Zhang, X.; Tang, L. Daily global plasmaspheric maps derived from cosmic GPS observations. *IEEE Trans. Geosci. Remote Sens.* **2014**, *52*, 6040–6046. [CrossRef]
15. Lin, J.; Yue, X.; Zhao, S. Estimation and analysis of GPS satellite DCB based on LEO observations. *GPS Solut.* **2016**, *20*, 251–258. [CrossRef]
16. Li, X.; Ma, T.; Xie, W.; Zhang, K.; Huang, J.; Ren, X. FY-3D and FY-3C onboard observations for differential code biases estimation. *GPS Solut.* **2019**, *23*, 1–14. [CrossRef]
17. Zhong, J.; Lei, J.; Dou, X.; Yue, X. Assessment of vertical TEC mapping functions for space-based GNSS observations. *GPS Solut.* **2016**, *20*, 353–362. [CrossRef]
18. Wautelet, G.; Loyer, S.; Mercier, F.; Perosanz, F. Computation of GPS P1–P2 differential code biases with Jason-2. *GPS Solut.* **2017**, *21*, 1619–1631. [CrossRef]
19. Li, W.; Li, M.; Shi, C.; Fang, R.; Zhao, Q.; Meng, X.; Yang, G.; Bai, W. GPS and BeiDou differential code bias estimation using Fengyun-3C satellite onboard GNSS observations. *Remote Sens.* **2017**, *9*, 1239. [CrossRef]
20. Yuan, L.; Jin, S.; Hoque, M. Estimation of LEO-GPS receiver differential code bias based on inequality constrained least square and multi-layer mapping function. *GPS Solut.* **2020**, *24*, 1–12. [CrossRef]
21. Yuan, L.; Hoque, M.; Jin, S. A new method to estimate GPS satellite and receiver differential code biases using a network of LEO satellites. *GPS Solut.* **2021**, *25*, 1–12. [CrossRef]
22. Heise, S.; Jakowski, N.; Wehrenpfennig, A.; Reigber, C.; Lühr, H. Sounding of the topside ionosphere/plasmasphere based on GPS measurements from CHAMP: Initial results. *Geophys. Res. Lett.* **2002**, *29*, 44-1–44-4. [CrossRef]
23. Spencer, P.S.; Mitchell, C.N. Imaging of 3-D plasmaspheric electron density using GPS to LEO satellite differential phase observations. *Radio Sci.* **2011**, *46*, 1–6. [CrossRef]
24. Wu, M.; Guo, P.; Xu, T.; Fu, N.; Xu, X.; Jin, H.; Hu, X. Data assimilation of plasmasphere and upper ionosphere using COSMIC/GPS slant TEC measurements. *Radio Sci.* **2015**, *50*, 1131–1140. [CrossRef]
25. Kim, E.; Kim, Y.; Jee, G.; Ssessanga, N. Reconstruction of plasmaspheric density distributions by applying a tomography technique to jason-1 plasmaspheric TEC measurements. *Radio Sci.* **2018**, *53*, 866–873. [CrossRef]
26. Prol, F.S.; Hoque, M.M. A Tomographic Method for the Reconstruction of the Plasmasphere Based on COSMIC/FORMOSAT-3 Data. *IEEE J. Sel. Top. Appl. Earth Obs. Remote Sens.* **2022**, *15*, 2197–2208. [CrossRef]
27. Fu, N.; Guo, P.; Wu, M.; Huang, Y.; Hu, X.; Hong, Z. The two-parts step-by-step ionospheric assimilation based on ground-based/spaceborne observations and its verification. *Remote Sens.* **2019**, *11*, 1172. [CrossRef]
28. Hernández-Pajares, M.; Juan, J.; Sanz, J. Neural network modeling of the ionospheric electron content at global scale using GPS data. *Radio Sci.* **1997**, *32*, 1081–1089. [CrossRef]

29. Hernández-Pajares, M.; Juan, J.; Sanz, J. New approaches in global ionospheric determination using ground GPS data. *J. Atmos. Sol.-Terr. Phys.* **1999**, *61*, 1237–1247. [CrossRef]
30. Orús, R.; Hernández-Pajares, M.; Juan, J.; Sanz, J. Improvement of global ionospheric VTEC maps by using kriging interpolation technique. *J. Atmos. Sol.-Terr. Phys.* **2005**, *67*, 1598–1609. [CrossRef]
31. Roma Dollase, D. Global Ionospheric Maps: Estimation and Assessment in Post-Processing and Real-Time. Ph.D. dissertations, Universitat Politècnica de Catalunya, Barcelona, Spain, 2019.
32. Hernández-Pajares, M.; Juan, J.M.; Sanz, J.; Aragón-Àngel, À.; García-Rigo, A.; Salazar, D.; Escudero, M. The ionosphere: Effects, GPS modeling and the benefits for space geodetic techniques. *J. Geod.* **2011**, *85*, 887–907. [CrossRef]
33. Hernández-Pajares, M.; Juan, J.; Sanz, J.; Orus, R.; García-Rigo, A.; Feltens, J.; Komjathy, A.; Schaer, S.; Krankowski, A. The IGS VTEC maps: A reliable source of ionospheric information since 1998. *J. Geod.* **2009**, *83*, 263–275. [CrossRef]
34. Hernández-Pajares, M.; Roma-Dollase, D.; Krankowski, A.; García-Rigo, A.; Orús-Pérez, R. Methodology and consistency of slant and vertical assessments for ionospheric electron content models. *J. Geod.* **2017**, *91*, 1405–1414 [CrossRef]
35. Roma-Dollase, D.; Hernández-Pajares, M.; Krankowski, A.; Kotulak, K.; Ghoddousi-Fard, R.; Yuan, Y.; Li, Z.; Zhang, H.; Shi, C.; Wang, C.; et al. Consistency of seven different GNSS global ionospheric mapping techniques during one solar cycle. *J. Geod.* **2018**, *92*, 691–706. [CrossRef]
36. Hernández-Pajares, M.; Lyu, H.; Aragón-Àngel, À.; Monte-Moreno, E.; Liu, J.; An, J.; Jiang, H. Polar Electron Content From GPS Data-Based Global Ionospheric Maps: Assessment, Case Studies and Climatology. *J. Geophys. Res. Space Phys.* **2020**, *125*, e2019JA027677. [CrossRef]
37. Liu, Q.; Hernández-Pajares, M.; Lyu, H.; Nishioka, M.; Yang, H.; Monte-Moreno, E.; Gulyaeva, T.; Béniguel, Y.; Wilken, V.; Olivares-Pulido, G.; et al. Ionospheric Storm Scale Index Based on High Time Resolution UPC-IonSAT Global Ionospheric Maps (IsUG). *Space Weather* **2021**, *19*, e2021SW002853. [CrossRef]
38. Hernández-Pajares, M.; Lyu, H.; Garcia-Fernandez, M.; Orus-Perez, R. A new way of improving global ionospheric maps by ionospheric tomography: Consistent combination of multi-GNSS and multi-space geodetic dual-frequency measurements gathered from vessel-, LEO-and ground-based receivers. *J. Geod.* **2020**, *94*, 1–16. [CrossRef]
39. Kotov, D.; Richards, P.G.; Truhlík, V.; Bogomaz, O.; Shulha, M.; Maruyama, N.; Hairston, M.; Miyoshi, Y.; Kasahara, Y.; Kumamoto, A.; et al. Coincident observations by the Kharkiv IS radar and ionosonde, DMSP and Arase (ERG) satellites and FLIP model simulations: Implications for the NRLMSISE-00 hydrogen density, plasmasphere and ionosphere. *Geophys. Res. Lett.* **2018**, *45*, 8062–8071. [CrossRef]
40. Kotov, D.; Richards, P.G.; Truhlík, V.; Maruyama, N.; Fedrizzi, M.; Shulha, M.; Bogomaz, O.; Lichtenberger, J.; Hernández-Pajares, M.; Chernogor, L.; et al. Weak magnetic storms can modulate ionosphere-plasmasphere interaction significantly: Mechanisms and manifestations at mid-latitudes. *J. Geophys. Res. Space Phys.* **2019**, *124*, 9665–9675. [CrossRef]
41. Olivares-Pulido, G.; Hernández-Pajares, M.; Lyu, H.; Gu, S.; García-Rigo, A.; Graffigna, V.; Tomaszewski, D.; Wielgosz, P.; Rapiński, J.; Krypia-Gregorczyk, A.; et al. Ionospheric tomographic common clock model of undifferenced uncombined GNSS measurements. *J. Geod.* **2021**, *95*, 1–13. [CrossRef]
42. Lyu, H.; Hernández-Pajares, M.; Nohutcu, M.; García-Rigo, A.; Zhang, H.; Liu, J. The Barcelona ionospheric mapping function (BIMF) and its application to northern mid-latitudes. *GPS Solut.* **2018**, *22*, 1–13. [CrossRef]
43. Blewitt, G. An automatic editing algorithm for GPS data. *Geophys. Res. Lett.* **1990**, *17*, 199–202. [CrossRef]
44. Jiahao, Z.; Lei, J.; Yue, X.; Dou, X. Determination of differential code bias of GNSS receiver onboard low Earth orbit satellite. *IEEE Trans. Geosci. Remote Sens.* **2016**, *54*, 4896–4905.
45. Montenbruck, O.; Andres, Y.; Bock, H.; van Helleputte, T.; van den Ijssel, J.; Loiselet, M.; Marquardt, C.; Silvestrin, P.; Visser, P.; Yoon, Y. Tracking and orbit determination performance of the GRAS instrument on MetOp-A. *GPS Solut.* **2008**, *12*, 289–299. [CrossRef]

MDPI AG
Grosspeteranlage 5
4052 Basel
Switzerland
Tel.: +41 61 683 77 34

Remote Sensing Editorial Office
E-mail: remotesensing@mdpi.com
www.mdpi.com/journal/remotesensing

www.ingramcontent.com/pod-product-compliance
Lightning Source LLC
Chambersburg PA
CBHW061451210926
51850CB00008B/1683
* 9 7 8 3 7 2 5 8 6 0 9 9 9 *